Stem Cell Biology and Regenerative Medicine

Series Editor
Kursad Turksen, Ph.D.
kursadturksen@gmail.com

More information about this series at http://www.springer.com/series/7896

Babak Arjmand
Editor

Perinatal Tissue-Derived Stem Cells

Alternative Sources of Fetal Stem Cells

Humana Press

Editor
Babak Arjmand
Endocrinology and Metabolism Research Center
Endocrinology and Metabolism Clinical Sciences Institute
Tehran University of Medical Sciences &
Brain and Spinal Cord Injury Research Center
Tehran University of Medical Sciences
Tehran, Iran

ISSN 2196-8985 ISSN 2196-8993 (electronic)
Stem Cell Biology and Regenerative Medicine
ISBN 978-3-319-83514-3 ISBN 978-3-319-46410-7 (eBook)
DOI 10.1007/978-3-319-46410-7

Printed on acid-free paper

This Humana Press imprint is published by Springer Nature
The registered company is Springer International Publishing AG
The registered company address is: Gewerbestrasse 11, 6330 Cham, Switzerland

*This book is dedicated to my **parents**,*
who have been my support during my life,
*to my loving wife **Parisa** who has always*
believed in me.
Also I want to dedicate it to my daughter
***Rasta** and my brothers, **Afshin** and **Farzin**.*

Preface

Nowadays, stem cells have a crucial role in tissue engineering and regenerative medicine. Also, basic researches in this field and its clinical applications are growing rapidly. Therefore, investigators, clinicians, and other relevant professionals have focused on regenerative medicine as a multidisciplinary area of interest.

Because of the unique biological characteristics and therapeutic potentials of perinatal tissue-derived stem cells, they are frequently suggested as an invaluable source for cellular therapy and regenerative medicine. In this volume, we address different types and properties of perinatal stem cells and also ethical considerations of their use in regenerative medicine. Furthermore, a brief review of their multi- or pluripotent and immunomodulatory properties, regenerative capacity, and therapeutic potentials is presented in this work. Additionally, we talk about the cGMP facility design and GMP-compliant manufacturing of perinatal stem cells for clinical translation.

It is my pleasure having the collaboration of prominent contributors in this volume, which could be valuable to both basic and clinical investigators who are interested in regenerative medicine.

I would like to acknowledge **Dr. Kursad Turksen**, Editor in Chief of the *Stem Cell Biology and Regenerative Medicine*, for his advice and support.

I also thank **Aleta Kalkstein**, Senior Editor, Hard Sciences, Cell Biology Stem Cell Research, and **Joseph Quatela**, Production Coordinator, at Springer for their continuous help and kind support to get the volume to the print stage.

Tehran, Iran Babak Arjmand

Contents

The original version of this book was revised. An erratum to this book can be found at DOI 10.1007/978-3-319-46410-7_12

Contributors

Leila Afshar Medical Ethics Department, Shahid Beheshti University of Medical Sciences, Tehran, Iran

Hamid Reza Aghayan Chronic Diseases Research Center, Endocrinology and Metabolism Population Sciences Institute, Tehran University of Medical Sciences, Tehran, Iran

Jafar Ai Department of Tissue Engineering and Applied Cell Sciences, Faculty of Advanced Technologies in Medicine, Tehran University of Medical Sciences, Tehran, Iran

Mehdi Aleahmad Department of Immunology, School of Public Health, Tehran University of Medical Sciences, Tehran, Iran

Shaghayegh Arasteh Reproductive Biotechnology Research Center, Avicenna Research Institute, ACECR, Tehran, Iran

Babak Arjmand Endocrinology and Metabolism Research Center, Endocrinology and Metabolism Clinical Sciences Institute, Tehran University of Medical Sciences, Tehran, Iran

Brain and Spinal Cord Injury Research Center, Tehran University of Medical Sciences, Tehran, Iran

Zohreh Bagher ENT and Head & Neck Research Center and Department, Hazrat Rasoul Akram Hospital, Iran University of Medical Sciences, Tehran, Iran

Scott R. Burger Advanced Cell & Gene Therapy, LLC, Chapel Hill, North Carolina, USA

Mohammad Reza Bolouri Immunology Research Center, Iran University of Medical Sciences, Tehran, Iran

Somayeh Ebrahimi-Barough Department of Tissue Engineering and Applied Cell Sciences, Faculty of Advanced Technologies in Medicine, Tehran University of Medical Sciences, Tehran, Iran

Khadijeh Falahzadeh Chronic Diseases Research Center, Endocrinology and Metabolism Population Sciences Institute, Tehran University of Medical Sciences, Tehran, Iran

Behrouz Farhadihosseinabad Biotechnology Department, School of Advanced Technologies in Medicine, Shahid Beheshti University of Medical Sciences, Tehran, Iran

Mazaher Gholipourmalekabadi Cellular and Molecular Research Center, Iran University of Medical Sciences, Tehran, Iran

Department of Tissue Engineering & Regenerative Medicine, Bhupal Nobles Post Graduate (B.N.P.G.) College, Tehran, Iran

Parisa Goodarzi Brain and Spinal Cord Injury Research Center, Tehran University of Medical Sciences, Tehran, Iran

Seyed Mahmoud Hashemi Department of Immunology, School of Medicine, Shahid Beheshti University of Medical Sciences, Tehran, Iran

Elham Hoveizi Department of Biology, Faculty of Sciences, Shahid Chamran University of Ahvaz, Ahvaz, Iran

Çiğdem İnci Centre for Regenerative Medicine and Stem Cell Manufacturing, LivMedCell, Liv Hospital, Istanbul, Turkey

Erdal Karaöz Centre for Regenerative Medicine and Stem Cell Manufacturing, LivMedCell, Liv Hospital, Istanbul, Turkey

Somaieh Kazemnejad Reproductive Biotechnology Research Center, Avicenna Research Institute, ACECR, Tehran, Iran

Sayeh Khanjani Reproductive Biotechnology Research Center, Avicenna Research Institute, ACECR, Tehran, Iran

Manijeh Khanmohammadi Reproductive Biotechnology Research Center, Avicenna Research Institute, ACECR, Tehran, Iran

Kiarash Khosrotehrani UQ Centre for Clinical Research, The University of Queensland, Brisbane, QLD, Australia

UQ Diamantina Institute, Translational Research Institute, The University of Queensland, Brisbane, QLD, Australia

Bagher Larijani Endocrinology and Metabolism Research Center, Endocrinology and Metabolism Clinical Sciences Institute, Tehran University of Medical Sciences, Tehran, Iran

Toshio Miki Department of Biochemistry and Molecular Biology, Keck School of Medicine, University of Southern California, Los Angeles, CA, USA

Fereshteh Mohamadi-Jahani Brain and Spinal Cord Injury Research Center, Tehran University of Medical Sciences, Tehran, Iran

Narendra Pal Singh Chauhan Department of Chemistry, Faculty of Advanced Technologies in Medicine, Iran University of Medical Sciences, Udaipur, Rajasthan, India

Reza Rahbarghazi Stem Cell Research Center, Tabriz University of Medical Sciences, Tabriz, Iran

Department of Applied Cell Sciences, Faculty of Advanced Medical Sciences, Tabriz University of Medical Sciences, Tabriz, Iran

Fakher Rahim Institute of Health Research, Thalassemia and Hemoglobinopathies Research Center, Ahvaz Jundishapur University of Medical Sciences, Ahvaz, Iran

Ali Samadikuchaksaraei Cellular and Molecular Research Center, Iran University of Medical Sciences, Tehran, Iran

Department of Tissue Engineering & Regenerative Medicine, Faculty of Advanced Technologies in Medicine, Iran University of Medical Sciences, Tehran, Iran

Department of Medical Biotechnology, Faculty of Allied Medicine, Iran University of Medical Sciences, Tehran, Iran

Abbas Shafiee UQ Centre for Clinical Research, The University of Queensland, Brisbane, QLD, Australia

Institute of Health and Biomedical Innovation, Queensland University of Technology, Brisbane, QLD, Australia

Abolfazl Shirazi Reproductive Biotechnology Research Center, Avicenna Research Institute, ACECR, Tehran, Iran

Sara Soudi Department of Immunology, Faculty of Medical Sciences, Tarbiat Modares University, Tehran, Iran

Fabio Triolo Department of Pediatric Surgery, McGovern Medical School, University of Texas Health Science Center at Houston, Houston, TX, USA

Amir-Hassan Zarnani Department of Immunology, School of Public Health, Tehran University of Medical Sciences, Tehran, Iran

Immunology Research Center, Iran University of Medical Sciences, Tehran, Iran

About the Editor

Babak Arjmand received his M.D. from Iran University of Medical Sciences and Ph.D. from Tehran University of Medical Sciences, Tehran, Iran. He is currently vice chancellor for research at Metabolic Disorders Research Center, Endocrinology and Metabolism Cellular-Molecular Sciences Institute in Tehran, Iran. His research focuses on the stem cell translation from the basic to the clinic and GMP-compliant manufacturing for cellular therapy. Dr. Arjmand is member of different cell and tissue committees and groups such as National Committee of Tissue, Cell and Gene Therapy at Iran Food and Drug Organization (MOH), Iranian Council of Stem Cell Technologies, and National Working Group for providing National Guideline on Stem Cell therapy.

Functional Dualism of Perinatal Stem Cells

Toshio Miki and Fabio Triolo

1 Introduction

Stem cell-based therapies hold the potential of alleviating the burden of many serious diseases. These promising stem cell-based approaches for patients with unmet medical needs rely mainly on two unique properties of stem cells: their differentiation capability to all three germ layers (pluripotency) and their immunomodulatory function. The pluripotency makes the stem cells able to generate desired types of cells for cell replacement therapies. The immunomodulatory properties can be utilized to control immunoreaction and subsequent pathological events. Traditionally, pluripotency has been considered a character of embryonic stem cells, and immunomodulatory properties one of mesenchymal stem cells from adult somatic tissues. During the last decade, however, many studies revealed that some perinatal stem cells represent a novel class of stem cells with intermediate characteristics of both pluripotent/embryonic and adult stem cells, as they possess the pluripotent stem cell-like differentiation potential and immunomodulatory effects similar to mesenchymal stem cells in vitro and in vivo. In addition, these perinatal stem cells are as genetically stable as adult stem cells. These unique characteristics, together with the absence of ethical issues concerning their procurement, attract many researchers in search of practical stem cells for prompt clinical translation.

In this chapter, we describe two types of perinatal stem cells: amniotic epithelial stem cells and Wharton's Jelly-derived stem cells, both of which possess embryonic cell-like differentiation properties and adult stem cell-like immunomodulatory

T. Miki, M.D., Ph.D. (✉)
Department of Biochemistry and Molecular Biology, Keck School of Medicine, University of Southern California, Los Angeles, CA, USA
e-mail: toshiomi@usc.edu

F. Triolo
Department of Pediatric Surgery, McGovern Medical School, University of Texas Health Science Center at Houston, Houston, TX, USA

B. Arjmand (ed.), *Perinatal Tissue-Derived Stem Cells*, Stem Cell Biology and Regenerative Medicine, DOI 10.1007/978-3-319-46410-7_1

properties. Unlike other types of placental stem cells, both of these cell types share a unique developmental origin. First, we outline various perinatal stem cells, their origin, and discuss their classification and advantages over other types of stem cells when considering the clinical application. Studies that indicate the beneficial properties of these cells are discussed in the following section. Lastly, preclinical studies with these placental stem cells are summarized to determine a strategic approach to translating the findings into clinical therapies.

2 Definition and Advantages of Perinatal Stem Cells

The placenta is the very first organ developed in our life, protecting the fetus by providing a stable environment and absorbing physical shocks throughout the gestation period (Donnelly and Campling 2014). All eutherian placenta provide common functions to their fetus with variations in the shape and microscopic structure between species. Human placenta consists of the umbilical cord, placental membrane, and a discoid shape placental body approximately 15–20 cm in diameter and 2–3 cm in thickness. Anatomically, the membrane and the discoid tissue are composed of three parts: the amnion, the chorion, and the decidua.

Although the placenta is normally considered postdelivery waste, it has been used as a nutrient source (placentophagy) or for medical purposes since ancient times. Over the last decade, there has been a growing interest in the placenta for its unique and rich source of perinatal stem cells. A number of studies have been conducted in order to identify and characterize perinatal stem cells from placenta tissue. The research publication trend from 1969 to 2014 reveals an emerging interest in stem cells isolated from placenta (Fig. 1). However, the generic placental/perinatal

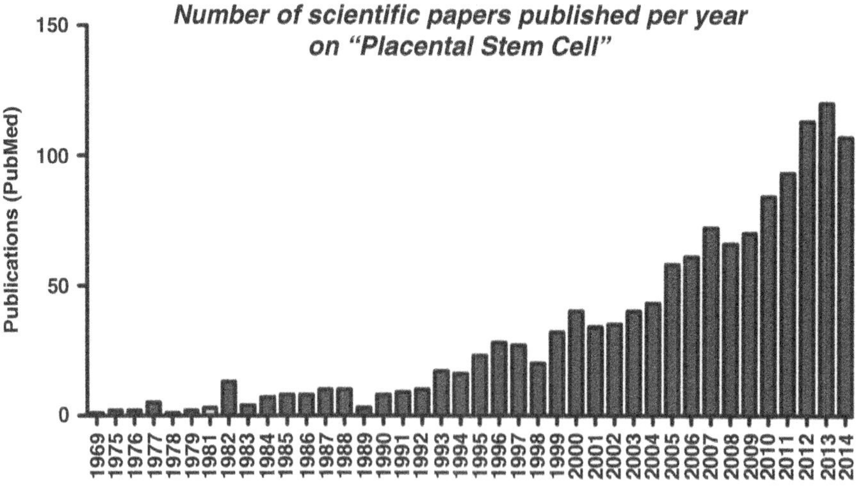

Fig. 1 The number of scientific papers published by year from 1969 to 2014 was searched in PubMed using the keywords "Placental Stem Cell." The publication trend reveals an emerging interest in stem cells isolated from placenta

stem cell terms do not specify the cell types and their use often generates confusion. Here, we clarify the type of stem cells isolated from human placenta and discuss the common and unique advantages of these cells.

Stem cells are traditionally classified as either embryonic or adult, based on the developmental stage they belong to. To this extent, placental stem cells are categorized as "perinatal" (i.e., around the time of birth), despite technically encompassing the time period ranging from the twentieth week of gestation through the first 28 days of life (neonatal period) (Cetrulo and Cetrulo 2009). Therefore, perinatal stem cells include cells isolated from amniotic fluid, umbilical cord blood, and postgestational maternal peripheral blood, as well as cells from the placenta (Fig. 2).

Regardless of the type of placental stem cells, there are common advantages over other types of stem cells.

First, placenta tissues are often discarded in the clinic, therefore, they are a readily available cell source, and the cells could be harvested with relatively simple procedures in a noninvasive manner. In addition, there are fewer limitations and ethical concerns compared to the use of other types of stem cells. Second, the placenta is a neonatal organ, thus the derived cells have less age- and environment-acquired DNA damage compared to stem cells from adult tissues and long-term cultured pluripotent stem cells.

It must be noted that perinatal stem cells are divided into two groups: placental and nonplacental stem cells. The cord blood is commonly obtained by irrigating a placenta via the umbilical vein. Although isolated from the placenta, neonatal hematopoietic stem cells in the cord blood are considered nonplacental stem cells. Fetal hematopoietic and mesenchymal stem cells have been identified in postgestational maternal peripheral blood and are also considered nonplacental stem cells (Khosrotehrani and Bianchi 2005; Nguyen Huu et al. 2006; O'Donoghue et al. 2003). The regenerative contribution of postgestational maternal peripheral blood-derived stem cells has been reported. The fetal cells in the maternal circulation system selectively home onto injured maternal hearts and undergo differentiation into diverse cardiac lineages in a fusion-independent manner (Kara et al. 2012). However, due to the rarity of these cells, postgestational maternal peripheral blood-derived stem cells are not ideal for future therapeutic applications. Amniotic fluid (AF) cells are a mixed population from multiple fetal tissues including the epithelium of the fetal skin, respiratory organs, and gastrointestinal tract. The heterogeneous nature of AF cells requires a definition of amniotic fluid stem cells and a standardized protocol for its isolation. Although some research only relied on the mesenchymal stem cell-like characteristics, most reliable studies were conducted with a subpopulation of AF cells, which express specific cell surface markers (e.g., c-kit). Based on their origin, AF stem cells are classified as nonplacental.

On the other hand, the cells isolated from placenta tissue are considered placental stem cells (Fig. 3) and are further classified based on their tissue of origin. As previously mentioned, the placental tissue includes both a discoid and a membranous component. The former consists of a fetal component known as the chorionic plate, which is derived from the trophoblast and the extraembryonic mesoderm. The chorionic plate/villi contains mesenchyme-derived cells, known as chorionic stem cells, which possess mesenchymal stem cell-like characteristics (Igura et al. 2004; Castrechini et al. 2010; Kim et al. 2011). Recently, Farmer's group successfully

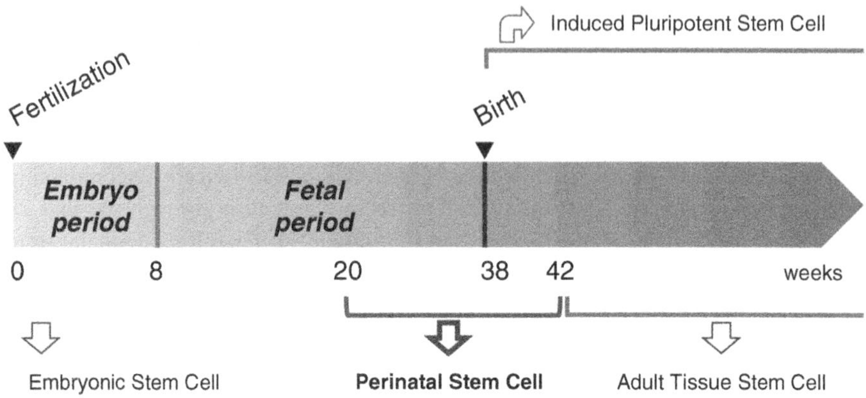

<table>
<tr><td rowspan="9">Perinatal Stem Cell</td><td rowspan="6">Placental Stem Cell</td><td rowspan="2">Amniotic Stem Cell</td><td>Amniotic Epithelial Stem Cell</td></tr>
<tr><td>Amniotic mesenchymal Stem Cell</td></tr>
<tr><td rowspan="2">Chorionic Stem Cell</td><td>Trophoblast Stem Cell</td></tr>
<tr><td>Chorionic Mesenchymal Stem Cell</td></tr>
<tr><td rowspan="2">Wharton's jelly Stem Cell</td><td>Wharton's jelly Mesenchymal Stem cell</td></tr>
<tr><td>Endothelial Progenitor cell</td></tr>
<tr><td rowspan="3">Non-Perinatal Stem Cell</td><td colspan="2">Amniotic Fluid Stem Cell</td></tr>
<tr><td colspan="2">Cord blood Stem Cell</td></tr>
<tr><td colspan="2">Postgestational maternal peripheral blood</td></tr>
</table>

Fig. 2 Human perinatal stem cell classification based on the origin

demonstrated the therapeutic efficacy of chorionic villus-derived cells on myelomeningocele in a large animal (fetal sheep) model (Wang et al. 2015). The therapeutic effect was mediated by neurotrophic factors that were secreted from the chorionic villus-derived cells.

The membranous component of the placenta includes the amnion, the chorion, and the decidua capsularis. The amnion contains the epithelial, compact, amniotic mesoderm, and spongy layer cells. Beneath the amniotic epithelium is a connective

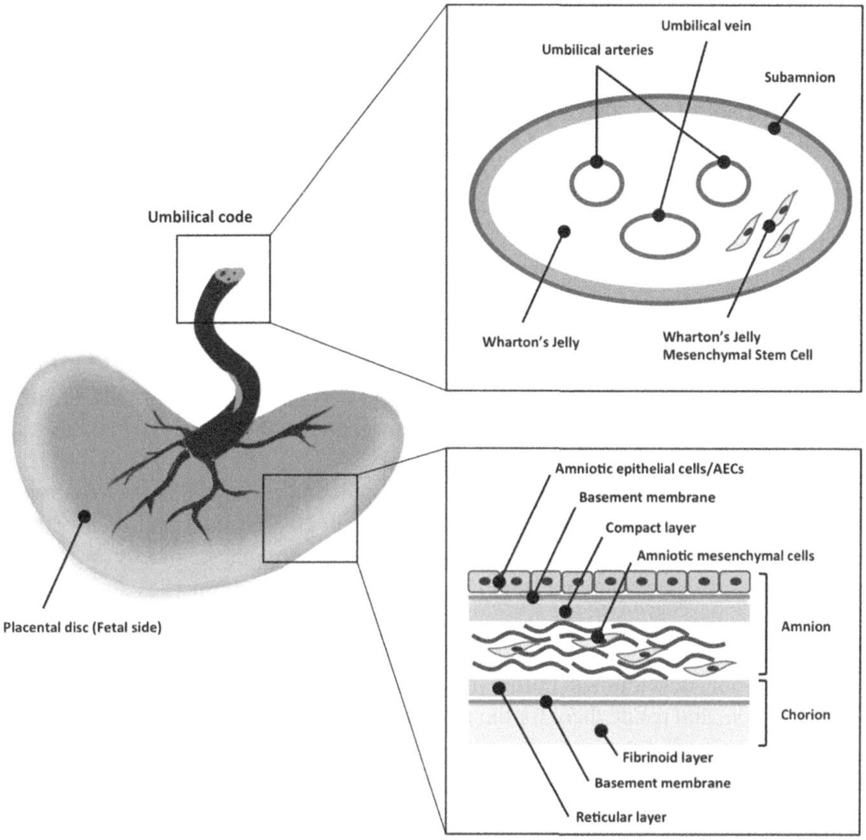

Fig. 3 Anatomical illustration of human placenta and the composition of stem cells

tissue (amniotic mesoderm layer) consisting of fetal fibroblasts, named amniotic mesenchymal stromal cells (AMSCs). These mesenchymal cells can be isolated by sequential trypsin-collagenase enzymatic digestion (Marongiu et al. 2010) and present mesenchymal cell characteristics, including specific cell surface marker (e.g., CD105, CD72, CD90) expression. Compared to mesenchymal stem cells derived from adult tissues, human AMSCs possess a greater expansion capacity (Lange-Consiglio et al. 2013). Furthermore, previous studies have shown the potential differentiation of human AMSCs into not only adipogenic, osteogenic, and chondrogenic lineages, but also cardiomyogenic (Zhao et al. 2005), angiogenic, neurogenic (Tamagawa et al. 2008), and pancreatic (Tamagawa et al. 2009) lineages.

The amniotic epithelium is the innermost extraembryonic layer facing the fetus and is uniquely derived from the epiblast before gastrulation. At day 8–9 after fertilization, the inner cell mass differentiates into two layers: the epiblast and the hypoblast. Subsequently, small cells (amnioblasts) appear from the epiblast cluster and migrate into the space between the trophoblast and the embryonic disc that will become the amniotic cavity. As the cavity grows, spatial segregation isolates amnioblasts from the epicenter of organogenesis, the embryo. It is therefore speculated that some epiblast-like stem cell characteristics are retained in human amnioblast-derived cells (hAECs), through the gestation period (Miki 2011). Some of the isolated hAECs have been found to express stem cell-specific surface markers similar to those of human embryonic stem cells (hESCs) including stage-specific embryonic antigens (SSEAs) 3 and 4, as well as tumor rejection antigens 1–60 and 1–81. Along with surface markers, hAECs express molecular markers of pluripotent stem cells, including OCT-4, NANOG, SOX-2, Lefty-A, FGF-4, REX-1, and TDGF-1 (Miki et al. 2005). Although pluripotency of a single AE cell has not yet been identified, exposure of hAEC cultures to various growth factors and chemicals in vitro yield morphological changes, demonstrating plasticity. Under appropriate culture conditions, hAECs exhibit the capability of differentiating into all three germ layers (Miki et al. 2005; Ilancheran et al. 2007). Reports have indicated the potential for hAECs to differentiate into various cell types including neurogenic, cardiomyogenic, hepatic, and pancreatic cell lineages (Miki and Strom 2006).

It has been reported that a potent stem cell population exists within the Wharton's Jelly of the human umbilical cord (Taghizadeh et al. 2011; Troyer and Weiss 2008; Wang et al. 2004). Interestingly, Wharton's Jelly mesenchymal stem cells (WJMSC) not only possess mesenchymal stem cell properties but also properties attributed to ESCs (Fong et al. 2007, 2011). It is speculated that WJMSCs are trapped within the mucous connective tissue of the umbilical cord (Wharton's Jelly) between day 4 and 12 of embryonic development during the process of hematopoietic and mesenchymal cell migration, and reside there for the duration of gestation (Taghizadeh et al. 2011).

In summary, various types of perinatal stem cells can be isolated from the placenta tissue: placental and nonplacental stem cells. Among the former, amniotic epithelial stem cells and Wharton's Jelly-derived stem cells both maintain embryonic or early fetal cell-like characteristics along with mesenchymal stem cell-like properties.

3 The Bifacial Properties of Amniotic Epithelial Stem Cells

hAECs possess bifacial properties: embryonic stem cell-like plasticity (differentiation capability) and mesenchymal stem cell-like immunomodulatory properties. This unique characteristic provides hAECs with a wider range of clinical applications when compared to other types of stem cells. In addition, hAECs possess low immunogenicity and are rich in lysosomes. Based on these four therapeutically advantageous properties, different strategic approaches can be envisioned for clinical cell therapy use (Miki 2016). One such approach takes advantage of the differentiation potential of hAECs to generate the desired types of functional cells for use in replacing injured or defective cells in order to restore organ function. The mesenchymal stem cell-like properties allow us to utilize the hAECs for: (1) modulating immunological conflict and easing the recipient's immunoreaction and (2) secreting appropriate trophic factors which encourage the recovery of the patient's damaged cells. It must be noted that hAECs are genetically stable (nontumorigenic) unlike other pluripotent stem cells. This nontumorigenic property allows to consider transplanting primary AECs with minimal concerns regarding tumor development.

3.1 *In Vivo Differentiation Potential*

The differentiation potential of AECs for several tissue types has been demonstrated in various in vivo models. Classically, the AEC is designated as an extraembryonic ectoderm lineage cell, and as such, the ectodermal characteristics of the primary AEC might lead researchers to induce neural and neuronal differentiation. Okawa et al. demonstrated that rat AECs are neural stem cell marker (neurofilament microtubule-associated protein 2 and nestin) positive and differentiate into neuron-like cells upon transplantation into the ischemic hippocampus of adult gerbils. The engrafted rat AE-derived neural cells survived for more than 5 weeks (Okawa et al. 2001). Similar neural differentiation and therapeutic efficacy of AECs were demonstrated in a stroke model in which intravenously transplanted hAECs migrated to the intracerebral ischemic area, reduced infarct volume, and improved behavioral function (Kranz et al. 2010). When rat AECs were transplanted into telencephalic ventricles of the developing rat brain at embryonic day (ED) 15.5, the transplanted cells migrated out from the injection site and integrated into the recipients' brain structures. The postnatal brain samples showed clear neural differentiation of rat AECs to neuronal cells (Marcus et al. 2008). Human AECs are not only able to differentiate into neural cells in the rodent brain, but also restore memory function in a transgenic mouse model of Alzheimer's disease (Xue et al. 2012). It is speculated that the signals and mechanisms which induce neural differentiation are provided from the microenvironment surrounding the engrafted hAEC. These data suggest the potential for clinical applications using hAECs as a source for cell replacement therapies aimed at neurodegenerative diseases.

The capability of AECs to differentiate into hepatocytes has also been demonstrated. LacZ-labeled rat AECs were engrafted in syngeneic rat liver, survived for more than 30 days after transplantation and integrated into the native liver plate structure (Nakajima et al. 2001). Likewise, lacZ-labeled rat AECs transplanted into fetal rat liver formed a hepatocyte-like cellular mass and survived for up to 14 days after birth (Takahashi et al. 2002). Marongiu et al. transplanted wild-type rat AECs into DPPIV- F344 rat livers and demonstrated that the transplanted cells integrated and formed clusters of mature hepatocytes in the recipients' livers (Marongiu et al. 2011). Human AECs demonstrated in vitro albumin production, glycogen storage, and albumin secretion, consistent with the hepatocyte gene expression profile. When human amnion tissue was implanted into the abdominal cavity of SCID mice, human albumin was detected in the mouse serum. The data indicated that the human amnion tissue survived and possessed hepatocyte-related functions in the peritoneal cavity of recipient mice (Takashima et al. 2004). Transcriptional analysis of hAECs transplanted SCID/Beige mouse livers indicated that transplanted hAECs terminally differentiated into mature hepatocytes in mouse liver and expressed functional marker genes including cytochrome P450 genes at equivalent levels to primaryhuman hepatocytes (Marongiu et al. 2011).

These data indicate that rodent and human AECs are able to differentiate into endoderm lineage tissues. It has been shown that hAECs express PDX-1, insulin, and glucagon mRNA after 2 weeks of culture under pancreatic differentiation conditions (Miki et al. 2005). Moreover, in vivo experiments demonstrated that nicotinamide-stimulated hAECs normalized the recipients' blood glucose levels for up to 4 weeks after being transplanted into the spleen or under the kidney capsule of streptozotocin-induced diabetic mice (Wei et al. 2003).

Human AECs also formed 3D "alveoli-like" structures when cultured in the small airway epithelial cell-specific medium, demonstrating differentiation potential into lung alveolar cells. The hAEC-derived alveoli-like cells expressed functional cystic fibrosis transmembrane conductance regulator (CFTR) and iodide/chloride (I2/Cl2) ion channels on the luminal side of those structures (Murphy et al. 2012). When hAECs were injected into bleomycin-induced mouse lung injury model lungs, the AECs produced surfactant proteins and remained engrafted for over 4 weeks (Moodley et al. 2013). The study also suggested that hAECs exert a wide range of anti-inflammatory effects by several possible mechanisms including reduced TGF-b, increased MMP-9 activity, GM-CSF secretion, and induction of IL-1RA.

3.2 Immunomodulatory Effect

It must be emphasized that, in general, hypoimmunogenicity and immunomodulation are different properties. In hAEC, however, these properties synergistically interact. It has been shown that hAECs express low MHC class I antigens and do not express MHC class II antigens on their cell surface. An immunotype mismatched human amniotic membrane, when transplanted under volunteers' skin, showed no rejection for more than 7 weeks (Akle et al. 1981). Low or no HLA antigen expression and the secretion of soluble HLA-G explained the immunotolerance of hAECs. It is known that HLA-G protects the fetus from being rejected by the maternal immune system (Hunt et al. 2005). The soluble HLA-G triggers maternal NK and T cells apoptosis through binding to the CD8 receptors/immunoglobulin-like transcript receptors (ILTR) (Banas et al. 2008). Levels of HLA-G on cultured AECs are upregulated by exposure to IFN-γ. The available data indicate that HLA-G molecule expression on the cell surface plays a major role in conferring low immunogenicity to hAECs. The secreted soluble HLA-G form also contributes to the immunomodulatory properties of hAECs.

Secreting immunomodulatory proteins is one of the mechanisms of AEC-mediated immunomodulation. It has been shown that AECs produce high levels of the Th2-related cytokines CCL2, CXCL8, and interleukin 6 (IL-6), which inhibit CD34+

cells and monocyte differentiation to DCs (Magatti et al. 2008). Additionally, AECs block the production of inflammatory cytokines TNF-α, CXCL10, CXCL9, and CCL5 in DC differentiation cultures (Magatti et al. 2009). These factors contribute to immunomodulatory effects of hAECs through (1) the suppression of proinflammatory cytokines, (2) the regulation of macrophage recruitment, and (3) the secretion of factors that inhibit the chemotactic activity of neutrophils and macrophages.

Immunomodulatory effects have also been demonstrated to function through different mechanisms. It has been shown that human AECs increased engraftment in a murine bone marrow transplantation model by facilitating microchimerism and research data suggest that the immunotolerant effect was mediated by direct cell-to-cell contact mechanisms. It is hypothesized that the direct interaction of membrane CD152 (CTLA4) with CD80 and CD86 on hAECs may activate Indoleamine 2,3-dioxygenase (IDO)-induced T cell suppression (Anam et al. 2013). Wolbank et al. demonstrated a dose-dependent inhibition of peripheral blood mononuclear cell (PBMC) immune responses in mixed lymphocyte reactions (up to 66–93 % inhibition) and phytohemagglutinin activation assays (up to 67–96 % inhibition). Due to the lack of the suppression effect in a transwell system, the investigators concluded that the AEC-mediated inhibition mechanism is dependent on cell-to-cell contact.

Based on these in vitro data, a number of preclinical studies were conducted to evaluate the immunomodulatory effects of AECs in different organs, including spinal cord, liver, and lung. After transplanting hAECs into damaged spinal cords in rat or monkey models, the AECs survived over 8 weeks and provided neurotrophic factors to induce axon growth and new collateral sprouting (Sankar and Muthusamy 2003; Wu et al. 2006). Transplantation of hAECs also exerts beneficial effects in a rat stroke model (Liu et al. 2008). McDonald et al. reported that intraperitoneal injection of hAECs suppressed symptoms of multiple sclerosis and decreased central nervous system inflammation, demyelination, and axonal degeneration in the mouse brain (McDonald et al. 2011). Systemic transplantation of hAECs in a carbon tetrachloride (CCl_4)-induced liver fibrosis murine model induced reduction of hepatocyte apoptosis and decrease of inflammation and fibrosis (Manuelpillai et al. 2010).

These results suggest that hAECs have the capacity to strongly suppress immune responses, potentially induce peripheral tolerance, and reverse the ongoing inflammatory damage. It seems that in these cell transplantation studies the hAECs transiently engrafted but did not transdifferentiate in the targeted cell types and it is believed that the therapeutic effects of AECs depend on the release of trophic and anti-inflammatory molecules. However, as mentioned earlier, these cells possess differentiation potential. Due to the limitation of using animal disease models, xenogeneic incompatibility or rejection might cause transient engraftment of hAECs. Although further research is required to identify the therapeutic mechanisms of AEC transplantation, there is a possibility that future clinical applications may benefit from a synergistic effect from both immunomodulation and differentiation capabilities of hAECs.

4 The Properties of Wharton's Jelly-Derived Stem Cells

During gestation, the fetus is connected to the placenta through the umbilical cord, a tubular structure covered by amniotic epithelium that protects a vein and two arteries from compression, torsion, and bending during pregnancy and delivery by embedding them in a viscoelastic mucoid tissue matrix, called Wharton's Jelly (WJ) (Wharton 1656). WJ is composed of hyaluronic acid and proteoglycans in an aqueous solution of salts, metabolites, and plasma proteins distributed in a fine homogeneous network of microfibrils formed by various collagen isoforms. It is also a rich source of perinatal stem cells, a bridge between embryonic and adult stem cells without the limitations of either. Since WJ is typically discarded as postdelivery medical waste, the use of its cells does not pose ethical concerns. In addition, it provides an ample supply of stem cells, with up to 10^6 cells per cm/g of umbilical cord (Lu et al. 2006; Seshareddy et al. 2008; Schugar et al. 2009; Tong et al. 2011; Li and Cai 2012; Christodoulou et al. 2013; Koliakos et al. 2011) having been isolated from various regions of the cord. Among them, WJMSC—Wharton's Jelly Mesenchymal Stem Cells (also reported as UCMSCs—Umbilical Cord Matrix Stem Cells) contained in the subamniotic and intervascular matrix (McElreavey et al. 1991) are particularly appealing for cell-based therapeutic use as they are safe (Matsuzuka et al. 2010), easy to isolate and expand, and they exhibit high plasticity and proliferative ability as well as immune-evasion and -regulation capacities and antitumoral properties.

4.1 Differentiation and Clinical Potential

WJMSCs meet the minimal criteria of MSCs according to the International Society for Cellular Therapy (Dominici et al. 2006) in terms of plastic adhesion, phenotype, and tri-lineage differentiation potential. They also express several pluripotency markers such as Oct-3/4, SSEA1, SSEA3, SSEA4, nucleostemin, SOX-2, KLF4, c-MYC, LIN28, POUF1, CRYPTO, REX1, and NANOG (Fong et al. 2011; Greco et al. 2007; Conconi et al. 2011; Hsieh et al. 2010; Gao et al. 2013), albeit at a much lower level compared to ESCs, which together with high expression of several tumor suppressor genes, might explain why they are not tumorigenic. Compared to stem cells from other regions of the umbilical cord and to adult MSCs, WJMSCs show greater differentiation potential and have been shown to differentiate into osteocytes, chondrocytes, adipocytes, endothelial cells, retinal cells, skeletal and cardiac muscle cells, neurons, glia, oligodendrocytes, hepatocytes, insulin-producing cells, and germ cells (Conconi et al. 2006, 2011; Mitchell et al. 2003; Zhang et al. 2010; Campard et al. 2008; Amidi et al. 2015; Bhandari et al. 2011).

Thanks to properties such as low immunogenicity and particularly, chondrogenic and osteogenic differentiation potential, WJMSCs promise to be an interesting cell source for cartilage, bone, and osteochondral tissue engineering. Wang et al. described an elegant WJMSC-based method to fabricate an integrated construct successfully mimicking native osteochondral tissue by sandwiching a layer of undifferentiated WJMSCs between two WJMSC-derived chondrogenic and osteogenic constructs (Wang et al. 2011). More recently, transplantation of WJMSCs in a canine model of intervertebral disc (IVD) degeneration significantly increased the expression of disc ECM components such as aggrecan and type II collagen, decelerating progressive IVD degeneration (Zhang et al. 2015). Given that WJMSCs can differentiate into nucleus pulposus (NP)-like cells by coculturing them with the disc NP cells in vitro (Ruan et al. 2012), it is likely that the transplanted WJMSCs differentiated into NP-like cells in the stromal microenvironment of discs in vivo. Furthermore, the hypoimmunogenic and immunomodulatory nature of WJMSCs makes them an attractive cell source to test in the treatment of degenerative or autoimmune joint diseases such as osteoarthritis and rheumatoid arthritis (Saulnier et al. 2015; Paz-Rodriguez 2016).

WJMSCs are also of great interest for the development of regenerative approaches to liver disease. They express liver-specific markers such as albumin, alpha-fetoprotein, and glucose-6-phosphatase, as well as liver progenitor markers (e.g., DKK1, DPP4, DSG2, CX43, and CK19) and transcription factors involved in liver development (e.g., GATA4, GATA6, SOX9, and SOX17) (Buyl et al. 2014; Khodabandeh et al. 2016). WJMSCs can be differentiated in vitro into hepatocyte-like cells and retain their hypoimmunogenicity, as the differentiative process doesn't change the immunological features of the cells (Zhao et al. 2009). Transplantation of undifferentiated WJMSCs in the liver of SCID mice with partial hepatectomy induces the engrafted cells to express albumin and alpha-fetoprotein after transplantation (Campard et al. 2008). Moreover, administration of undifferentiated WJMSCs in murine models of liver fibrosis can rescue the injured livers through transdifferentiation of WJMSCs in liver-like cells, as well as through the secretion of bioactive factors and/or cytokines by undifferentiated WJMSCs to support liver repair (Tsai et al. 2009; Hammam et al. 2016). The initial reports of the clinical application of WJMSCs to liver disease appear very promising and show that the use of these cells is safe and clinically feasible. In particular, WJMSC transfusion in patients with primary biliary cirrhosis with an incomplete response to ursodeoxycholic acid improves liver function, clinical symptoms, and the quality of life of patients (Wang et al. 2013). Furthermore, WJMSC treatment improves hepatic function and ascites in decompensated liver cirrhosis patients (Zhang et al. 2012) and enhances liver function, decreases MELD score, and increases survival rates in HBV-associated acute-on-chronic liver failure patients (Shi et al. 2012).

WJMSCs are very interesting candidates for use in cardiovascular disease treatment, as well. They exhibit high expression of early cardiac transcription factors, such as Flk-1, Isl-1, and Nkx2.5 (Gao et al. 2013), they can differentiate in vitro into connexin 43-, α-cardiac actin- and Troponin T-expressing cardiomyocyte-like cells (Nartprayut et al. 2013; Lupu et al. 2011) and are able to integrate into cardiac tissue (Lin et al. 2016). Undifferentiated WJMSCs transplanted in a porcine model of acute myocardial infarction (AMI) support cardiac regeneration by transdifferentiating in cardiomyocytes and vascular endothelial cells, as well as promoting recruitment and differentiation of resident cardiac stem cells in neonatal cardiomyocytes, reducing fibrosis and apoptosis and improving ventricular remodeling and function (Zhang et al. 2013). Intracoronary administration of allogeneic WJMSCs in humans with AMI within 7 days of reperfusion therapy reduces infarct size, improves heart function, and prevents left ventricular remodeling, with no signs of immune response or tumor formation caused by the transplanted cells (Gao et al. 2015; Musialek et al. 2015). Furthermore, successful WJMSC-based heart valve leaflet fabrication was achieved and validated in large animal models (Semenov and Breymann 2011; Lanuti et al. 2015), confirming that WJMSCs have great potential for cardiovascular tissue engineering applications.

Perhaps, one of the most promising applications of WJMSCs is diabetes treatment. WJMSCs are able to differentiate into insulin producing islet-like clusters in vitro that lead to the functional recovery in diabetic rats (Chao et al. 2008; Yu et al. 2015). Undifferentiated WJMSCs achieve normoglycemia within 1 week of injection in NOD mice and are also able to prevent or delay the onset of diabetes (Hu et al. 2014). Most importantly, in addition to controlling hyperglycemia, the immunoregulatory properties of WJMSCs are able to suppress T-cell mediated autoimmune attacks on the pancreas, allowing the regeneration of the islets (Hu et al. 2014; Tsai et al. 2015). It has been shown that the use of WJMSCs for the treatment of Type 1 and Type 2 Diabetes in humans is safe, effective, and can restore islet function (Hu et al. 2013; Liu et al. 2014). Interestingly, WJMSCs also enhance healing of excisional and diabetic wounds via differentiation into keratinocytes and release of important wound healing molecules (Fong et al. 2014).

Importantly, WJMSCs have a spontaneous tendency toward a neural lineage differentiation commitment, are able to differentiate into neural and glial-like cell types in vitro, and their potential use for nerve repair in clinical applications is currently being investigated. For example, when WJMSCs were tested in vivo in the treatment of sciatic nerve axonotmesis and neurotmesis injuries using the rat model, the experiments revealed that both undifferentiated and glial-like differentiated WJMSCs boosted the recovery of sensory and motor function of the rat sciatic nerve, indicating that Wharton's Jelly may be a valuable cell source for the repair of peripheral nerve damage (Ribeiro et al. 2013). Furthermore, intrathecal injection of WJMSCs has been shown to be safe and able to delay the progression of neurologic deficits for spinocerebellar ataxia and multiple system atrophy-cerebellar type in humans, indicating their potential for the treatment of neurodegenerative disorders (Dongmei et al. 2011).

Taken together, the available data show that WJMSCs are primitive stem cells with properties bridging those of adult MSCs and ESCs, without the technical and biological limitations of the former, as well as the ethical and tumorigenic limitations of the latter. Furthermore, their ample accessibility, with over 130 million annual births worldwide, coupled with their noninvasive isolation, high expansion potential, great plasticity, nontumorigenic nature, and potent immune-privileged status, make them an ideal source for both autologous and allogeneic use in regenerative medicine applications.

4.2 Immune Properties

WJMSCs express low levels of normal (HLA-A,B and C) as well as noncanonical (HLA-E, F, and G) MHC class I antigens, and lack MHC class II (HLA-DR, DP, and DQ) as well as CD40/CD40L, CD80, CD86, and B7-DC costimulatory antigens implicated in the activation of T and B cell responses (Chen et al. 2012). Lack of expression of these molecules is probably responsible for the nonimmunogenic nature of these cells, while the expression of B7-H1, a negative regulatory molecule, contributes to suppressing T-cell proliferation. WJMSCs also suppress monocyte to dendritic cell differentiation and maturation in a contact-dependent manner, as well as induction of regulatory T cell generation, confirming the strong immunomodulatory functions of these cells (Tipnis et al. 2010). Nonclassical HLA molecules such as HLA-E and HLA-G have been associated with the induction of tolerance of NK-cells toward self-cells, as well as with the maternal tolerance to the semiallogeneic embryo. In particular, HLA-G can inhibit T-cell activation and is one of the main molecules responsible for the immune-avoidance mechanisms during embryo implantation and fetus development (Rouas-Freiss et al. 1997; Fanchin et al. 2007). Furthermore, HLA-E can inhibit NK cells and when expressed by pig cells, can alleviate human NK cell-mediated rejection of porcine xenografts (Crew et al. 2005). Mixed lymphocyte reaction experiments showed that WJMSCs do not induce, but inhibit, the proliferation of stimulated immune cells and that the inhibitory effect is dose dependent, showing that these cells have an immune suppression function. Furthermore, WJMSCs express several immunomodulatory genes, such as VEGF, TGFβ1, HGF, HMOX1, IL1β, IL-6, LIF, LGALS-1/3/8, Cox1/2, and PTGE at high levels (Chen et al. 2012). WJMSCs have more robust immunomodulatory properties than adult MSCs. For example, WJMSCs strongly attenuate mitogen-driven T-cell responses at a much lower dose range than bone marrow-derived MSCs (BMMSCs), and suppress alloantigen-driven T-cells to a greater extent compared to BMMSCs or adipose tissue-derived MSCs (Prasanna and Jahnavi 2011). WJMSCs don't elicit any immune response under xeno-transplant settings even in the absence of immune suppression. For example, human WJMSCs survived for 4 months after being transplanted in an immune competent rat model of spinal cord injury in the absence of immune suppression (Yang et al. 2008). Despite being an extremely promising cellular source for the development of allogeneic clinical applications, under certain circumstances, such as injection in an inflamed region, repeated injections in the same region, or stimulation with IFN-gamma prior to injection, WJMSCs can elicit an immune response (Cho et al. 2008), so these aspects must be carefully evaluated when considering a therapeutic application.

The ability to modulate immunological responses ranks WJMSCs as an important stem cell source for allogeneic applications, possibly not requiring HLA matching before transplantation. However,what sets these cells apart from other promising sources is their ability to maintain a sort of positional memory that allows them to continue expressing molecules with immune-modulating activity after they are extracted from WJ and expanded ex vivo (La Rocca et al. 2012). Most importantly, they are able to pass on this ability to their differentiated progeny. In fact, by contrast with BMMSCs, the immunogenicity of which increases upon differentiation, cells differentiated from WJMSCs not only retain their hypoimmunogenic status, but have been shown to express high levels of potent inhibitors of the immune response, such as IDO, HLA-G, and PGE2. Taken together, these data indicate that the immunoprivileged status of WJMSCs remains stable even after multidirectional differentiation (Zhao et al. 2009; Kalaszczynska and Ferdyn 2015).

5 Conclusion

Placenta-derived stem cells are easily accessible and do not have the limitations and ethical concerns associated with the clinical use of other pluripotent stem cells. In this chapter, we classified placenta-derived stem cells into placental and nonplacental stem cells. Among them, we focused on amniotic epithelial stem cells and Wharton's Jelly-derived stem cells and discussed their unique functional dualism.

The abundance of human placenta and the multiple therapeutic capabilities of placental stem cells make them one of the best cell sources for practical clinical translation.

Acknowledgements This work was supported by California Institute for Regenerative Medicine (CIRM) grant TR3-05488 (TM).

Disclosure of Potential Conflicts of Interest. T.M. owns stock in Stemnion, LLC. The authors have received no payment for the preparation of this manuscript and state no other financial and nonfinancial conflict of interests.

References

Akle CA, Adinolfi M, Welsh KI, Leibowitz S, McColl I (1981) Immunogenicity of human amniotic epithelial cells after transplantation into volunteers. Lancet 2:1003–1005. doi:10.1016/S0140-6736(81)91212-5

Amidi F, Ataie Nejad N, Agha Hoseini M, Nayernia K, Mazaheri Z, Yamini N, Saeednia S (2015) In vitro differentiation process of human Wharton's jelly mesenchymal stem cells to male germ cells in the presence of gonadal and non-gonadal conditioned media with retinoic acid. In Vitro Cell Dev Biol Anim 51:1093–1101. doi:10.1007/s11626-015-9929-4

Anam K, Lazdun Y, Davis PM, Banas RA, Elster EA, Davis TA (2013) Amnion-derived multipotent progenitor cells support allograft tolerance induction. Am J Transplant 13:1416–1428. doi:10.1111/ajt.12252

Banas RA, Trumpower C, Bentlejewski C, Marshall V, Sing G, Zeevi A (2008) Immunogenicity and immunomodulatory effects of amnion-derived multipotent progenitor cells. Hum Immunol 69:321–328. doi:10.1016/j.humimm.2008.04.007

Bhandari DR, Seo KW, Sun B, Seo MS, Kim HS, Seo YJ, Marcin J, Forraz N, Le Roy H, Larry D, Colin M, Kang KS (2011) The simplest method for in vitro β-cell production from human adult stem cells. Differentiation 82:144–152. doi:10.1016/j.diff.2011.06.003

Buyl K, De Kock J, Najar M, Lagneaux L, Branson S, Rogiers V, Vanhaecke T (2014) Characterization of hepatic markers in human Wharton's Jelly-derived mesenchymal stem cells. Toxicol In Vitro 28:113–119. doi:10.1016/j.tiv.2013.06.014

Campard D, Lysy PA, Najimi M, Sokal EM (2008) Native umbilical cord matrix stem cells express hepatic markers and differentiate into hepatocyte-like cells. Gastroenterology 134:833–848. doi:10.1053/j.gastro.2007.12.024

Castrechini NM, Murthi P, Gude NM, Erwich JJHM, Gronthos S, Zannettino A, Brennecke SP, Kalionis B (2010) Mesenchymal stem cells in human placental chorionic villi reside in a vascular Niche. Placenta 31:203–212. doi:10.1016/j.placenta.2009.12.006

Cetrulo CL, Cetrulo KJ (2009) Perinatal stem cells. Wiley, New York

Chao KC, Chao KF, Fu YS, Liu SH (2008) Islet-like clusters derived from mesenchymal stem cells in Wharton's jelly of the human umbilical cord for transplantation to control type 1 diabetes. PLoS One. doi:10.1371/journal.pone.0001451

Chen H, Zhang N, Li T, Guo J, Wang Z, Yang M, Gao L (2012) Human umbilical cord Wharton's jelly stem cells: immune property genes assay and effect of transplantation on the immune cells of heart failure patients. Cell Immunol 276:83–90. doi:10.1016/j.cellimm.2012.03.012

Cho PS, Messina DJ, Hirsh EL, Chi N, Goldman SN, Lo DP, Harris IR, Popma SH, Sachs DH, Huang CA (2008) Immunogenicity of umbilical cord tissue derived cells. Blood 111:430–438. doi:10.1182/blood-2007-03-078774

Christodoulou I, Kolisis FN, Papaevangeliou D, Zoumpourlis V (2013) Comparative evaluation of human mesenchymal stem cells of fetal (Wharton's Jelly) and adult (adipose tissue) origin during prolonged in vitro expansion: considerations for cytotherapy. Stem Cells Int. doi:10.1155/2013/246134

Conconi MT, Burra P, Di Liddo R, Calore C, Turetta M, Bellini S, Bo P, Nussdorfer GG, Parnigotto PP (2006) CD105(+) cells from Wharton's jelly show in vitro and in vivo myogenic differentiative potential. Int J Mol Med 18:1089–1096

Conconi MT, Di Liddo R, Tommasini M, Calore C, Parnigotto PP (2011) Phenotype and differentiation potential of stromal populations obtained from various zones of human umbilical cord: an overview. Open Tissue Eng Regen Med J 4:6–20. doi:10.2174/1875043501104010006

Crew MD, Cannon MJ, Phanavanh B, Garcia-Borges CN (2005) An HLA-E single chain trimer inhibits human NK cell reactivity towards porcine cells. Mol Immunol 42:1205–1214. doi:10.1016/j.molimm.2004.11.013

Dominici M, Le Blanc K, Mueller I, Slaper-Cortenbach I, Marini F, Krause D, Deans R, Keating A, Prockop D, Horwitz E (2006) Minimal criteria for defining multipotent mesenchymal stromal cells. The International Society for Cellular Therapy position statement. Cytotherapy 8:315–317. doi:10.1080/14653240600855905

Dongmei H, Jing L, Mei X, Ling Z, Hongmin Y, Zhidong W, Li D, Zikuan G, Hengxiang W (2011) Clinical analysis of the treatment of spinocerebellar ataxia and multiple system atrophy-cerebellar type with umbilical cord mesenchymal stromal cells. Cytotherapy 13:913–917. doi:10.3109/14653249.2011.579958

Donnelly L, Campling G (2014) Functions of the placenta. Anaesth Intensive Care Med 15:136–139. doi:10.1016/j.mpaic.2014.01.004

Fanchin R, Gallot V, Rouas-Freiss N, Frydman R, Carosella ED (2007) Implication of HLA-G in human embryo implantation. Hum Immunol 68:259–263. doi:10.1016/j.humimm.2006.11.002

Fong CY, Richards M, Manasi N, Biswas A, Bongso A (2007) Comparative growth behaviour and characterization of stem cells from human Wharton's jelly. Reprod Biomed Online 15:708–718. doi:10.1016/S1472-6483(10)60539-1

Fong CY, Chak LL, Biswas A, Tan JH, Gauthaman K, Chan WK, Bongso A (2011) Human Wharton's jelly stem cells have unique transcriptome profiles compared to human embryonic stem cells and other mesenchymal stem cells. Stem Cell Rev 7:1–16. doi:10.1007/s12015-010-9166-x

Fong CY, Tam K, Cheyyatraivendran S, Gan SU, Gauthaman K, Armugam A, Jeyaseelan K, Choolani M, Biswas A, Bongso A (2014) Human Wharton's jelly stem cells and its conditioned medium enhance healing of excisional and diabetic wounds. J Cell Biochem 115:290–302. doi:10.1002/jcb.24661

Gao LR, Zhang NK, Ding QA, Chen HY, Hu X, Jiang S, Li TC, Chen Y, Wang ZG, Ye Y, Zhu ZM (2013) Common expression of stemness molecular markers and early cardiac transcription factors in human Wharton's jelly-derived mesenchymal stem cells and embryonic stem cells. Cell Transpl 22:1883–1900. doi:10.3727/096368912X662444

Gao LR, Chen Y, Zhang NK, Yang XL, Liu HL, Wang ZG, Yan XY, Wang Y, Zhu ZM, Li TC, Wang LH, Chen HY, Chen YD, Huang CL, Qu P, Yao C, Wang B, Chen GH, Wang ZM, Xu ZY, Bai J, Lu D, Shen YH, Guo F, Liu MY, Yang Y, Ding YC, Yang Y, Tian HT, Ding QA, Li LN, Yang XC, Hu X (2015) Intracoronary infusion of Wharton's jelly-derived mesenchymal stem cells in acute myocardial infarction: double-blind, randomized controlled trial. BMC Med 13:162. doi:10.1186/s12916-015-0399-z

Greco SJ, Liu K, Rameshwar P (2007) Functional similarities among genes regulated by OCT4 in human mesenchymal and embryonic stem cells. Stem Cells 25:3143–3154. doi:10.1634/stemcells.2007-0351

Hammam OA, Elkhafif N, Attia YM, Mansour MT, Elmazar MM, Abdelsalam RM, Kenawy SA, El-Khatib AS (2016) Wharton's jelly-derived mesenchymal stem cells combined with praziquantel as a potential therapy for Schistosoma mansoni-induced liver fibrosis. Sci Rep 6:21005. doi:10.1038/srep21005

Hsieh J, Fu Y, Chang S, Tsuang Y, Wang H (2010) Functional module analysis reveals differential osteogenic and stemness potentials in human mesenchymal stem cells from bone marrow and Wharton's jelly of umbilical cord. Stem Cells Dev 19:1895–1910. doi:10.1089/scd.2009.0485

Hu J, Yu X, Wang Z, Wang F, Wang L, Gao H, Chen Y, Zhao W, Jia Z, Yan S, Wang Y (2013) Long term effects of the implantation of Wharton's jelly-derived mesenchymal stem cells from the umbilical cord for newly-onset type 1 diabetes mellitus. Endocr J 60:347–357. doi:10.1507/endocrj.EJ12-0343

Hu J, Wang Y, Wang F, Wang L, Yu X, Sun R, Wang Z, Wang L, Gao H, Fu Z, Zhao W, Yan S (2014) Effect and mechanisms of human Wharton's jelly-derived mesenchymal stem cells on type 1 diabetes in NOD model. Endocrine 48:124–134. doi:10.1007/s12020-014-0219-9

Hunt JS, Petroff MG, McIntire RH, Ober C (2005) HLA-G and immune tolerance in pregnancy. FASEB J 19:681–693. doi:10.1096/fj.04-2078rev

Igura K, Zhang X, Takahashi K, Mitsuru A, Yamaguchi S, Takashi TA (2004) Isolation and characterization of mesenchymal progenitor cells from chorionic villi of human placenta. Cytotherapy 6:543–553. doi:10.1080/14653240410005366

Ilancheran S, Michalska A, Peh G, Wallace EM, Pera M, Manuelpillai U (2007) Stem cells derived from human fetal membranes display multilineage differentiation potential. Biol Reprod 77:577–588. doi:10.1095/biolreprod.106.055244

Kalaszczynska I, Ferdyn K (2015) Wharton's jelly derived mesenchymal stem cells: future of regenerative medicine? Recent findings and clinical significance. Biomed Res Int. doi:10.1155/2015/430847

Kara RJ, Bolli P, Karakikes I, Matsunaga I, Tripodi J, Tanweer O, Altman P, Shachter NS, Nakano A, Najfeld V, Chaudhry HW (2012) Fetal cells traffic to injured maternal myocardium and undergo cardiac differentiation. Circ Res 110:82–93. doi:10.1161/CIRCRESAHA.111.249037

Khodabandeh Z, Vojdani Z, Talaei-Khozani T, Jaberipour M, Hosseini A, Bahmanpour S (2016) Comparison of the expression of hepatic genes by human Wharton's Jelly Mesenchymal stem cells cultured in 2D and 3D Collagen culture systems. Iran J Med Sci 41:28–36

Khosrotehrani K, Bianchi DW (2005) Multi-lineage potential of fetal cells in maternal tissue: a legacy in reverse. J Cell Sci 118:1559–1563. doi:10.1242/jcs.02332

Kim MJ, Shin KS, Jeon JH, Lee DR, Shim SH, Kim JK, Cha DH, Yoon TK, Kim GJ (2011) Human chorionic-plate-derived mesenchymal stem cells and Wharton's jelly-derived mesenchymal stem cells: a comparative analysis of their potential as placenta-derived stem cells. Cell Tissue Res 346:53–64. doi:10.1007/s00441-011-1249-8

Koliakos I, Tsagias N, Karagiannis V (2011) Mesenchymal cells isolation from Wharton's jelly, in perspective to clinical applications. J Biol Res 16:194–201

Kranz A, Wagner D-C, Kamprad M, Scholz M, Schmidt UR, Nitzsche F, Aberman Z, Emmrich F, Riegelsberger U-M, Boltze J (2010) Transplantation of placenta-derived mesenchymal stromal cells upon experimental stroke in rats. Brain Res 1315:128–136. doi:10.1016/j.brainres.2009.12.001

La Rocca G, Corrao S, Lo Iacono M, Corsello T, Farina F, Anzalone R (2012) Novel immunomodulatory markers expressed by human WJ-MSC: an updated review in regenerative and reparative medicine. Open Tissue Eng Regen Med J 5:50–58. doi:10.2174/1875043501205010050

Lange-Consiglio A, Tassan S, Corradetti B, Meucci A, Perego R, Bizzaro D, Cremonesi F (2013) Investigating the efficacy of amnion-derived compared with bone marrow-derived mesenchymal stromal cells in equine tendon and ligament injuries. Cytotherapy 15:1011–1020. doi:10.1016/j.jcyt.2013.03.002

Lanuti P, Serafini F, Pierdomenico L, Simeone P, Bologna G, Ercolino E, Di Silvestre S, Guarnieri S, Canosa C, Impicciatore GG, Chiarini S, Magnacca F, Mariggiò MA, Pandolfi A, Marchisio M, Di Giammarco G, Miscia S (2015) Human mesenchymal stem cells reendothelialize porcine heart valve scaffolds: novel perspectives in heart valve tissue engineering. Biores Open Access 4:288–297. doi:10.1089/biores.2015.0019

Li DR, Cai JH (2012) Methods of isolation, expansion, differentiating induction and preservation of human umbilical cord mesenchymal stem cells. Chin Med J (Engl) 125:4504–4510. doi:10.3760/cma.j.issn.0366-6999.2012.24.032

Lin YL, Chen CP, Lo CM, Wang HS (2016) Stiffness-controlled three-dimensional collagen scaffolds for differentiation of human Wharton's jelly mesenchymal stem cells into cardiac progenitor cells. J Biomed Mater Res A. doi:10.1002/jbm.a.35762

Liu T, Wu J, Huang Q, Hou Y, Jiang Z, Zang S, Guo L (2008) Human amniotic epithelial cells ameliorate behavioral dysfunction and reduce infarct size in the rat middle cerebral artery occlusion model. Shock 29:603–611. doi:10.1097/SHK.0b013e318157e845

Liu X, Zheng P, Wang X, Dai G, Cheng H, Zhang Z, Hua R, Niu X, Shi J, An Y (2014) A preliminary evaluation of efficacy and safety of Wharton's jelly mesenchymal stem cell transplantation in patients with type 2 diabetes mellitus. Stem Cell Res Ther 5:57. doi:10.1186/scrt446

Lu LL, Liu YJ, Yang SG, Zhao QJ, Wang X, Gong W, Han ZB, Xu ZS, Lu YX, Liu D, Chen ZZ, Han ZC (2006) Isolation and characterization of human umbilical cord mesenchymal stem cells with hematopoiesis-supportive function and other potentials. Haematologica 91:1017–1026

Lupu M, Khalil M, Andrei E, Iordache F, Pfannkuche K, Neef K, Georgescu A, Buzila C, Brockmeier K, Maniu H, Hescheler J (2011) Integration properties of Wharton's jelly-derived novel mesenchymal stem cells into ventricular slices of murine hearts. Cell Physiol Biochem 28:63–76. doi:10.1159/000331714

Magatti M, De Munari S, Vertua E, Gibelli L, Wengler GS (2008) Human amnion mesenchyme harbors cells with allogeneic T-cell suppression and stimulation capabilities. Stem Cells 26:182–192. doi:10.1634/stemcells.2007-0491

Magatti M, De Munari S, Vertua E, Nassauto C, Albertini A, Wengler GS, Parolini O (2009) Amniotic mesenchymal tissue cells inhibit dendritic cell differentiation of peripheral blood and amnion resident monocytes. Cell Transplant 18:899–914. doi:10.3727/096368909X471314

Manuelpillai U, Tchongue J, Lourensz D, Vaghjiani V, Samuel CS, Liu A, Williams ED, Sievert W (2010) Transplantation of human amnion epithelial cells reduces hepatic fibrosis in immunocompetent CCl_4-treated mice. Cell Transplant 19:1157–1168. doi:10.3727/096368910X504496

Marcus AJ, Coyne TM, Black IB, Woodbury D (2008) Fate of amnion-derived stem cells transplanted to the fetal rat brain: migration, survival and differentiation. J Cell Mol Med 12:1256–1264. doi:10.1111/j.1582-4934.2008.00180.x

Marongiu F, Gramignoli R, Sun Q, Tahan V, Miki T, Dorko K, Ellis E, Strom SC (2010) Isolation of amniotic mesenchymal stem cells. Curr Protoc Stem Cell Biol Chapter 1:Unit 1E.5. doi:10.1002/9780470151808.sc01e05s12

Marongiu F, Gramignoli R, Dorko K, Miki T, Ranade AR, Paola Serra M, Doratiotto S, Sini M, Sharma S, Mitamura K, Sellaro TL, Tahan V, Skvorak KJ, Ellis ECS, Badylak SF, Davila JC, Hines R, Laconi E, Strom SC (2011) Hepatic differentiation of amniotic epithelial cells. Hepatology 53:1719–1729. doi:10.1002/hep.24255

Matsuzuka T, Rachakatla RS, Doi C, Maurya DK, Ohta N, Kawabata A, Pyle MM, Pickel L, Reischman J, Marini F, Troyer D, Tamura M (2010) Human umbilical cord matrix-derived stem cells expressing interferon-β gene significantly attenuate bronchioloalveolar carcinoma xenografts in SCID mice. Lung Cancer 70:28–36. doi:10.1016/j.lungcan.2010.01.003

McDonald C, Siatskas C, Bernard CAC (2011) The emergence of amnion epithelial stem cells for the treatment of Multiple Sclerosis. Inflamm Regen 31:256–271. doi:10.2492/inflammregen.31.256

McElreavey KD, Irvine AI, Ennis KT, McLean WH (1991) Isolation, culture and characterisation of fibroblast-like cells derived from the Wharton's jelly portion of human umbilical cord. Biochem Soc Trans 19:29S

Miki T (2011) Amnion-derived stem cells: in quest of clinical applications. Stem Cell Res Ther 2:25. doi:10.1186/scrt66

Miki T (2016) A rational strategy for the use of amniotic epithelial stem cell therapy for liver diseases. Stem Cells Transl Med 5:405–409. doi:10.5966/sctm.2015-0304

Miki T, Strom SC (2006) Amnion-derived pluripotent/multipotent stem cells. Stem Cell Rev 2:133–142. doi:10.1385/SCR:2:2:133

Miki T, Lehmann T, Cai H, Stolz DB, Strom SC (2005) Stem cell characteristics of amniotic epithelial cells. Stem Cells 23:1549–1559. doi:10.1634/stemcells.2004-0357

Mitchell KE, Weiss ML, Mitchell BM, Martin P, Davis D, Morales L, Helwig B, Beerenstrauch M, Abou-Easa K, Hildreth T, Troyer D, Medicetty S (2003) Matrix cells from Wharton's jelly form neurons and glia. Stem Cells 21:50–60. doi:10.1634/stemcells.21-1-50

Moodley Y, Vaghjiani V, Chan J, Baltic S, Ryan M, Tchongue J, Samuel CS, Murthi P, Parolini O, Manuelpillai U (2013) Anti-inflammatory effects of adult stem cells in sustained lung injury: a comparative study. PLoS One 8:e69299. doi:10.1371/journal.pone.0069299

Murphy SV, Lim R, Heraud P, Cholewa M, Le Gros M, de Jonge MD, Howard DL, Paterson D, McDonald C, Atala A, Jenkin G, Wallace EM (2012) Human amnion epithelial cells induced to express functional cystic fibrosis transmembrane conductance regulator. PLoS One. doi:10.1371/journal.pone.0046533

Musialek P, Mazurek A, Jarocha D, Tekieli L, Szot W, Kostkiewicz M, Banys RP, Urbanczyk M, Kadzielski A, Trystula M, Kijowski J, Zmudka K, Podolec P, Majka M (2015) Myocardial regeneration strategy using Wharton's jelly mesenchymal stem cells as an off-the-shelf "unlimited" therapeutic agent: results from the Acute Myocardial Infarction First-in-Man Study. Postepy Kardiol Interwencyjnej 11:100–107. doi:10.5114/pwki.2015.52282

Nakajima T, Enosawa S, Mitani T, Li XK, Suzuki S, Amemiya H, Koiwai O, Sakuragawa N (2001) Cytological examination of rat amniotic epithelial cells and cell transplantation to the liver. Cell Transplant 10:423–427

Nartprayut K, U-Pratya Y, Kheolamai P, Manochantr S, Chayosumrit M, Issaragrisil S, Supokawej A (2013) Cardiomyocyte differentiation of perinatally-derived mesenchymal stem cells. Mol Med Rep 7:1465–1469. doi:10.3892/mmr.2013.1356

Nguyen Huu S, Dubernard G, Aractingi S, Khosrotehrani K (2006) Feto-maternal cell trafficking: a transfer of pregnancy associated progenitor cells. Stem Cell Rev 2:111–116. doi:10.1385/SCR:2:2:111

O'Donoghue K, Choolani M, Chan J, de la Fuente J, Kumar S, Campagnoli C, Bennett PR, Roberts IAG, Fisk NM (2003) Identification of fetal mesenchymal stem cells in maternal blood: implications for non-invasive prenatal diagnosis. Mol Hum Reprod 9:497–502. doi:10.1093/molehr/gag063

Okawa H, Okuda O, Arai H, Sakuragawa N, Sato K (2001) Amniotic epithelial cells transform into neuron-like cells in the ischemic brain. Neuroreport 12:4003–4007

Paz-Rodriguez J (2016) Feasibility Study of Umbilical Cord Tissue Derived Mesenchymal Stem Cells (UC-MSC) in Disease Modifying Anti-Rheumatic Drugs (DMARD) Resistant Rheumatoid Arthritis. ClinicalTrials.gov Identifier: NCT01985464

Prasanna SJ, Jahnavi VS (2011) Wharton's jelly mesenchymal stem cells as off-the-shelf cellular therapeutics: a closer look into their regenerative and immunomodulatory properties. Open Tissue Eng Regen Med J 4:28–38. doi:10.2174/1875043501104010028

Ribeiro J, Gartner A, Pereira T, Gomes R, Lopes MA, Gonçalves C, Varejão A, Luís AL, Maurício AC (2013) Perspectives of employing mesenchymal stem cells from the Wharton's jelly of

the umbilical cord for peripheral nerve repair. Int Rev Neurobiol 108:79–119. doi:10.1016/B978-0-12-410499-0.00004-6
Rouas-Freiss N, Gonçalves RM, Menier C, Dausset J, Carosella ED (1997) Direct evidence to support the role of HLA-G in protecting the fetus from maternal uterine natural killer cytolysis. Proc Natl Acad Sci U S A 94:11520–11525. doi:10.1073/pnas.94.21.11520
Ruan D, Zhang Y, Wang D, Zhang C, Wu J, Wang C, Shi Z, Xin H, Xu C, Li H, He Q (2012) Differentiation of human Wharton's jelly cells toward nucleus pulposus-like cells after coculture with nucleus pulposus cells in vitro. Tissue Eng Part A 18:167–175. doi:10.1089/ten.TEA.2011.0186
Sankar V, Muthusamy R (2003) Role of human amniotic epithelial cell transplantation in spinal cord injury repair research. Neuroscience 118:11–17. doi:10.1016/S0306-4522(02)00929-6
Saulnier N, Viguier E, Perrier-Groult E, Chenu C, Pillet E, Roger T, Maddens S, Boulocher C (2015) Intra-articular administration of xenogeneic neonatal Mesenchymal Stromal Cells early after meniscal injury down-regulates metalloproteinase gene expression in synovium and prevents cartilage degradation in a rabbit model of osteoarthritis. Osteoarthr Cartil 23:122–133. doi:10.1016/j.joca.2014.09.007
Schugar RC, Chirieleison SM, Wescoe KE, Schmidt BT, Askew Y, Nance JJ, Evron JM, Peault B, Deasy BM (2009) High harvest yield, high expansion, and phenotype stability of CD146 mesenchymal stromal cells from whole primitive human umbilical cord tissue. J Biomed Biotechnol. doi:10.1155/2009/789526
Semenov OV, Breymann C (2011) Mesenchymal stem cells derived from Wharton's Jelly and their potential for cardio-vascular tissue engineering. Open Tissue Eng Regen Med J 4:64–71. doi:10.2174/1875043501104010064
Seshareddy K, Troyer D, Weiss ML (2008) Method to isolate mesenchymal-like cells from wharton's jelly of umbilical cord. Methods Cell Biol 86:101–119. doi:10.1016/S0091-679X(08)00006-X
Shi M, Zhang Z, Xu R, Lin H, Fu J, Zou Z, Zhang A, Shi J, Chen L, Lv S, He W, Geng H, Jin L, Liu Z, Wang FS (2012) Human mesenchymal stem cell transfusion is safe and improves liver function in acute-on-chronic liver failure patients. Stem Cells Transl Med 1:725–731. doi:10.5966/sctm.2012-0034
Taghizadeh RR, Cetrulo KJ, Cetrulo CL (2011) Wharton's Jelly stem cells: future clinical applications. Placenta. doi:10.1016/j.placenta.2011.06.010
Takahashi N, Enosawa S, Mitani T, Lu H, Suzuki S, Amemiya H, Amano T, Sakuragawa N (2002) Transplantation of amniotic epithelial cells into fetal rat liver by in utero manipulation. Cell Transplant 11:443–449
Takashima S, Ise H, Zhao P, Akaike T, Nikaido T (2004) Human amniotic epithelial cells possess hepatocyte-like characteristics and functions. Cell Struct Funct 29:73–84. doi:10.1247/csf.29.73
Tamagawa T, Ishiwata I, Ishikawa H, Nakamura Y (2008) Induced in-vitro differentiation of neural-like cells from human amnion-derived fibroblast-like cells. Hum Cell 21:38–45. doi:10.1111/j.1749-0774.2008.00049.x
Tamagawa T, Ishiwata I, Sato K, Nakamura Y (2009) Induced in vitro differentiation of pancreatic-like cells from human amnion-derived fibroblast-like cells. Hum Cell 22:55–63. doi:10.1111/j.1749-0774.2009.00069.x
Tipnis S, Viswanathan C, Majumdar AS (2010) Immunosuppressive properties of human umbilical cord-derived mesenchymal stem cells: role of B7-H1 and IDO. Immunol Cell Biol 88:795–806. doi:10.1038/icb.2010.47
Tong CK, Vellasamy S, Tan BC, Abdullah M, Vidyadaran S, Seow HF, Ramasamy R (2011) Generation of mesenchymal stem cell from human umbilical cord tissue using a combination enzymatic and mechanical disassociation method. Cell Biol Int 35:221–226. doi:10.1042/CBI20100326
Troyer DL, Weiss ML (2008) Wharton's jelly-derived cells are a primitive stromal cell population. Stem Cells 26:591–599. doi:10.1634/stemcells.2007-0439
Tsai PC, Fu TW, Chen YMA, Ko TL, Chen TH, Shih YH, Hung SC, Fu YS (2009) The therapeutic potential of human umbilical mesenchymal stem cells from Wharton's jelly in the treatment of rat liver fibrosis. Liver Transpl 15:484–495. doi:10.1002/lt.21715

Tsai PJ, Wang HS, Lin GJ, Chou SC, Chu TH, Chuan WT, Lu YJ, Weng ZC, Su CH, Hsieh PS, Sytwu HK, Lin CH, Chen TH, Shyu JF (2015) Undifferentiated Wharton's jelly mesenchymal stem cell transplantation induces insulin-producing cell differentiation and suppression of T-cell-mediated autoimmunity in nonobese diabetic mice. Cell Transplant 24:1555–1570. doi: 10.3727/096368914X683016

Wang H-S, Hung S-C, Peng S-T, Huang C-C, Wei H-M, Guo Y-J, Fu Y-S, Lai M-C, Chen C-C (2004) Mesenchymal stem cells in the Wharton's jelly of the human umbilical cord. Stem Cells 22:1330–1337. doi:10.1634/stemcells.2004-0013

Wang L, Zhao L, Detamore MS (2011) Human umbilical cord mesenchymal stromal cells in a sandwich approach for osteochondral tissue engineering. J Tissue Eng Regen Med 5:712–721. doi:10.1002/term.370

Wang L, Li J, Liu H, Li Y, Fu J, Sun Y, Xu R, Lin H, Wang S, Lv S, Chen L, Zou Z, Li B, Shi M, Zhang Z, Wang FS (2013) A pilot study of umbilical cord-derived mesenchymal stem cell transfusion in patients with primary biliary cirrhosis. J Gastroenterol Hepatol 28(Suppl 1):85–92. doi:10.1111/jgh.12029

Wang A, Brown EG, Lankford L, Keller BA, Pivetti CD, Sitkin NA, Beattie MS, Bresnahan JC, Farmer DL (2015) Placental mesenchymal stromal cells rescue ambulation in ovine myelomeningocele. Stem Cells Transl Med 4:659–669. doi:10.5966/sctm.2014-0296

Wei JP, Zhang TS, Kawa S, Aizawa T, Ota M, Akaike T, Kato K, Konishi I, Nikaido T (2003) Human amnion-isolated cells normalize blood glucose in streptozotocin-induced diabetic mice. Cell Transplant 12:545–552

Wharton T (1656) Adenographia: sive glandularum totius corporis descriptio. Typis J.G. Impensis Authoris, Londini

Wu Z, Hui G, Lu Y, Wu X, Guo L (2006) Transplantation of human amniotic epithelial cells improves hindlimb function in rats with spinal cord injury. Chin Med J (Engl) 119:2101–2107

Xue S, Chen C, Dong W, Hui G, Liu T, Guo L (2012) Therapeutic effects of human amniotic epithelial cell transplantation on double-transgenic mice co-expressing APPswe and PS1ΔE9-deleted genes. Sci China Life Sci 55:132–140. doi:10.1007/s11427-012-4283-1

Yang CC, Shih YH, Ko MH, Hsu SY, Cheng H, Fu YS (2008) Transplantation of human umbilical mesenchymal stem cells from Wharton's jelly after complete transection of the rat spinal cord. PLoS One. doi:10.1371/journal.pone.0003336

Yu YB, Bian JM, Gu DH (2015) Transplantation of insulin-producing cells to treat diabetic rats after 90% pancreatectomy. World J Gastroenterol 21:6582–6590

Zhang HT, Fan J, Cai YQ, Zhao SJ, Xue S, Lin JH, Jiang XD, Xu RX (2010) Human Wharton's jelly cells can be induced to differentiate into growth factor-secreting oligodendrocyte progenitor-like cells. Differentiation 79:15–20. doi:10.1016/j.diff.2009.09.002

Zhang Z, Lin H, Shi M, Xu R, Fu J, Lv J, Chen L, Lv S, Li Y, Yu S, Geng H, Jin L, Lau GKK, Wang FS (2012) Human umbilical cord mesenchymal stem cells improve liver function and ascites in decompensated liver cirrhosis patients. J Gastroenterol Hepatol 27(suppl 2):112–120. doi:10.1111/j.1440-1746.2011.07024.x

Zhang W, Liu XC, Yang L, Zhu DL, Zhang YD, Chen Y, Zhang HY (2013) Wharton's jelly-derived mesenchymal stem cells promote myocardial regeneration and cardiac repair after miniswine acute myocardial infarction. Coron Artery Dis 24:549–558. doi:10.1097/MCA.0b013e3283640f00

Zhang Y, Tao H, Gu T, Zhou M, Jia Z, Jiang G, Chen C, Han Z, Xu C, Wang D, He Q, Ruan D (2015) The effects of human Wharton's jelly cell transplantation on the intervertebral disc in a canine disc degeneration model. Stem Cell Res Ther 6:154. doi:10.1186/s13287-015-0132-z

Zhao P, Ise H, Hongo M, Ota M, Konishi I, Nikaido T (2005) Human amniotic mesenchymal cells have some characteristics of cardiomyocytes. Transplantation 79:528–535

Zhao Q, Ren H, Li X, Chen Z, Zhang X, Gong W, Liu Y, Pang T, Han ZC (2009) Differentiation of human umbilical cord mesenchymal stromal cells into low immunogenic hepatocyte-like cells. Cytotherapy 11:414–426. doi:10.1080/14653240902849754

Immunomodulatory Properties of Perinatal Tissue-Derived Mesenchymal Stem Cells

Seyed Mahmoud Hashemi and Sara Soudi

1 Immunogenicity and Immunomodulatory Properties of Wharton's Jelly-Derived Mesenchymal Stem Cells (WJ-MSCs)

The umbilical cord contains two arteries and a vein and a mucilaginous proteoglycan-rich connective tissue known as Wharton's Jelly that surrounds the umbilical vessels and covered by amniotic epithelium (Taghizadeh et al. 2011). MSCs can be isolated from the different compartments of the umbilical cord (Karahuseyinoglu et al. 2007; Troyer and Weiss 2008). Stem cells have been reported in umbilical cord blood, the Wharton's jelly, subendothelial layer of the umbilical vein, and in other layers of umbilical vessels' perivascular region (Fong et al. 2007). WJ-MSCs are primitive mesenchymal cells that trapped in the connective tissue matrix through the developing cord, during embryogenesis (Taghizadeh et al. 2011). WJ-MSCs have been isolated from different regions: the perivascular compartment surrounding the blood vessels, the intervascular zone, and the subamnion (Bongso and Fong 2013). However, derivation protocol for WJ-MSCs has not been standardized.

The phenotype of WJ-MSCs appears to be similar to bone marrow stromal and other MSCs. WJ-MSCs are negative for CD34, CD45, CD14, CD33, CD56, CD31, and

---

The original version of this chapter was revised. An erratum to this chapter can be found at DOI 10.1007/978-3-319-46410-7_12

S.M. Hashemi (✉)
Department of Immunology, School of Medicine, Shahid Beheshti University of Medical Sciences, Tehran, Iran
e-mail: smmhashemi@sbmu.ac.ir

S. Soudi (✉)
Department of Immunology, Faculty of Medical Sciences, Tarbiat Modares University, Tehran, Iran
e-mail: soudi@modares.ac.ir

B. Arjmand (ed.), *Perinatal Tissue-Derived Stem Cells*, Stem Cell Biology and Regenerative Medicine, DOI 10.1007/978-3-319-46410-7_2

human leukocyte antigen (HLA) class II and positive for CD73, CD90, CD105, CD10, CD13, CD29, CD44, CD146, CD271, and HLA-class I (Wang et al. 2004; Weiss et al. 2008; Subramanian et al. 2015). Immunogenicity WJ-MSCs have been characterized both in vitro and in vivo. Human UC-MSCs as well as WJ-MSCs do not express (HLA)-DR and the co-stimulatory molecules, CD40, CD80, and CD86 that are required for T cell activation (Weiss et al. 2008; Tipnis et al. 2010). HLA-DR expression increased after in vitro interferon-γ (IFN-γ) treatment. However, no significant change in the expression of co-stimulatory molecules was observed (Tipnis et al. 2010). Immunogenicity of human WJ-MSCs has been assessed by in vitro assays including mixed lymphocyte reaction (MLR). The results of Weiss et al. who assessed the effect of WJ-MSCs on one- and two-way MLR assays showed that they do not stimulate T cell proliferation in a one-way MLR, and that they inhibit the proliferation of stimulated T cells in a two-way MLR (Weiss et al. 2008). The immunogenicity of human WJ-MSCs has been reported to be lower than human BM-MSCs. In vitro activation of allogeneic lymphocytes or peripheral blood by human BM-MSCs was significantly stronger than WJ-MSCs (Prasanna et al. 2010; Deuse et al. 2011). In vivo immunogenicity of WJ-MSCs has been assessed by allogeneic and xenogeneic transplantation. WJ-MSCs has been reported to survive in vivo after xenogeneic and allogeneic transplantation.

It has been reported that xenogeneic in vivo immune activation of BM-MSCs was significantly stronger than WJ-MSCs. Although both BM-MSCs and umbilical cord lining MSCs are recognized by allogeneic and xenogeneic lymphocytes, umbilical cord lining MSCs are less immunogenic and were more slowly rejected in immunocompetent mice (Deuse et al. 2011).

After xenotransplantation of pig umbilical cord matrix MSCs into rat brain the cells engraft and proliferate without requiring immune suppression (Medicetty et al. 2004). In another study, human WJ-MSCs survived for 16 weeks in the spinal cord of immune competent rats in the absence of any immune suppressive drugs (Yang et al. 2008). In a recent study, the effects of intra-hippocampal transplantation of human WJ-MSCs on rat pilocarpine-induced epilepsy was evaluated (Huang et al. 2015). In addition to their effects in the central nervous system, xenotransplantation of human WJ-MSCs was reported in rat models of peritoneal fibrosis (Fan et al. 2016) and carbon tetrachloride (CCl4)-induced liver fibrosis (Tsai et al. 2009). These results indicate that human WJ-MSCs are a good stem cell source for xenotransplantation.

WJ-MSCs are also capable of immune suppression and immune avoidance similar to other types of MSCs. Immunomodulatory properties of WJ-MSCs are mediated by soluble factors such as cytokines and immunosuppressive molecules. It has been shown that WJ-MSCs secreted a number of soluble suppressive cytokines such as transforming growth factor-beta (TGF-β), insulin like growth factor (IGF), platelet-derived growth factor (PDGF), epidermal growth factor (EGF), hepatocyte growth factor (HGF), leukemia inhibitory factor (LIF), and interleukin (IL)-10 (Liu et al. 2012; Wang et al. 2010a; Choi et al. 2013a). WJ-MSCs compared with MSC from other sources produce large amounts of IL-10, higher levels of TGF-β and HLA-G.

In addition, PGE2, indoleamine 2, 3-dioxygenase (IDO), and NO have been reported to have immunoregulative functions in different types of MSCs. However, blocking experiment indicated that PGE2 was more effective than TGF-β, IDO, and

NO in immunosuppressive effects of WJ-MSCs (Chen et al. 2010; Wang et al. 2010b; Choi et al. 2013b; Donders et al. 2015). In addition to immunomodulatory agents, WJ-MSCs have been reported to secrete angiogenic and wound healing promoting factors such as TGF-β, vascular endothelial growth factor (VEGF), PDGF, IGF, IL-6, and IL-8 (Choi et al. 2013b). Furthermore, WJ-MSCs express vascular endothelial growth factor (VEGF) and IL-6, which have been shown to modulate differentiation of lymphoid precursors and differentiation of monocytes to dendritic cells (Weiss et al. 2008). Recent studies suggest that immunomodulatory properties of WJ-MSCs were enhanced upon stimulation with proinflammatory cytokines, IFN-γ, TNF-α, and IL-1b (Donders et al. 2015; Prasanna et al. 2010; Tipnis et al. 2010). Moreover, IFN-γ and IL-1β produced by activated peripheral blood mononuclear cell (PBMC) upregulated the expression of cyclooxygenase-2 (COX-2) and the production of PGE2 by human umbilical cord mesenchymal stem cells (hUCMSCs) (Chen et al. 2010). WJ-MSCs has suppressive effects on differentiation, proliferation, and function of immune cells such as T cells, dendritic cells (DC), and NK cells via contact-dependent mechanisms as well as through soluble molecules. WJ-MSCs are able to inhibit polyclonal T cell proliferation. They can functionally inhibit IFN-γ production by activated T cells and induce IL-10 secretion as well as induction of regulatory T cell (Treg) generation (Donders et al. 2015; Tipnis et al. 2010; Chen et al. 2010; Zhou et al. 2011).

It has been reported that MSCs induced CD4+ CD25+ FOXP3+ regulatory T cells after in vitro coculture with naïve T cells (Yousefi et al. 2016) and cell contact is more effective than soluble mediators.

Moreover, in vivo studies reveal that WJ-MSCs increasing the frequency of Treg cells (Tregs) and reestablishing the balance between Th1/Th2 and Th17/Treg-related cytokines (Alunno et al. 2014; Sun et al. 2010). Several studies reported that WJ-MSCs inhibit differentiation, maturation, and functionality of DCs. WJ-MSCs reduce the expression of HLA-DR, CD80, and CD83 and resulted in impaired allostimulatory ability of DCs (Donders et al. 2015; Saeidi et al. 2013; Tipnis et al. 2010). In several experimental models, such as type 1 diabetes, myocardial infarction, and Parkinson's disease, severe and refractory systemic lupus erythematosus, in vivo immunomodulatory, and anti-inflammatory effects of WJ-MSCs have been investigated (Chao et al. 2008; López et al. 2013; Wu et al. 2007; Sun et al. 2010). Low immunogenicity and immunomodulatory properties of WJ-MSCs make it promising to use in allogeneic clinical applications in inflammatory and autoimmune diseases (Fig. 1).

2 Immunomodulatory Properties of Umbilical Cord Blood-Derived Mesenchymal Stem Cells (UCB-MSCs)

Umbilical cord blood has been accepted as a well-established source for hematopoietic stem cells. However, it is still controversial whether MSCs can be isolated from cord blood.

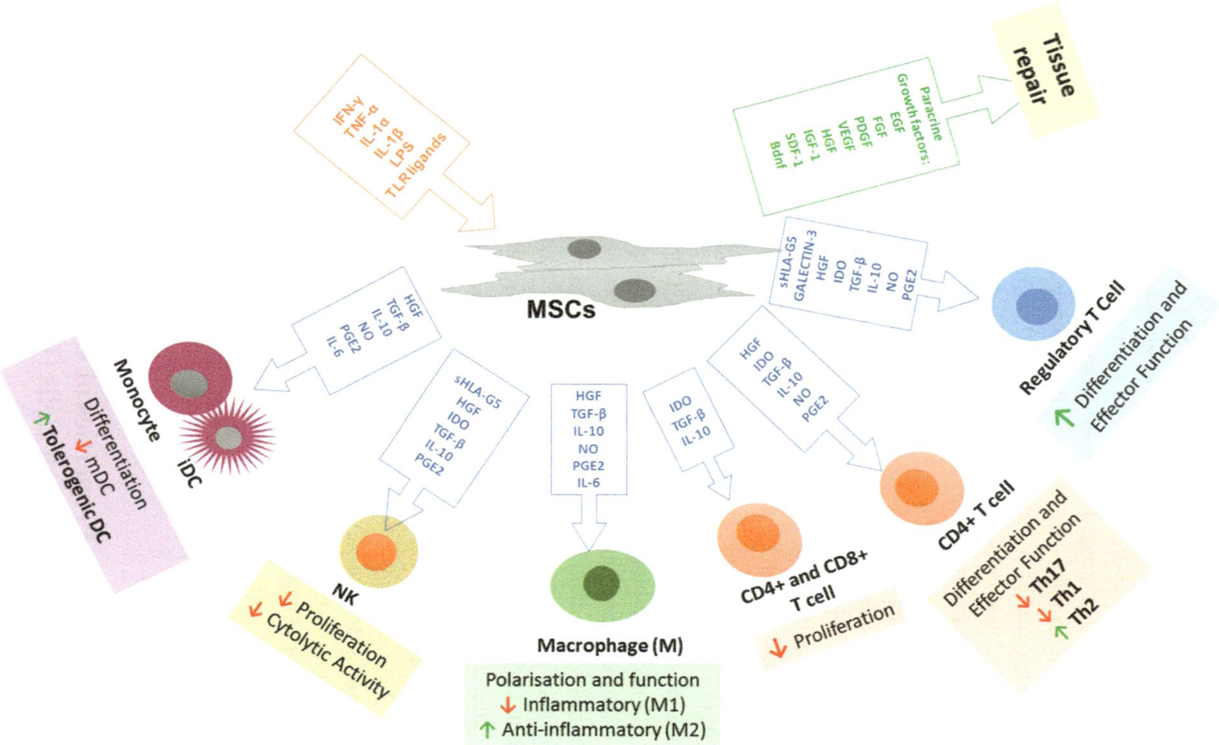

Fig. 1 Immunomodulatory effects of mesenchymal stem cells on differentiation and function of different types of immune cells such as helper T cell subsets, dendritic cells (DCs), regulatory T cells (T), natural killer (NK) cells, monocytes, and macrophages. Immunomodulatory effects of MSCs is dependent on cell–cell contact and soluble factors released by MSCs. *HGF* hepatocyte growth factor. *iDC* immature dendritic cell, *IDO* indoleamine 2, 3-dioxygenase, *IL* interleukin, *mDC* mature dendritic cell, *NO* nitric oxide, *PGE2* prostaglandin E2, *TGF-β* transforming growth factor β

The presence of MSCs in UCB has been reported in some studies (Lee et al. 2004; Bieback et al. 2004; Heo et al. 2016) whereas others have suggested that the UCB is not a rich source of MSCs due to very low frequency and MSC isolation protocol (Flynn et al. 2007; Secco et al. 2008; Mareschi et al. 2001; Wexler et al. 2003).

The cells were positive for CD29, CD44, CD73, CD90, and CD10 whereas MSCs were negative for CD14, CD31, CD34, CD45, and CD106, which are known markers of hematopoietic and endothelial cells (Lee et al. 2004; Bieback et al. 2004; Liu and Hwang 2005; Heo et al. 2016).

The cytokine expression profile of UCB-MSCs has been reported to be similar to that of BM-MSCs, except that UCB-MSCs expressed IL-12 but not G-CSF (Liu and Hwang 2005).

Using cytokine protein array Liu and Hwang et al. reported that UCB-MSCs produced cytokines including proinflammatory: IL-1b, IL-6; anti-inflammatory: TGF-β2, TGF-β3, MIF, LIF; growth factor: GM-CSF, VEGF, FGF-4, FGF-7, FGF-9, PIGF, oncostatin M; growth factor receptor: IGFBP-1, IGFBP-2, IGFBP-3, IGFBP-4; chemokines: GRO, IL-8, MCP-1, MIP-3a, PARC, IP10, ENA-78, GCP-2, osteoprotegerin; TIMP-1, TIMP-2 the natural inhibitors of matrix metalloproteinases (Liu and Hwang 2005). IL-6, IL-8 and TIMP-1, TIMP-2 are abundant CB-MSCs cytokines (Hwang et al. 2009; Flynn et al. 2007).

3 Immunogenicity and Immunomodulatory Properties of Amniotic Membrane-Derived MSCs (AM-MSCs)

The innermost layer that surrounded the embryo is amniotic membrane (AM) that is a fetal component of extra embryonic membranes. This multilayer membrane with 0.02–0.5 mm thickness has diverse clinical application because of both its physical and cellular structure. The physical aspect of AM application is related to an integrated translucent avascular membrane which provides a permeable barrier with high elasticity that resists against proteolytic factors and fractional forces. Basement membrane proteoglycans, laminins, different types of collagens, and cytoskeletal proteins are responsible for these physical properties. With regards to these characteristics AM is used in general surgery for treatment of corneal, conjunctival and limbal lesions, reconstitution of burned skins, and wound healing. The second aspect of AM application is its cellular component composed of two main cellular compartments which are separated by basement membrane. The inner layer adjacent to amniotic fluid (Rennie et al. 2012a; Mamede et al. 2012; Danforth and Hull 1958) is amniotic epithelial cells (AEC) and the outer layer of AM is amniotic mesenchymal stromal cells that according to agreement of "International Placenta Stem Cell Society" called amniotic mesenchymal stem cell (AMSC) (Parolini et al. 2008, 2009). Both of them are categorized as stem cells because of their ability to self-renewal and differentiation to other lineages (Insausti et al. 2010). Amniotic epithelial cells express pluripotency transcription factors such as Oct-4, Sox-2, Nanog, and Rex-1 (Parolini et al. 2008; Insausti et al. 2010) and can differentiate to the three germinal layers: ectoderm, mesoderm, and endoderm (Tamagawa et al. 2004; Miki and Strom 2006). In addition, AEC show pluripotent cell surface markers such as SSEA-3 and SSEA-4 (stage-specific embryonic antigen 3 and 4), TRA 1-60 and TRA 1-81 (tumor rejection antigen 1-60 and 1-81) that is associated with embryonic stem cells. They also express cell–cell interaction molecules such as E-Cadherin, CD9, CD29, CD104, CD49e, CD49f, CD49d, and CD44 (Roubelakis et al. 2012; Insausti et al. 2010). Amniotic mesenchymal stem cells have adipogenic, chondrogenic, osteogenic, and angiogenic differentiation potential (Ilancheran et al. 2009; Alviano et al. 2007; Ilancheran et al. 2007; In't Anker et al. 2004) although hepatic, neurogenic, and myogenic differentiated lines have been reported too (Portmann-Lanz et al. 2006). Similar to bone marrow and other adult tissue isolated mesenchymal cells, they highly express CD90, CD73, CD105, and CD29 cell surface markers and do not express hematopoietic cell surface markers such as CD45, CD34, CD14, CD11b, and CD19 (Ilancheran et al. 2007; Roubelakis et al. 2012; Parolini et al. 2008). Human amniotic epithelial cells (hAEC) are separated by trypsin digestion of amniotic membrane which is mechanically separated from chorion. More enzymatic digestion with collagenase will terminate to complete isolation of AM-MSCs (In't Anker et al. 2004; Miki and Strom 2006; Wei et al. 2009; Bilic et al. 2004; Soncini et al. 2007). AM isolated stem cells are cultured in DMEM or α-MEM medium supplemented with fetal bovine serum (FBS) and epidermal growth factor (EGF) with or without leukemia inhibitory factor (LIF) according to the laboratory setup (Manochantr et al. 2010; Lisi et al. 2012; Tamagawa et al. 2007). The significant high ratio of stem cell to naïve population of AM (5–50 %) compared to others somatic tissues (0.01–0.1 %) is one of the main feature of AM for clinical application, as an average of 5×10^8 AM-MSCs (Bilic et al. 2008; Parolini et al. 2008) and 100×10^6 hAECs are obtained from one AM (Lagasse et al. 2000; Miki 2011).

To behave as an immunomodulatory agent, isolated cells should express and secrete immunoregulatory molecules, sense inflammatory and anti-inflammatory conditions, and interact with immune cells. AM-derived stem cells do not express polymorphic HLA-A, B, C, and DR antigens that demonstrated their low immunogenicity after allo- or xeno-transplantation (Li et al. 2005). Transplantation of a monolayer of human amniotic epithelial cells can survive for a long time without induction of any acute immune responses against transplant (Akle et al. 1981). Xenograft amniotic membranes transplanted to the limbal area, intracorneal space, and under the kidney capsule show no or low host cell infiltration and few host vessels formation (Kubo et al. 2001). Amnion-derived MSCs not only are low immunogenic but also are immunosuppressive. According to this feature, co-transplantation in conjunction with umbilical cord blood-derived hematopoietic stem cells reduces potential of graft-versus-host disease in recipients (Li et al. 2007). Although we do not know the complete immunosuppressive mechanisms of AM-MSCs, following inhibitory molecules are suggested as important ones. AM-MSCs express high level of IL-10 and IL-1 receptor agonists at transcriptional level and release the proteins to amnion where counteracts with inflammatory cytokine products such as TNF-α, IL-1, IL-8, and IL-6 and suppress their more production by immune cells. In addition, AM-MSCs exert their inhibitory effect on T lymphocyte proliferation through self-secreted IL-10 or induction of IL-10 producing immune cells by its other inhibitory molecules like (Indoleamine 2,3-dioxygenase) IDO (Yang et al. 2009). (IDO) enzyme which catalyzes essential amino acid tryptophan through kynurenine pathway works as an immunoregulatory molecule through inhibition of T lymphocyte and NK cell populations' growth and activity (Kang et al. 2012; Spaggiari et al. 2008). Co-culturing of AM-MSCs with PBMC and other inhibitory molecules like Prostaglandin E (PGE) augmented IDO production (Kang et al. 2012). Prostaglandin E2 (PGE2) is an anti-inflammatory lipid mediator that is produced by arachidonic acid processing by COX-1 and COX-2 enzyme (Smith et al. 1996). PGE2 constitutively produced by AM-MSCs and increased when AM-MSCs were co-cultured with PBMCs (Kang et al. 2012). PGE2 are dominant immunosuppressive molecule of AM-MSCs, because it enhances its own production that result in complete suppression of surrounded inflammatory molecules (Kalinski 2012) PGE2 use different ways to play its immunosuppressive role on T lymphocytes. T lymphocyte proliferation and activation were suppressed by inhibition of IL-2 production and induction of cAMP production by PGE2, respectively (Walker et al. 1983). PGE2 promoted FOXP3$^+$CD4$^+$CD25$^+$ regulatory T cell differentiation and affected T helper cell polarization to the benefit of Th2 subtype through induction of IL-10 and IL-4 cytokine production and inhibition of IL-12 and IL-2 production (Mahic et al. 2006; Demeure et al. 1997). PGE2 inhibit inflammatory cell migration and induce regulatory cell maintenance by modulation of chemokine production. In addition, it interacts with dendritic cells and suppressed DC-mediated T cell activation by suppression of antigen presentation and can inhibit activation of macrophages and NK cells (Yañez et al. 2010; Sreeramkumar et al. 2012).

Transforming growth factor beta (TGF-β) family are consisted from highly similar three isoforms (TGF-β 1, TGF-β 2, and TGF-β 3) that secreted in the inactive latent form to extracellular matrix. *Activation of* TGF-β will be primed by proteolytic function of matrix metalloproteases and reactive oxygen species (Barcellos-Hoff and Dix 1996; Yu and Stamenkovic 2000). Activated TGF-β interacts with TGF-β receptors on immune cells and triggers both anti-inflammatory and proinflammatory functions in the context-dependent manner. TGF-β is abundantly secreted by all types of MSCs like AM-MSCs. Amnion-derived MSCs exerted the most parts of its immunomodulatory effects through TGF-β that was abrogated with anti-TGF-β antibody. Different studies show increased level of TGF-β expression in AM-MSCs at both mRNA and protein level after co-culturing with immune cells (Kang et al. 2012; Chen et al. 2011a). Secreted TGF-β will suppress immune cell proliferation through cell cycle blocking. It ligates to TGF-β receptors on B lymphocytes and induces apoptosis (Spender et al. 2009). In addition, TGF-β suppressed B cell activation by inhibition of NF-kB and cytokine production and interfere with antibody production (Cazac and Roes 2000). AM-MSCs can direct T helper cell differentiation to regulatory T cells or Th17 subtype through TGF-β1. This induction suppresses Th1 or Th2 differentiation of helper T cells (Li and Flavell 2008), inhibit T cell proliferation, and suppress cytotoxic T cell activity by inhibition of expression of cytolytic gene products. TGF-B production by AM-MSCs suppressed inflammatory cytokine production by classical macrophages and promotes alternative macrophage activation which secretes anti-inflammatory cytokines and help to tissue repair (Gong et al. 2012).

Beside the soluble factors, AM-MSCs express cell membrane bond suppressive molecules. Induction of nonclassical class I HLA-G molecules on the surface of AM-MSCs is among the immunosuppressive mechanisms. Physiological expression of HLA-G is restricted to AM and thymus in the body (Lefebvre et al. 2000). HLA-G transcripts can be alternatively spliced to membrane bound and soluble proteins. Cell bond HLA-G induces tolerance in natural killer cells especially through activation of killing inhibitory receptor ILT (*Ig-like transcript*) pathway. Soluble HLA-G interaction with CD8+ marker on T and NK cells upregulate FasL expression and induce apoptosis (Contini et al. 2003). In addition, soluble HLA-G redirect helper T lymphocyte to regulatory phenotype (Lila et al. 2001) and exerts immunosuppressive effect on DC maturation that in consequence terminated to less activation of NK cells and T lymphocytes (Gros et al. 2008). Increase in immunosuppressive cytokine production by mononuclear cells is another effect of soluble HLA-G on immune cells (Hunt et al. 2006).

Programmed death-ligand 1 (PD-L1) or B7 homolog 1 (B7-H1) is another transmembrane protein expressed on the AM-MSCs. This regulatory molecule interacts with PD-1 on T lymphocyte and disturbs TCR signaling pathway through attenuation of NF-Kb and AP-1 activation (Sheppard et al. 2004). This attenuation results in IL-2 reduction and suppression of T lymphocyte proliferation.

Fas ligand as a member of the tumor necrosis factor (TNF) family are located in transmembrane part of AM-MSCs and interacts with Fas(CD95) receptors on immune cells. Induction of apoptosis in Fas-expressing T lymphocytes is an immunoregulatory way that suppresses cytotoxic T cell function (Mazar et al. 2009). Uptake of apoptotic T cell particles by macrophages turn them to alternatively activated macrophages with high TGF-β production and tolerogenic function (Akiyama et al. 2012).

Although there is no doubt on immunoregulatory function of AM-MSCs that exerted by its membrane bond or soluble factors, different studies demonstrated that they are not spontaneous suppressors and should be excited under inflammatory condition (Shi et al. 2012). Inflammation may provide MSCs migration and homing to injured site. MSCs produce growth factors, chemokines, chemokine receptors, and other cell adhesion molecules in response to TNF-α, IL-1β, and other inflammatory cytokines secreted by immune cells at inflammation site (Ullah et al. 2015). There are also reports that show the production of immunosuppressive molecules of AM-MSCs needs stimulation especially by IFN-γ or microbial ligands (Chang et al. 2006) (Nurmenniemi et al. 2010). Matrix metallo-proteases (MMPs) and chemokine receptors, chemokine receptor type 4 (CXCR4), are the main factors for MSCs migration to and homing in injured site (Ries et al. 2007). After migration, resident MSCs secrete chemokines (CCL2, CXCL9, CXCL10, and CXCL11) and express cell adhesion molecules like intercellular adhesion molecules (ICAM)-1 and vascular cell adhesion molecules (VCAM)-1 which attracted immune cells and facilitate close contact with them at inflammation site (Ren et al. 2010; Shi et al. 2012).

AM-MSCs apply all mentioned inhibitory mechanisms in direct interaction with innate and adaptive immune cells, to suppress their function (Insausti et al. 2014). According to Magatti et al. reports, AM-MSCs block DC maturation and differentiation from monocytes through inhibition of CD80, CD86, and HLA-DR expression and induction of cell cycle arrest at G0 phase (Magatti et al. 2009) AM-MSCs induce tolerogenic dendritic cells and macrophages by their soluble factors and direct transmembrane HLA-G interaction with ILT receptors which terminated to differentiation of regulatory T cells (LeMaoult et al. 2007). Natural killer cells are innate lymphoid cells that patrol the body and screen tumor, microbial infected or foreign cells ligands that interacted with activating NK receptors. Following activation, NK cells release the content of cytolytic granules including perforins and granzymes and kill involved cells (Vivier et al. 2008). However, NK cells express killing inhibitory receptors including KIR, NKG2A/CD94, ILT2, and so on, that recognize MHC class I (HLA-A, -B or -C) molecules on every normal cells in the body and tolerate them (Campbell and Purdy 2011). Because of the absence of MHC class I molecules on AM-MSCs, they can be killed by active NK cells while inhibiting their cytotoxic effect on other cells. PGE2 and IDO) production by AM-MSCs downregulate NK cell killer activating receptors and inhibit their proliferation (Spaggiari et al. 2008). AM-MSCs interrupt NK cell communication with other immune cells via soluble or membrane bond HLA-G which ligated to the killing inhibitory receptors on dendritic cells, T and B lymphocytes and affected their cytokine production and ligand-receptor engagement with NK cells (Gros et al. 2008). T lymphocytes as the main player of adaptive immunity, respond to environmental stimulus after antigen recognition by their antigen-specific T cell receptors. Antigen-specific T lymphocytes are divided to two main categories according to how they act; 1) cytotoxic T lymphocyte which destroy and kill the cells who introduced antigens by class I MHC molecules in the cell–cell contact manner, and 2) helper T lymphocyte which recognize antigens on class II MHC molecules and produced the wide range of cytokines from regulatory to inflammatory and anti-inflammatory properties. AM-MSCs do not express MHC molecules and escape from T lymphocytes recognition system, however have reciprocal effect on each other. Kang et al. showed that AM-MSCs produce increased level of IL-10, TGF-β, hepatic growth factor (HGF), IDO), and COX-2 in co-culture with PBMCs or in the presence of PBMC supernatant (Kang et al. 2012). In the reciprocal interaction, AM-MSCs secreted factors that inhibit T cell proliferation in response to phytohemagglutinin or allogeneic stimulation in the dose-dependent manner (Li et al. 2007) (Banas et al. 2008). Researchers demonstrated that cytokine production of mitogen-stimulated T lymphocytes will be affected in the presence of AM-MSCs in the culture. According to analysis of cytokine level in the supernatant of AM-MSCs – PBMC co-culture, changes in level of IL-2, IL-4, IL-7, IL-10, IL-15, TGF-β, and IFN-γ production were observed while IL-10 and TGF-β had the significant increased level and IFN-γ showed the decreased level compared to PBMC culture alone (Li et al. 2007; Roelen et al. 2009). Differentiation to different subtypes of helper T lymphocytes is dependent on surrounded cytokines, so AM-MSCs can trigger TH2 and regulatory subtypes and suppress TH1 differentiation, because of augmentation of IL-10 and TGF-β production.

Different reports confirmed that placental MSCs support FOXP3+ regulatory T cell induction and proliferation through induction of tolerogenic antigen-presenting cells or regulatory cytokines (LeMaoult et al. 2007; Chen et al. 2011b).

4 Immunomodulatory Properties of Amniotic Fluid Mesenchymal Stem Cells (AF-Mscs)

Amniotic fluid (AF) is the secretion of chorio-amniotic membrane and fetal skin that provide water and nutrients in the amniotic bag to create a safe environment for embryo development (Ganatra 2003). Amniotic fluid volume increases during pregnancy as a result of active transport of sodium and chloride that induces water transport across membrane. Electrolytes, protein, lipid, carbohydrate, and embryo produced urine and respiratory fluid are soluble components of amniotic fluid (Zhao 2015; Westgren et al. 1995). Most of these soluble components are secreted by diverse cell population that separated from different tissues of developing embryo and immersed in amniotic fluid. The cells are derived from placenta, skin, digestive, urinary, and respiratory tracts of embryo and are used for prenatal genetic diagnosis by amniocentesis (Siegel et al. 2007). However, they have pluripotent and multipotent stem cell characteristics and are considered in clinics for their tissue regeneration and immunomodulatory properties (Rennie et al. 2012b).

Human amniotic fluid-derived MSCs (AF-MSCs) can be isolated from amnion fluid of pregnant woman at 16–20 weeks of gestation. This adherent fibroblastic-like cells is expanded in culture media containing 89 % DMEM-High Glucose or α-MEM, 10 %FBS, 1 % penicillin–streptomycin supplemented with/without 4–10 ng/ml bFGF (Liu et al. 2009; Li et al. 2015). AF-MSCs are not tumorigenic after injection to nude mice, however well growing AF-MSCs can be cultured up to more than 20 passages. The population doubling time will increase from 36 h at first passage to 48, 55, and 97 h for P5, P10, and P20, respectively (Li et al. 2015), so usually the cells are used up to 4–8 passages for experimental use. Immunophenotype analysis showed that AF-MSCs represented high expression of CD73, CD105, CD90, CD166, and HLA-ABC, while are negative for CD45,CD34, CD14, and HLA-DR cell surface markers (Parolini et al. 2009; Li et al. 2015). AF-MSCs express pluripotency markers of Oct-4, Nanog, Sox-2, and Rex-1 in different gestational age (Tsai et al. 2004). Although AF-MSCs have diverse differentiation potential to different cells like alveolar epithelial cells and hepatocytes (Li et al. 2014; Zheng et al. 2008), they are characterized by differentiation to adipogenic, chondrogenic, and osteogenic cells after in vitro culture in the presence of specific differentiation promoting media (Li et al. 2015). AF-MSCs create an immunoprivileged status in the amniotic cavity to protect fetus from rejection by mother immune system because of their low immunogenicity and immunosuppressive activity. Low immunogenicity of AF-MSCs is related to the absence of HLA-DR and positive co-stimulatory molecules of CD40, CD80, and CD86. In addition, they express high level of negative co-stimulatory molecules of B7H1, B7H2, B7H3, B7H4, and BTLA in the cell surface (Moorefield et al. 2011). Low immunogenicity of AF-MSCs introduced them as a source of allogeneic MSC transplantation, while cell surface expression of HLA-ABC and probable low level expression of class II HLA molecules promoted allo-antibody production (Schu et al. 2012). So it seems that they are suitable for autologous not allogeneic transplantation. AF-MSCs like other MSCs take part in immunosuppressive processes through secretion of anti-inflammatory molecules that are one of the main soluble components of amniotic fluid. IL-10, IL-1 receptor agonists, and other inhibitory secretions counteract with inflammatory functions of immune cells that causes inhibition of neutrophil infiltration to damaged site (Cargnoni et al. 2009) or production of proinflammatory cytokines of TNF-α and C-X-C motif chemokine ligand 10 (CXCL10) by activated dendritic cells (Magatti et al. 2009). AF-MSCs can suppress T lymphocyte proliferation and activation by PGE2 and IDO) as discussed earlier (Kang et al. 2012). AF-MSCs are one of the complex components of amniotic fluid contributed to wound healing and tissue regeneration (Silini et al. 2013). Fibroblast proliferation and differentiation to myofibroblasts are the primary step of wound repair that terminated to regeneration of epithelium, connective tissue, and vasculature. AF-MSCs have paracrine role by secretion of different growth factors containing vascular endothelial growth factor (VEGF), epithelial growth factor (EGF), basic fibroblasts growth factors (bFGF), members of the insulin growth factor-binding protein (IGFBP) superfamily, and transforming growth factor beta (TGF-ß) in wound repair (Skardal 2014; Sorrell and Caplan 2010). In addition, MSCs can differentiate directly to myofibroblasts and augment vascularization (Yamaguchi et al. 2005). AF-MSCs are also involved in the last step of wound repair that was accompanied by increase in matrix metalloproteases (MMPs) and decrease in TGF-β that terminated to collagen degradation, fibroblast apoptosis, and tissue-specific cell proliferation (Darby and Hewitson 2007). MSCs can bound MMPs at the cell surface and activate exogenous pro-MMPs which may further participate in extracellular matrix degradation and tissue remodeling (Lozito et al. 2014).

5 Immunomodulatory Properties of Chorion-Derived Mesenchymal Stem Cells (CMSCs)

Chorion is the outer layer of fetal part extraembryonic membrane that is connected to decidua as maternal part of placenta. Both decidua and chorion form the placenta membrane that separates maternal from fetal blood (Witkowska-Zimny and Wrobel 2011). Chorion is composed of chorionic plate and chorionic villi that are a rich source of mesenchymal stem cells that is known as CP-MSC and CV-CMSCs, respectively (Soncini et al. 2007; Jones et al. 2002). These fetal tissue-isolated MSCs are primitive than adult MSCs and have greater life span and self-renewal capacity. However, different studies demonstrated that maternal part isolated MSCs like decidua (D-MSC) have a greater life span than CP-MSC and CV-MSC (Soncini et al.

2007; Fukuchi et al. 2004). They showed an intermediate phenotype of adult MSCs and pluripotent stem cells and can differentiate into cells developed from three germ layers (Wang et al. 2014; Abumaree et al. 2013; Chang et al. 2007). Although preterm-isolated chorion-derived mesenchymal stem cells (CMSCs) show higher stemness and expression level of NANOG, SOX2, c-MYC, and KLF4 and generates better embryoid body rather than term-isolated CMSCs, their application technically are impossible because of ethical problem (Jones et al. 2012). Although MSCs obtained from different fetal or adult tissues contribute the same phenotype and immunomodulatory properties, they have differences in magnitude and quality of these characteristics according to species origin, tissue source, and localization (Hass et al. 2011; Hashemi et al. 2013). CP-MSC and CV-MSC are negative for hematopoietic cell surface markers and express MSC-specific markers of CD105, CD73, CD90, and CD29. Their low immunogenicity and immunoprivileged phenotype are related to low or negative expression of HLA-DR that may be converted to immunogenic after differentiation or stimulation with IFN-γ (Huang et al. 2010; Chan et al. 2008). Expression of HLA-ABC and HLA-G is higher in CP-MSC and CV-MSC compared to adult MSCs. Strongly HLA-G positive CMSCs reflected their immunosuppressive role in pregnancy and their potential in graft tolerance (Hunt et al. 2005; Menier et al. 2010). As Bailo et al. demonstrated that engraftment of chorion-derived cells can be successfully transplanted into neonatal swine and rats and create tissues with human microchimerism (Bailo et al. 2004). Moreover, higher expression of HLA-G on human placenta-derived MSCs (hP-MSC) compared to adult MSCs makes them resistant to NK cytotoxicity and suppressed NK cells efficiently. Different studies demonstrated that hP-MSC suppress allogeneic T cell proliferation and activation through IL-10 and TGF-β production (Li et al. 2007) and induction of Treg cell increase (Chang et al. 2006). Recent study showed that CV-MSC have two subpopulation according to CD106 (VCAM-1) expression on the cell surface that affected their immunomodulatory capacity and biological activity. CD^{106+}CV-MSC demonstrated low colony forming capacity and proliferation potential compared to CD^{106-}CV-MSC, while exerts higher immunosuppressive activity (Yang et al. 2013). CD^{106+}CV-MSC have augmented inhibitory activity on T cell function through complete suppression of IFN-γ secretion by PHA-activated T lymphocyte and suppression of Tbet expression that directed Th1 polarization (Yang et al. 2013). Moreover, increased expression of COX-2, IL-1a, IL-1b, IL-6, and IL-8 appeared in CD^{106+}CV-MSC compared to CD^{106-}CV-MSC (Yang et al. 2013). CP-MSCs are also the active immuromodulator of T cell responses, as they suppress IFN-γ production and induce IL-4, IL-13, IL-2, and GM-CSF production if co-cultured with activated T cells in the dose-dependent manner (Lee et al. 2012). Like AF-MSC, CP-MSC has antifibrotic effect. They counteract with TGF-β in wound healing process and suppress collagen formation by production and activation of MMPs (Lee et al. 2010). However, different comparative studies on immunomodulatory function of F-MSCs and adult MSCs demonstrated the superior immunoregulatory function of F-MSCs beside their low immunogenicity (Lee et al. 2012; Chen et al. 2011a). In addition, F-MSCs function is different from adult MSCs in response to IFN-γ and TNF-α stimulation. IFN-γ stimulation will turn adult MSCs to active antigen-presenting cells (APCs) by upregulation of MHC class II molecules while F-MSCs behave as poor APCs (Chang et al. 2006; Chan et al. 2008; Stagg et al. 2006). Finally, because of F-MSCs isolation has no ethical problem; they are an available source for therapeutic use in tissue regeneration and immunosuppression aspects.

References

Abumaree MH et al (2013) Phenotypic and functional characterization of mesenchymal stem cells from chorionic villi of human term placenta. Stem Cell Rev 9(1):16–31

Akiyama K et al (2012) Mesenchymal-stem-cell-induced immunoregulation involves FAS-ligand-/FAS-mediated T cell apoptosis. Cell Stem Cell 10:544–555

Akle CA et al (1981) Immunogenicity of human amniotic epithelial cells after transplantation into volunteers. Lancet 318(8254):1003–1005

Alunno A et al (2014) In vitro immunomodulatory effects of microencapsulated umbilical cord Wharton jelly-derived mesenchymal stem cells in primary Sjögren's syndrome. Rheumatology (United Kingdom) 54(1):163–168, http://www.rheumatology.oxfordjournals.org/cgi/doi/10.1093/rheumatology/keu292

Alviano F et al (2007) Term Amniotic membrane is a high throughput source for multipotent Mesenchymal Stem Cells with the ability to differentiate into endothelial cells in vitro. BMC Dev Biol 7:11, http://www.pubmedcentral.nih.gov/articlerender.fcgi?artid=1810523&tool=pmcentrez&rendertype=abstract

Bailo M et al (2004) Engraftment potential of human amnion and chorion cells derived from term placenta. Transplantation 78(10):1439–1448, http://www.ncbi.nlm.nih.gov/entrez/query.fcgi?cmd=Retrieve&db=PubMed&dopt=Citation&list_uids=15599307

Banas RA et al (2008) Immunogenicity and immunomodulatory effects of amnion-derived multipotent progenitor cells. Hum Immunol 69(6):321–328

Barcellos-Hoff MH, Dix TA (1996) Redox-mediated activation of latent transforming growth factor-beta 1. Mol Endocrinol (Baltimore, MD) 10(9):1077–1083

Bieback K et al (2004) Critical parameters for the isolation of mesenchymal stem cells from umbilical cord blood. Stem Cells (Dayton, Ohio) 22(4):625–634, http://www.ncbi.nlm.nih.gov/pubmed/15277708. Accessed 22 Apr 2016

Bilic G et al (2004) In vitro lesion repair by human amnion epithelial and mesenchymal cells. Am J Obstet Gynecol 190(1):87–92

Bilic G et al (2008) Comparative characterization of cultured human term amnion epithelial and mesenchymal stromal cells for application in cell therapy. Cell Transplant 17(8):955–968

Bongso A, Fong C (2013) The therapeutic potential, challenges and future clinical directions of stem cells from the Wharton's Jelly of the human umbilical cord. Stem Cell Rev Rep 9(2):226–240, http://link.springer.com/10.1007/s12015-012-9418-z

Campbell KS, Purdy AK (2011) Structure/function of human killer cell immunoglobulin-like receptors: lessons from polymorphisms, evolution, crystal structures and mutations. Immunology 132(3):315–325

Cargnoni A et al (2009) Transplantation of allogeneic and xenogeneic placenta-derived cells reduces bleomycin-induced lung fibrosis. Cell Transplant 18(4):405–422

Cazac BB, Roes J (2000) TGF-β receptor controls B cell responsiveness and induction of IgA in vivo. Immunity 13(4):443–451, http://www.cell.com/article/S1074761300000443/fulltext

Chan WK et al (2008) MHC expression kinetics and immunogenicity of mesenchymal stromal cells after short-term IFN-γ challenge. Exp Hematol 36(11):1551–1561

Chang C-J et al (2006) Placenta-derived multipotent cells exhibit immunosuppressive properties that are enhanced in the presence of interferon-gamma. Stem Cells 24(11):2466–2477

Chang CM et al (2007) Placenta-derived multipotent stem cells induced to differentiate into insulin-positive cells. Biochem Biophys Res Commun 357(2):414–420

Chao KC et al (2008) Islet-like clusters derived from mesenchymal stem cells in Wharton's jelly of the human umbilical cord for transplantation to control type 1 diabetes. PLoS One 3(1):e1451

Chen K et al (2010) Human umbilical cord mesenchymal stem cells hUC-MSCs exert immunosuppressive activities through a PGE2-dependent mechanism. Clin Immunol 135(3):448–458, http://dx.doi.org/10.1016/j.clim.2010.01.015

Chen P-M et al (2011a) Immunomodulatory properties of human adult and fetal multipotent mesenchymal stem cells. J Biomed Sci 18(1):49, http://www.jbiomedsci.com/content/18/1/49

Chen P-M et al (2011b) Immunomodulatory properties of human adult and fetal multipotent mesenchymal stem cells. J Biomed Sci 18(1):49, http://www.jbiomedsci.com/content/18/1/49

Choi M et al (2013a) Proangiogenic features of Wharton's jelly-derived mesenchymal stromal/stem cells and their ability to form functional vessels. Int J Biochem Cell Biol 45(3):560–570, http://dx.doi.org/10.1016/j.biocel.2012.12.001

Choi M et al (2013b) Proangiogenic features of Wharton's jelly-derived mesenchymal stromal/stem cells and their ability to form functional vessels. Int J Biochem Cell Biol 45(3):560–570

Contini P et al (2003) Soluble HLA-A,-B,-C and -G molecules induce apoptosis in T and NK CD8+ cells and inhibit cytotoxic T cell activity through CD8 ligation. Eur J Immunol 33(1):125–134, http://www.ncbi.nlm.nih.gov/pubmed/12594841. Accessed 21 Jun 2016

Danforth D, Hull RW (1958) The microscopic anatomy of the fetal membranes with particular reference to the detailed structure of the amnion. Am J Obstet Gynecol 75(3):536–547, discussion 548–550. http://www.ncbi.nlm.nih.gov/pubmed/13508744. Accessed 17 Jun 2016

Darby IA, Hewitson TD (2007) Fibroblast differentiation in wound healing and fibrosis. Int Rev Cytol 257:143–179

Demeure CE et al (1997) Prostaglandin E2 primes naive T cells for the production of anti-inflammatory cytokines. Eur J Immunol 27:3526–3531

Deuse T et al (2011) Immunogenicity and immunomodulatory properties of umbilical cord lining mesenchymal stem cells. Cell Transplant 20(5):655–667

Donders R et al (2015) Human Wharton's Jelly-derived stem cells display immunomodulatory properties and transiently improve rat experimental autoimmune encephalomyelitis. Cell Transplant 24(10):2077–2098, http://openurl.ingenta.com/content/xref?genre=article&issn=0963-6897&volume=24&issue=10&spage=2077

Fan Y-P et al (2016) The therapeutic potential of human umbilical mesenchymal stem cells from Whartons Jelly in the treatment of rat peritoneal dialysis-induced fibrosis. Stem Cells Transl Med 5(2):235–247, http://stemcellstm.alphamedpress.org/cgi/doi/10.5966/sctm.2015-0001

Flynn A, Barry F, O'Brien T (2007) UC blood-derived mesenchymal stromal cells: an overview. Cytotherapy 9(8):717–726, http://linkinghub.elsevier.com/retrieve/pii/S1465324907701408

Fong CY et al (2007) Comparative growth behaviour and characterization of stem cells from human Wharton's jelly. Reprod Biomed Online 15(6):708–718, http://linkinghub.elsevier.com/retrieve/pii/S1472648310605391

Fukuchi Y et al (2004) Human placenta-derived cells have mesenchymal stem/progenitor cell potential. Stem Cells 22:649–658

Ganatra MA (2003) Amniotic membrane in surgery. J Pak Med Assoc 53(1):29–32

Gong D et al (2012) TGFβ signaling plays a critical role in promoting alternative macrophage activation. BMC Immunol 13:31

Gros F et al (2008) Soluble HLA-G molecules impair natural killer/dendritic cell crosstalk via inhibition of dendritic cells. Eur J Immunol 38(3):742–749

Hashemi SM et al (2013) Comparative immunomodulatory properties of adipose-derived mesenchymal stem cells conditioned media from BALB/c, C57BL/6, and DBA mouse strains. J Cell Biochem 114(4):955–965, http://www.ncbi.nlm.nih.gov/pubmed/23225199. Accessed 11 Aug 2014

Hass R et al (2011) Different populations and sources of human mesenchymal stem cells (MSC): a comparison of adult and neonatal tissue-derived MSC. Cell Commun Signal 9(1):12, http://www.biosignaling.com/content/9/1/12

Heo JS et al (2016) Comparison of molecular profiles of human mesenchymal stem cells derived from bone marrow, umbilical cord blood, placenta and adipose tissue. Int J Mol Med 37(1):115–125, http://www.pubmedcentral.nih.gov/articlerender.fcgi?artid=4687432&tool=pmcentrez&rendertype=abstract

Huang XP et al (2010) Differentiation of allogeneic mesenchymal stem cells induces immunogenicity and limits their long-term benefits for myocardial repair. Circulation 122(23):2419–2429

Huang PY et al (2015) Xenograft of human umbilical mesenchymal stem cells from Wharton's jelly as a potential therapy for rat pilocarpine-induced epilepsy. Brain Behav Immun 54:45–58, http://dx.doi.org/10.1016/j.bbi.2015.12.021

Hunt JS et al (2005) HLA-G and immune tolerance in pregnancy. FASEB J 19(7):681–693

Hunt JS et al (2006) The role of HLA-G in human pregnancy. Reprod Biol Endocrinol 4(Suppl 1):S10

Hwang JH et al (2009) Comparison of cytokine expression in mesenchymal stem cells from human placenta, cord blood, and bone marrow. J Korean Med Sci 24(4):547–554, http://jkms.org/DOIx.php?id=10.3346/jkms.2009.24.4.547. Accessed 22 Apr 2016

Ilancheran S et al (2007) Stem cells derived from human fetal membranes display multilineage differentiation potential. Biol Reprod 77(3):577–588, http://www.ncbi.nlm.nih.gov/pubmed/17494917

Ilancheran S, Moodley Y, Manuelpillai U (2009) Human fetal membranes: a source of stem cells for tissue regeneration and repair? Placenta 30(1):2–10

In't Anker PS et al (2004) Isolation of mesenchymal stem cells of fetal or maternal origin from human placenta. Stem Cells (Dayton, Ohio) 22(7):1338–1345, http://www.ncbi.nlm.nih.gov/pubmed/15579651

Insausti CL et al (2010) The amniotic membrane as a source of stem cells. Histol Histopathol 25(1):91–98

Insausti CL et al (2014) Amniotic membrane-derived stem cells: immunomodulatory properties and potential clinical application. Stem Cells Cloning 7(1):53–63

Jones EA et al (2002) Isolation and characterization of bone marrow multipotential mesenchymal progenitor cells. Arthritis Rheum 46:3349–3360

Jones GN et al (2012) Ontological differences in first compared to third trimester human fetal placental chorionic stem cells. PLoS One 7(9):e43395

Kalinski P (2012) Regulation of immune responses by prostaglandin E2. J Immunol (Baltimore, MD: 1950) 188(1):21–28, http://www.pubmedcentral.nih.gov/articlerender.fcgi?artid=3249979&tool=pmcentrez&rendertype=abstract\http://www.jimmunol.org/content/188/1/21.short

Kang JW et al (2012) Immunomodulatory effects of human amniotic membrane-derived mesenchymal stem cells. J Vet Sci 13(1):23–31

Karahuseyinoglu S et al (2007) Biology of stem cells in human umbilical cord stroma: in situ and in vitro surveys. Stem Cells 25(2):319–331, http://www.ncbi.nlm.nih.gov/pubmed/17053211

Kubo M et al (2001) Immunogenicity of human amniotic membrane in experimental xenotransplantation. Investig Ophthalmol Vis Sci 42(7):1539–1546

Lagasse E et al (2000) Purified hematopoietic stem cells can differentiate into hepatocytes in vivo. Nat Med 6(11):1229–1234

Lee OK et al (2004) Isolation of multipotent mesenchymal stem cells from umbilical cord blood. Blood 103(5):1669–1675, http://www.bloodjournal.org/content/103/5/1669.abstract. Accessed 22 Apr 2016

Lee M-J et al (2010) Anti-fibrotic effect of chorionic plate-derived mesenchymal stem cells isolated from human placenta in a rat model of CCl(4)-injured liver: potential application to the treatment of hepatic diseases. J Cell Biochem 111(6):1453–1463, http://www.ncbi.nlm.nih.gov/pubmed/20830742. Accessed 25 Jun 2016

Lee JM et al (2012) Comparison of immunomodulatory effects of placenta mesenchymal stem cells with bone marrow and adipose mesenchymal stem cells. Int Immunopharmacol 13(2):219–224, http://linkinghub.elsevier.com/retrieve/pii/S1567576912000975

Lefebvre S et al (2000) Modulation of HLA-G expression in human thymic and amniotic epithelial cells. Hum Immunol 61(11):1095–1101

LeMaoult J et al (2007) Immune regulation by pretenders: cell-to-cell transfers of HLA-G make effector T cells act as regulatory cells. Blood 109(5):2040–2048
Li MO, Flavell RA (2008) TGF-??: a master of all T cell trades. Cell 134(3):392–404
Li H et al (2005) Immunosuppressive factors secreted by human amniotic epithelial cells. Investig Ophthalmol Vis Sci 46(3):900–907
Li C et al (2007) Human-placenta-derived mesenchymal stem cells inhibit proliferation and function of allogeneic immune cells. Cell Tissue Res 330(3):437–446
Li Y et al (2014) Differentiation of human amniotic fluid-derived mesenchymal stem cells into type II alveolar epithelial cells in vitro. Int J Mol Med 33(6):1507–1513
Li L et al (2015) Characteristics of human amniotic fluid mesenchymal stem cells and their tropism to human ovarian cancer. PLoS One 10(4):e0123350
Lila N et al (2001) Soluble HLA-G protein secreted by allo-specific CD4+ T cells suppresses the allo-proliferative response: a CD4+ T cell regulatory mechanism. Proc Natl Acad Sci U S A 98(21):12150–12155
Lisi A et al (2012) A combined synthetic-fibrin scaffold supports growth and cardiomyogenic commitment of human placental derived stem cells. PLoS One 7(4):e34284
Liu CH, Hwang SM (2005) Cytokine interactions in mesenchymal stem cells from cord blood. Cytokine 32(6):270–279
Liu H et al (2009) Effects of different culture conditions on isolation and expansion of stem cells from second-trimester amniotic fluids. Zhonghua Fu Chan Ke Za Zhi 44(4):241–245
Liu K-J et al (2011) Surface expression of HLA-G is involved in mediating immunomodulatory effects of placenta-derived multipotent cells (PDMCs) towards natural killer lymphocytes. Cell Transplant 20(11–12):1721–1730
Liu S et al (2012) Immune characterization of mesenchymal stem cells in human umbilical cord Wharton's jelly and derived cartilage cells. Cell Immunol 278(1-2):35–44
López Y et al (2013) Wharton's jelly or bone marrow mesenchymal stromal cells improve cardiac function following myocardial infarction for more than 32 weeks in a rat model: a preliminary report. Curr Stem Cell Res Ther 8:46–59, http://www.ncbi.nlm.nih.gov/pubmed/23270633\nC:\Users\Flo\AppData\Local\MendeleyLtd.\Mendeley Desktop\Downloaded\López et al. - 2013 - Wharton's jelly or bone marrow mesenchymal stromal cells improve cardiac function following myocardial infarction.pdf
Lozito TP et al (2014) Human mesenchymal stem cells generate a distinct pericellular zone of MMP activities via binding of MMPs and secretion of high levels of TIMPs. Matrix Biol 34:132–143
Magatti M et al (2009) Amniotic mesenchymal tissue cells inhibit dendritic cell differentiation of peripheral blood and amnion resident monocytes. Cell Transplant 18(8):899–914
Mahic M et al (2006) FOXP3 + CD4 + CD25+ adaptive regulatory T cells express cyclooxygenase-2 and suppress effector T cells by a prostaglandin E2-dependent mechanism. J Immunol 177(1):246–254
Mamede A et al (2012) Amniotic membrane: from structure and functions to clinical applications. Cell Tissue Res 349(2):447–458
Manochantr S et al (2010) Isolation, characterization and neural differentiation potential of amnion derived mesenchymal stem cells. J Med Assoc Thai 93(Suppl 7):S183–S191, http://www.ncbi.nlm.nih.gov/pubmed/21294413
Mareschi K et al (2001) Isolation of human mesenchymal stem cells: bone marrow versus umbilical cord blood. Haematologica 86(10):1099–1100, http://www.ncbi.nlm.nih.gov/pubmed/11602418. Accessed 4 Mar 2016
Mazar J et al (2009) Cytotoxicity mediated by the Fas ligand (FasL)-activated apoptotic pathway in stem cells. J Biol Chem 284(33):22022–22028
Medicetty S et al (2004) Transplantation of pig stem cells into rat brain: proliferation during the first 8 weeks. Exp Neurol 190(1):32–41
Menier C et al (2010) Recent advances on the non-classical major histocompatibility complex class i HLA-G molecule. Tissue Antigens 75(3):201–206

Miki T (2011) Amnion-derived stem cells: in quest of clinical applications. Stem Cell Res Ther 2(3):25, http://www.pubmedcentral.nih.gov/articlerender.fcgi?artid=3152995&tool=pmcentrez&rendertype=abstract

Miki T, Strom SC (2006) Amnion-derived pluripotent/multipotent stem cells. Stem Cell Rev 2:133–142

Moorefield EC et al (2011) Cloned, CD117 selected human amniotic fluid stem cells are capable of modulating the immune response. PLoS One 6(10)

Nurmenniemi S et al (2010) Toll-like receptor 9 ligands enhance mesenchymal stem cell invasion and expression of matrix metalloprotease-13. Exp Cell Res 316(16):2676–2682

Parolini O et al (2008) Concise review: isolation and characterization of cells from human term placenta: outcome of the first international Workshop on Placenta Derived Stem Cells. Stem Cells (Dayton, Ohio) 26(2):300–311, http://www.ncbi.nlm.nih.gov/pubmed/17975221

Parolini O et al (2009) Amniotic membrane and amniotic fluid-derived cells: potential tools for regenerative medicine? Regen Med 4(2):275–291

Portmann-Lanz CB et al (2006) Placental mesenchymal stem cells as potential autologous graft for pre- and perinatal neuroregeneration. Am J Obstet Gynecol 194(3):664–673

Prasanna SJ et al (2010) Pro-inflammatory cytokines, IFNγ and TNFα, influence immune properties of human bone marrow and Wharton jelly mesenchymal stem cells differentially. PLoS One 5(2):e9016, http://www.pubmedcentral.nih.gov/articlerender.fcgi?artid=2814860&tool=pmcentrez&rendertype=abstract

Ren G et al (2010) Inflammatory cytokine-induced intercellular adhesion molecule-1 and vascular cell adhesion molecule-1 in mesenchymal stem cells are critical for immunosuppression. J Immunol 184:2321–2328

Rennie K et al (2012a) Applications of amniotic membrane and fluid in stem cell biology and regenerative medicine. Stem Cells Int 2012. Article ID 721538, 13 pages

Rennie K et al (2012b) Applications of amniotic membrane and fluid in stem cell biology and regenerative medicine. Stem Cells Int 2012. Article ID 721538, 13 pages

Ries C et al (2007) MMP-2, MT1-MMP, and TIMP-2 are essential for the invasive capacity of human mesenchymal stem cells: differential regulation by inflammatory cytokines. Blood 109(9):4055–4063

Roelen DL et al (2009) Differential immunomodulatory effects of fetal versus maternal multipotent stromal cells. Hum Immunol 70(1):16–23

Roubelakis MG, Trohatou O, Anagnou NP (2012) Amniotic fluid and amniotic membrane stem cells: marker discovery. Stem Cells Int 2012, Article ID 107836, 9 pages

Saeidi M et al (2013) Immunomodulatory effects of human umbilical cord Wharton's jelly-derived mesenchymal stem cells on differentiation, maturation and endocytosis of monocyte-derived dendritic cells. Iran J Allergy Asthma Immunol 12(1):37–49, http://eutils.ncbi.nlm.nih.gov/entrez/eutils/elink.fcgi?dbfrom=pubmed&id=23454777&retmode=ref&cmd=prlinks. Accessed 3 Jun 2016

Schu S et al (2012) Immunogenicity of allogeneic mesenchymal stem cells. J Cell Mol Med 16(9):2094–2103

Secco M et al (2008) Multipotent stem cells from umbilical cord: cord is richer than blood! Stem Cells 26(1):146–150

Sheppard KA et al (2004) PD-1 inhibits T-cell receptor induced phosphorylation of the ZAP70/CD3?? Signalosome and downstream signaling to PKC?? FEBS Lett 574(1-3):37–41

Shi Y et al (2012) How mesenchymal stem cells interact with tissue immune responses. Trends Immunol 33(3):136–143, http://www.pubmedcentral.nih.gov/articlerender.fcgi?artid=3412175&tool=pmcentrez&rendertype=abstract. Accessed 4 Nov 2014

Siegel N et al (2007) Stem cells in amniotic fluid as new tools to study human genetic diseases. Stem Cell Rev 3(4):256–264

Silini A et al (2013) Soluble factors of amnion-derived cells in treatment of inflammatory and fibrotic pathologies. Curr Stem Cell Res Ther 8(1):6–14, http://www.ncbi.nlm.nih.gov/pubmed/23270631

Skardal A (2014) Amniotic fluid stem cells for wound healing. In: Atala A, Murphy SV (eds) Perinatal stem cells, vol XXIII. Springer, New York, pp 17–24
Smith WL, Garavito RM, DeWitt DL (1996) Prostaglandin endoperoxide H synthases (cyclooxygenases)-1 and -2. J Biol Chem 271(52):33157–33160
Soncini M et al (2007) Isolation and characterization of mesenchymal cells from human fetal membranes. J Tissue Eng Regen Med 1(4):296–305
Sorrell JM, Caplan AI (2010) Topical delivery of mesenchymal stem cells and their function in wounds. Stem Cell Res Ther 1(4):30
Spaggiari GM et al (2008) Mesenchymal stem cells inhibit natural killer-cell proliferation, cytotoxicity, and cytokine production: role of indoleamine 2,3-dioxygenase and prostaglandin E2. Blood 111:1327–1333
Spender LC et al (2009) TGF-beta induces apoptosis in human B cells by transcriptional regulation of BIK and BCL-XL. Cell Death Differ 16(4):593–602
Sreeramkumar V, Fresno M, Cuesta N (2012) Prostaglandin E2 and T cells: friends or foes? Immunol Cell Biol 90(6):579–586, http://www.nature.com/doifinder/10.1038/icb.2011.75\http://www.ncbi.nlm.nih.gov/pubmed/21946663\http://www.pubmedcentral.nih.gov/articlerender.fcgi?artid=PMC3389798
Stagg J et al (2006) Interferon-gamma-stimulated marrow stromal cells: a new type of nonhematopoietic antigen-presenting cell. Blood 107(6):2570–2577
Subramanian A et al (2015) Comparative characterization of cells from the various compartments of the human umbilical cord shows that the Wharton's Jelly compartment provides the best source of clinically utilizable mesenchymal stem cells. PLoS One 10(6):e0127992, http://www.pubmedcentral.nih.gov/articlerender.fcgi?artid=4464659&tool=pmcentrez&rendertype=abstract. Accessed 3 Jun 2016
Sun L et al (2010) Umbilical cord mesenchymal stem cell transplantation in severe and refractory systemic lupus erythematosus. Arthritis Rheum 62(8):2467–2475
Taghizadeh RR, Cetrulo KJ, Cetrulo CL (2011) Wharton's Jelly stem cells: future clinical applications. Placenta 32(Suppl 4):S311–S315
Tamagawa T, Ishiwata I, Saito S (2004) Establishment and characterization of a pluripotent stem cell line derived from human amniotic membranes and initiation of germ layers in vitro. Hum Cell 17(3):125–130
Tamagawa T et al (2007) Differentiation of mesenchymal cells derived from human amniotic membranes into hepatocyte-like cells in vitro. Hum Cell 20(3):77–84
Tipnis S, Viswanathan C, Majumdar AS (2010) Immunosuppressive properties of human umbilical cord-derived mesenchymal stem cells: role of B7–H1 and IDO. Immunol Cell Biol 88(8):795–806, http://dx.doi.org/10.1038/icb.2010.47
Troyer DL, Weiss ML (2008) Concise review: Wharton's Jelly-derived cells are a primitive stromal cell population. Stem Cells 26:591–599
Tsai MS et al (2004) Isolation of human multipotent mesenchymal stem cells from second-trimester amniotic fluid using a novel two-stage culture protocol. Hum Reprod 19(6):1450–1456
Tsai P et al (2009) The therapeutic potential of human umbilical mesenchymal stem cells from Wharton's jelly in the treatment of rat liver fibrosis. Liver Transpl 15(5):484–495, http://www.ncbi.nlm.nih.gov/pubmed/19399744
Ullah I, Baregundi Subbarao R, Rho G-J (2015) Human mesenchymal stem cells - current trends and future prospective. Biosci Rep 35(2):e00191, http://www.bioscirep.org/content/35/2/e00191.abstract
Vivier E et al (2008) Functions of natural killer cells. Nat Immunol 9(5):503–510
Walker C et al (1983) Lymphokine regulation of activated (G1) lymphocytes. 1. Prostaglandin E2-induced inhibition of interleukin 2 production. J Immunol (Baltimore, MD : 1950) 130(4):1770–1773
Wang HS et al (2004) Mesenchymal stem cells in the Wharton's jelly of the human umbilical cord. Stem Cells 22(7):1330–1337

Wang D et al (2010a) CD14+ monocytes promote the immunosuppressive effect of human umbilical cord matrix stem cells. Exp Cell Res 316(15):2414–2423, http://dx.doi.org/10.1016/j.yexcr.2010.04.018

Wang D et al (2010b) CD14+ monocytes promote the immunosuppressive effect of human umbilical cord matrix stem cells. Exp Cell Res 316(15):2414–2423

Wang J et al (2014) The subtype CD200-positive, chorionic mesenchymal stem cells from the placenta promote regeneration of human hepatocytes. Biotechnol Lett 36(6):1335–1341

Wei JP et al (2009) Human amniotic mesenchymal cells differentiate into chondrocytes. Cloning Stem Cells 11(1):19–26, http://www.ncbi.nlm.nih.gov/pubmed/19226212

Weiss ML et al (2008) Immune properties of human umbilical cord Wharton's jelly-derived cells. Stem Cells (Dayton, Ohio) 26(11):2865–2874, http://www.ncbi.nlm.nih.gov/pubmed/18703664

Westgren M et al (1995) Cytokines in fetal blood and amniotic fluid in Rh-immunized pregnancies. Obstet Gynecol 86(2):209–213

Wexler SA et al (2003) Adult bone marrow is a rich source of human mesenchymal "stem" cells but umbilical cord and mobilized adult blood are not. Br J Haematol 121(2):368–374, http://www.ncbi.nlm.nih.gov/pubmed/12694261. Accessed 22 Apr 2016

Witkowska-Zimny M, Wrobel E (2011) Perinatal sources of mesenchymal stem cells: Wharton's jelly, amnion and chorion. Cell Mol Biol Lett 16(3):493–514

Wu KH et al (2007) Therapeutic potential of human umbilical cord derived stem cells in a rat myocardial infarction model. Ann Thorac Surg 83(4):1491–1498, http://www.ncbi.nlm.nih.gov/pubmed/17383364

Yamaguchi Y et al (2005) Bone marrow cells differentiate into wound myofibroblasts and accelerate the healing of wounds with exposed bones when combined with an occlusive dressing. Br J Dermatol 152(4):616–622

Yañez R et al (2010) Prostaglandin E2 plays a key role in the immunosuppressive properties of adipose and bone marrow tissue-derived mesenchymal stromal cells. Exp Cell Res 316(19):3109–3123

Yang CC et al (2008) Transplantation of human umbilical mesenchymal stem cells from Wharton's jelly after complete transection of the rat spinal cord. PLoS One 3(10):e3336

Yang S-H et al (2009) Soluble mediators from mesenchymal stem cells suppress T cell proliferation by inducing IL-10. Exp Mol Med 41:315–324

Yang ZX et al (2013) CD106 identifies a subpopulation of mesenchymal stem cells with unique immunomodulatory properties. PLoS One 8(3):1–12

Yousefi F et al (2016) In vivo immunomodulatory effects of adipose-derived mesenchymal stem cells conditioned medium in experimental autoimmune encephalomyelitis. Immunol Lett 172:94–105, http://www.sciencedirect.com/science/article/pii/S0165247816300256. Accessed 2 Mar 2016

Yu Q, Stamenkovic I (2000) Cell surface-localized matrix mealloproteinase-9 proteolytically activates TGF-beta and promotes tumor invasion and angiogenesis. Genes Dev 14:163–176

Zhao RC (2015) Stem cells: basics and clinical translation. Springer, Netherlands

Zheng Y-B et al (2008) Characterization and hepatogenic differentiation of mesenchymal stem cells from human amniotic fluid and human bone marrow: a comparative study. Cell Biol Int 32(11):1439–1448

Zhou C et al (2011) Immunomodulatory effect of human umbilical cord Wharton's jelly-derived mesenchymal stem cells on lymphocytes. Cell Immunol 272(1):33–38, http://www.ncbi.nlm.nih.gov/pubmed/22004796

Umbilical Cord Tissue and Wharton's Jelly Mesenchymal Stem Cells Properties and Therapeutic Potentials

Erdal Karaöz and Çiğdem İnci

1 Umbilical Cord Tissue

Umbilical cord (UC) tissue is composed of connective tissue (Wharton's jelly), amniotic epithelium, two umbilical arteries, and an umbilical vein (Can and Karahuseyinoglu 2007). During pregnancy, UC transfers all the necessary nutrition and oxygen from mother's blood to the fetus through placenta. Wharton's jelly (WJ) supports and protects structure of umbilical arteries by covering them and provides good blood circulation (Nagamura-Inoue and He 2014).

It has long been known that UC, which is thrown away after birth, contains stem cells. The presence of hematopoietic stem cells in human cord blood was first reported in 1974 (Knudtzon 1974). It was followed by the first successful cord blood transplantation to the patient with Fanconi anemia (Broxmeyer et al. 1989). Since then, other clinical trials with cord blood-derived hematopoietic stem cells also reported with encouraging results (Gluckman and Rocha 2005).

UC tissue is also one of the most promising sources of mesenchymal stem cells (MSCs) with several advantages. The unusual fibroblasts in UC were reported for the first time in 1970 (Parry 1970). According to electron microscope observations, WJ cells were different from smooth muscle cells with their poor mitochondrial content and lack of regimented plasmalemmal vesicles. Then, McElreavey et al. isolated and cultured the fibroblast-like cells from WJ in 1991 (McElreavey et al. 1991). In the next years, MSCs isolated from different regions of UC such as amnion, perivascular area, arteries, vein, and WJ (Mennan et al. 2013; Subramanian et al. 2015).

E. Karaöz, Ph.D (✉) • Ç. İnci, M.Sc.
Centre for Regenerative Medicine and Stem Cell Manufacturing, LivMedCell, Liv Hospital, Ahmet Adnan Saygun Cd. Canan Sk. No:5, İstanbul, Turkey
e-mail: ekaraoz@hotmail.com; incicigdem@gmail.com

B. Arjmand (ed.), *Perinatal Tissue-Derived Stem Cells*, Stem Cell Biology and Regenerative Medicine, DOI 10.1007/978-3-319-46410-7_3

Large numbers of MSCs can be isolated from WJ because cord matrix has a wider surface area than other compartments of UC. Even though immunophenotypic properties of cells isolated from different regions do not show significant difference, WJ-derived cells have much lesser CD40+ cells which demonstrate non-MSCs contaminants (Wetzig et al. 2013). The telomerase levels of the cells are higher in WJ-MSCs at late passages, suggesting that WJ-MSCs could retain their immature phenotype during long-term ex vivo culture (Fig. 1). It has been also shown that WJ-MSCs do not show genetic instability or oncogene activation and retain their differentiation and proliferation abilities for long periods of time in culture conditions (Scheers et al. 2013). Although there is no significant difference between adipogenic differentiation potential of stem cells from different compartments of UC, WJ-MSCs show better osteogenic and chondrogenic differentiation potential (Subramanian et al. 2015).

Although embryonic stem cells (ESC) have high self-renewal capacity and ability to differentiate into three germ layers, they have limited clinical applications because of ethical considerations and technical difficulties. In contrast, UC-derived MSCs

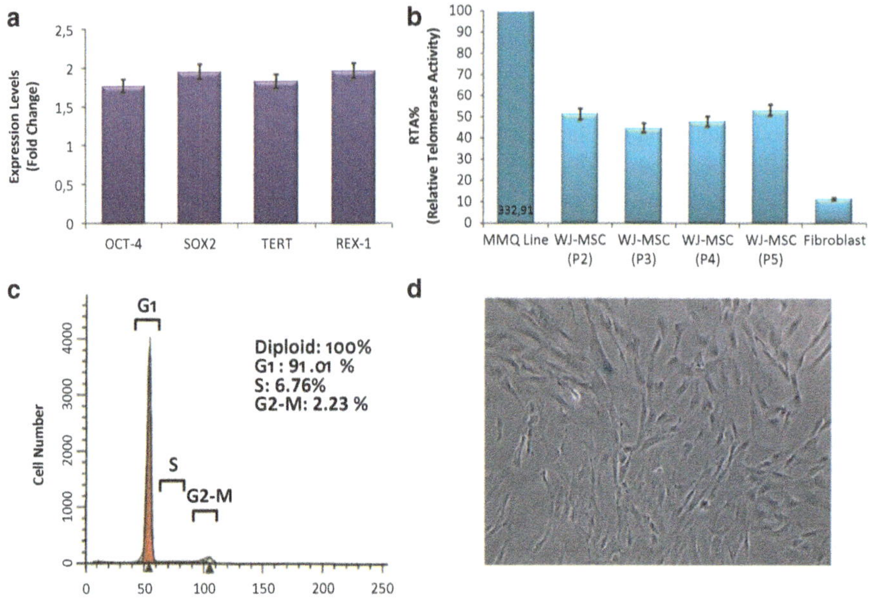

Fig 1 (**a**) Real-time PCR analyses of WJ-MSCs to evaluate the expressions of stemness genes, including octamer-binding transcription factor 4 (OCT-4), SRY box-2 (SOX2), telomerase reverse transcriptase (TERT), and reduced expression 1 (REX-1) in relation to the expression of a housekeeping gene GAPDH. (**b**) Relative telomerase activities of WJ-MSCs at passage 2 (WJ-MSC P2), passage 3 (WJ-MSC P3), passage 4 (WJ-MSC P4), passage 5 (WJ-MSC-P5), and MCF7 breast cancer cell line. (**c**) Cell cycle analysis of WJ-MSCs at P3, WJ-MSCs are under proliferative status. (**d**) Senescence assay results of WJ-MSCs at P3. β-Gal positive senescent cell rate is less than 1 % (Unpublished Data)

possess self-renewal and multipotency properties between adult and embryonic stem cells and collection procedure of UC is considered ethical (Fong et al. 2011). UC has been described as the preferential source of stem cells due to their large donor pool and reduced immunogenicity (Kalaszczynska and Ferdyn 2015). Another advantage of UC as a stem cell source, UC-derived stem cells do not form benign tumors (teratoma) in vivo in contrast to pluripotent stem cells (Hentze et al. 2009).

2 Wharton's Jelly-Derived Mesenchymal Stem Cells

2.1 Isolation Methods of WJ-MSCs

Cell culture techniques have an effect on the quality and quantity of the cells. There are various different techniques for isolation of adult stem cells from WJ. After removing vessels, WJ could be separated from the cord and for the next step there are two main procedures for isolation of WJ-MSCs: enzymatic and explant methods. Collagenase, hyaluronidase, dispase, and trypsin could be used for enzymatic digestion (Salehinejad et al. 2012; Nagamura-Inoue and He 2014). In the explant method, cord is minced into small pieces which are then placed into the bottom of the culture dishes and MSCs migrate from tissue fragments. But, commonly the cell recovery rate is poor because the cord fragments often float in the medium because of mechanical forces applied to the tissue fragments during culture processes.

Similarly to the conventional culture methods, which are involving the use of animal-derived supplements like fetal bovine serum (FBS), WJ-MSCs can also expand in large scales in *xeno*-free culture conditions (Corotchi et al. 2013). Therefore, isolation and production methods of WJ-MSCs can be adapted to clinical/good manufacturing practice (GMP) grade conditions. For instance, *xeno*-free culture media or human platelet lysates (HPL) can be alternatives to FBS, to be used as culture supplements for clinical applications. According to our laboratory experiences and (unpublished data) other researcher's reports, culture media supplemented with HPL enhances the proliferation and differentiation rate of WJ-MSCs (Jonsdottir-Buch et al. 2013; Antoninus et al. 2015) (Fig. 2).

2.2 Characterization of WJ-MSCs

Referring to the Mesenchymal and Tissue Stem Cell Committee of the International Society for Cellular Therapy, MSCs are plastic adherent when maintained in standard culture conditions and express CD73, CD90, and CD105, lack expression of CD45, CD34, CD14, CD79, and HLA-DR (Dominici et al. 2006). MSCs have capacity to differentiate in vitro toward osteoblasts, adipocytes, and chondrocytes.

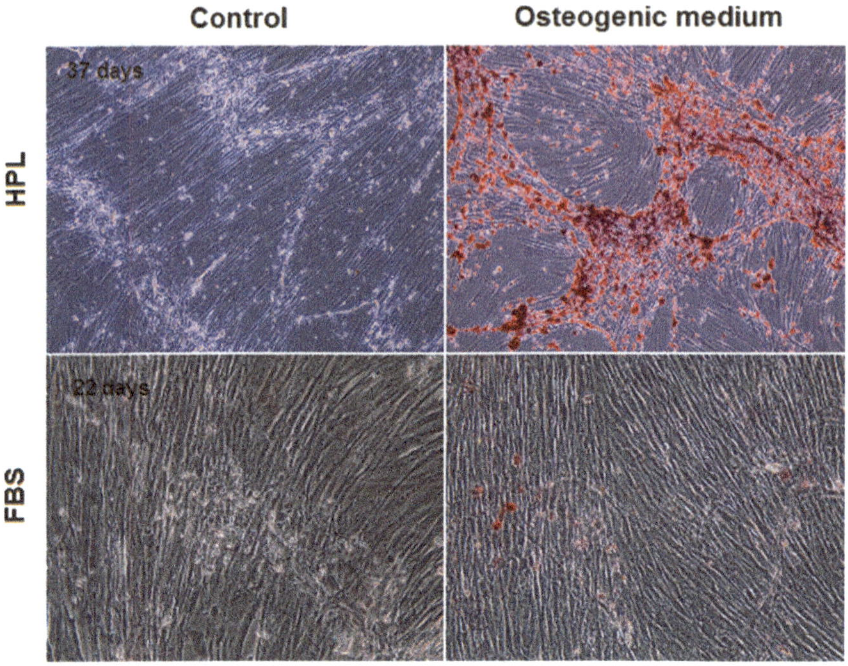

Fig. 2 Alizarin Red S Staining of WJ-MSCs cultured with fetal bovine serum (FBS) and human platelet lysates (HPL) after incubation in osteogenic differentiation medium (Unpublished Data). WJ-MSCs which cultured with HPL have more potential for differentiation to osteogenic lineages

WJ-MSCs express surface markers that are common in MSCs and do not express hematopoietic stem cell markers (Figs. 3, 4, and 5). They have trilineage—osteoblasts, adipocytes, and chondrocytes—differentiation capacity (Fong et al. 2007; Wang et al. 2009). WJ-MSCs also differentiate into hepatocyte-like cells (Borhani-Haghighi et al. 2015; Mortezaee et al. 2015; Zheng et al. 2015), insulin-producing cells (Kao et al. 2015), cardiomyocytes (Wang et al. 2004), neuron-like cells (Zhuang et al. 2015; Leite et al. 2014), muscle cells (Trivanović et al. 2013), fibroblasts (Han et al. 2011) in vitro.

WJ-MSCs have a higher telomerase enzyme activity level than somatic cells, while it is lower than cancer cell lines and ESCs. Moreover, their telomerase enzyme activity level stays stable for long times in vitro (Fig. 1). WJ-MSCs also maintain expression of pluripotency markers such as Sox2, Nanog, and Oct4 at low levels relative to ESCs (Carlin et al. 2006; Nekanti et al. 2010) (Fig. 1). During development, these factors regulate the expression levels of other genes. This situation suggests that WJ-MSCs are more immature than other sources of adult stem cells and their potential is closer to the ESCs. Expression of pluripotency markers at low levels and telomerase activity rates could be an explanation for why WJ-MSCs do not form teratomas.

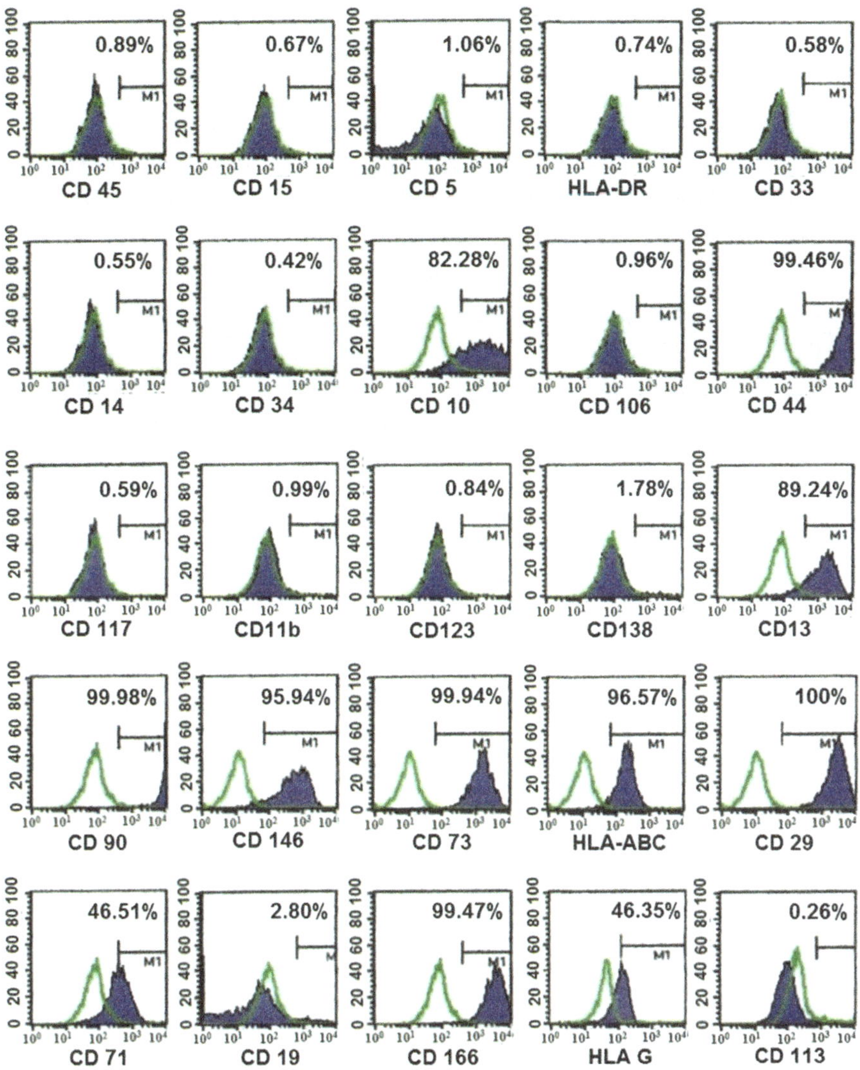

Fig. 3 Flow cytometric analysis of cell surface markers of WJ-MSCs at P3 (Unpublished Data)

2.3 *Immunomodulatory Properties of WJ-MSCs*

Use of allogeneic MSCs in clinical applications is considered safe and effective because of their immunomodulatory properties. Many researchers explain MSCs' immunomodulatory function by the lack of costimulatory molecules which are essential for T cell activation such as CD40, CD80, and CD86 expression and secretion of indoleamine-2,3-dioxygenase (IDO), transforming growth factor β (TGF-β), hepatocyte growth factor (HGF) and prostaglandin E2 (PGE2), nitric oxide (NO),

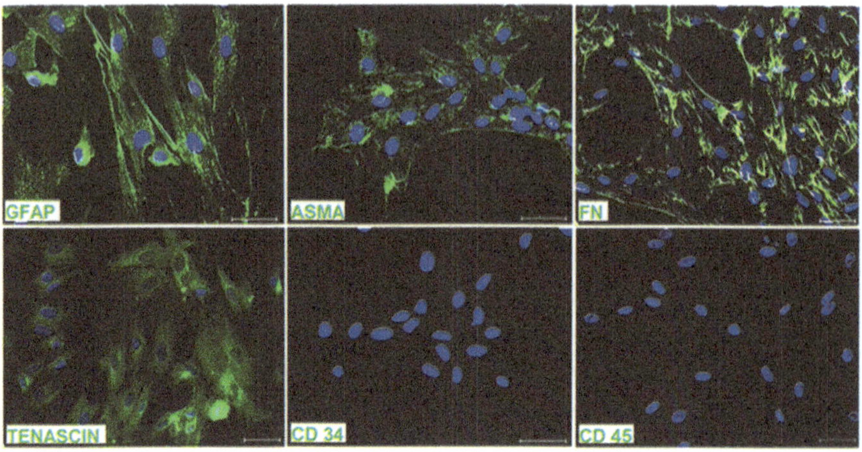

Fig. 4 Immunofluorescent staining for glial fibrillary acidic protein (GFAP), alpha smooth muscle actin (ASMA), Fibronectin (FN), Tenascin, CD 34, and CD 45. Nuclei were labeled with DAPI (*blue*) (Unpublished Data). The cells were negative for CD 34 and CD45; positive for GFAP, ASMA, FN, and Tenascin

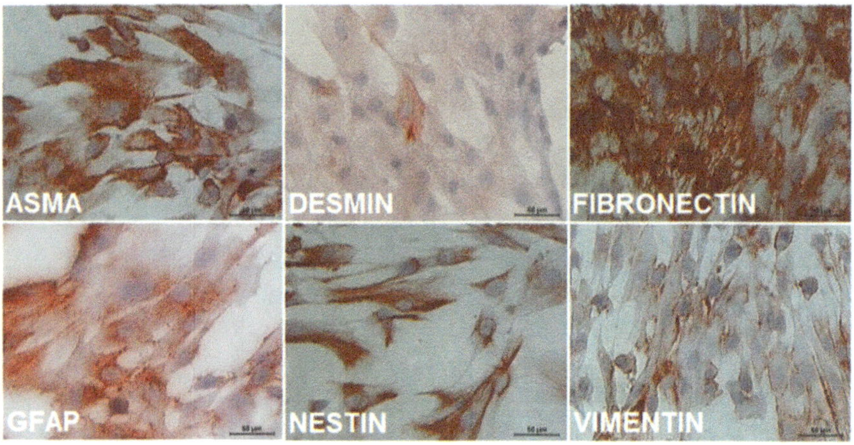

Fig. 5 Immunophenotype of WJ-MSCs. Staining patterns are shown for Asma, Desmin, Fibronectin, Glial fibrillary acidic protein (GFAP), Nestin, and Vimentin. Scale bars: 50 μm (Unpublished Data)

interleukin-6 (IL-6), and human leukocyte antigen-G (HLA-G) at high levels (Weiss et al. 2008; Kalaszczynska and Ferdyn 2015). IDO, PGE2, TGF-β, HGF, NO, IL-6, and HLA-G have been reported to be involved in T cell suppression (Nicola et al. 2002; Ren et al. 2008). It has been shown that retroviral transduction of MSCs with CD80 or CD86 did not result in lymphoproliferation (Klyushnenkova et al. 2005). Regarding MSCs' immunosuppressive effects on T cells, when they are separated by a semipermeable membrane, soluble factors play a crucial role in their low immunogenicity (Jacobs et al. 2013, Sarıboyacı et al. 2014). In contrast, some researchers suggested that cell–cell contact is more important and efficient than soluble factors in the immunosuppressive ability of MSCs (Xu et al. 2007; He et al. 2015). Whether because of cell–cell contact or through paracrine mechanisms, it can easily be said that WJ-MSCs possess high immunosuppressive properties.

Similar to other sources of MSCs, WJ-MSCs also express MHC class-I antigens at low levels but not class-II antigens and costimulatory antigens that are involved in activation of T and B cell responses (Weiss et al. 2008; Conconi et al. 2011; Nagamura-Inoue and He 2014; Kalaszczynska and Ferdyn 2015). Distinctly, WJ-MSCs express higher levels of HLA-G which plays an important role in avoiding maternal immunity against the fetus during pregnancy (Fig. 6) (Conconi et al. 2011). HLA-G exerts an immunosuppressive effect by inhibiting natural killer (NK) cells and T cell-mediated cytolysis through interactions with inhibitory receptors and inducing regulatory immunosuppressive cells (Nasef et al. 2007; Ding et al. 2015). HLA-G secretion is also associated with a better graft acceptance (Lila et al. 2000). HLA-G expression properties of WJ-MSCs make them natural inhibitors against cell rejection, creating a very suitable cell source for third party or allogeneic applications.

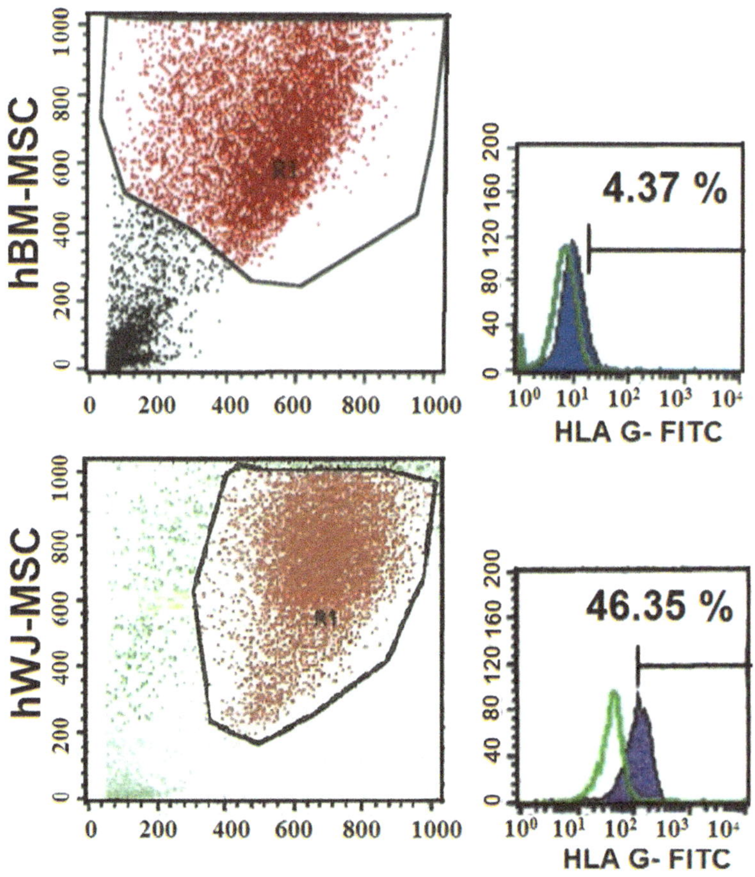

Fig. 6 Flow cytometric analysis of cell surface marker HLA-G of BM-MSC and WJ-MSC (Unpublished Data)

2.4 Homing

After WJ-MSCs are infused and joined to the circulation process, they tend to migrate and home toward tumor tissues or damaged areas mostly but not toward healthy tissues (Rachakatla et al. 2007; Doi et al. 2010). Similar to the migration of leukocytes to sites of inflammation, homing ability arises from the interaction between chemokines and receptor expression in stem cells (Matsuzuka et al. 2010; Kholodenko et al. 2013). MSCs migrate into the damaged tissues in response to the signals from injured areas. It has been shown that tumor-conditioned medium induced CXCR4 expression, which is involved in the migration mechanism of stem cells on bone marrow-derived MSCs (Song and Li 2011). These chemokine receptor expression profiles vary between different sources of the cells, age, isolation methods, and passage number of the cells (Kholodenko et al. 2013; Sohni and Verfaillie 2013). WJ-MSCs express stromal cell-derived factor-1 (SDF1), transforming growth factor beta receptor III (TGFBR3), and fibroblast growth factor receptor 2 (FGFR2) as chemokine receptors (Matsuzuka et al. 2010). SDF1 and its receptor CXCR4 are the most studied chemotactic factor pair in cell migration. SDF1 also plays role in hematopoietic cell homing, migration of mature lymphocytes, and primordial germ cells (Horuk and Peiper 1996; Doitsidou et al. 2002).

The homing ability of stem cells toward inflammatory tissues/tumor areas would make them important vehicles for targeted cell therapy. Ren et al. showed that MSCs can be genetically modified to express IFN-α, which has antitumor characteristics. Systemic administration of this modified stem cells increased apoptosis of cancer cells and decreased vasculature of tumor areas in a mouse melanoma lung metastasis model (Ren et al. 2008). A similar approach used for showing the antitumor and antimetastatic effect of NK-4 gene transduced MSCs in vivo. NK-4 expressing MSCs selectively migrated to the tumor tissues of the lung when they are injected intravenously and the cell therapy prolonged survival of the lung metastasis model (Kanehira et al. 2007). Nakamizo et al. engineered stem cells to release IFN-β to demonstrate that MSCs have a strong tropism for brain tumors. IFN-β expressing MSCs increased survival of the intracranial human glioma xenografts (Nakamizo et al. 2005). It has been documented that WJ-MSCs were localized in the periphery of tumor tissues after transplantation (Ayuzawa et al. 2009). As a result, WJ-MSCs may also serve as a vehicle for selective delivery of therapeutic reagents in tumors.

2.5 Antitumor Activity of WJ-MSCs

It has been shown that MSCs attenuate tumor cells (Khakoo et al. 2006; Qiao et al. 2008). Human WJ-MSCs inhibited MDA 231 cancer cell's proliferation by stimulating the intrinsic apoptosis pathway in vitro (Ayuzawa et al. 2009). Also *xeno*-transplantation of human WJ-MSCs intravenously or intratumorally attenuated metastatic tumor growth in breast carcinoma-induced SCID mice (Ayuzawa et al. 2009). In recent years, a lot of studies demonstrate that un-engineered WJ-MSCs can inhibit tumor growth of various types of cancers both in vivo and in vitro (Ayuzawa et al. 2009; Doi et al. 2010; Gauthaman et al. 2012). In contrast, there are also reports that MSCs can also form a cancer stem cell niche and support cancer cells' growth (Ramasamy et al. 2006; Patel et al. 2010). Although some in vitro and in vivo studies have reported a positive relationship between MSCs and cancer, there aren't any studies reporting certain evidences of cancer formation and progression after MSC infusion to human until today.

On the other hand, WJ-MSCs can be engineered to suppress tumorigenicity. Wang et al. showed that artificial fusion of human WJ-MSCs with esophageal carcinoma cells induced apoptosis of cancer cells (Wang et al. 2011). IFN-β which is a potent inhibitor of proliferation of cancer cell lines, transfected WJ-MSCs could cause death of bronchioloalveolar carcinoma cells both in vivo and in vitro (Matsuzuka et al. 2010). WJ-MSCs can potentially be used for targeted delivery of cancer therapeutics through their migration ability to tumor area.

3 In Vitro Studies

There are several *in vitro* studies to examine the potential impact of WJ-MSCs on different disorders such as Alzheimer's disease (AD), bone, and cartilage defects.

WJ-MSCs have a potential to improve neurological function. Lee et al. investigated the effect of WJ-MSCs on AD by coculturing stem cells with hippocampal neurons induced with beta amyloid peptide (Aβ) (Lee et al. 2010). Incubation of hippocampal neurons with Aβ induced apoptosis whereas coculturing with WJ-MSCs ameliorated the apoptosis rates. According to these data, researchers experienced WJ-MSC transplantation to AD mouse model, finalized with promising results supporting in vitro studies.

Using matrices/scaffolds with cells is one of the tissue engineering strategies which hold the potential to be a successful therapy for incurable end-stage diseases. The procedure involves harvesting cells from donor biopsy, expanding these cells on scaffolds in vitro, and implantation of scaffolds with cells to damaged area. Different cell sources are used for tissue engineering experiments such as chondrocytes and keratinocytes (Bailey et al. 2007; Killat et al. 2013). MSCs could also be used in rebuilding damaged or diseased tissues through their regenerative capacity.

Allogeneic use of MSCs is a significant alternative for end-stage organ failure that a tissue biopsy may not yield enough cells for seeding or the cases that could not get a biopsy such as myocardium, cartilage. The viability of using umbilical cord-derived cells for the engineering of cardiovascular constructs has been described in previous studies (Hoerstrup et al. 2002; Schmidt et al. 2006a). Living blood vessels generated from umbilical cord-derived progenitor cells involving WJ-MSCs, with a three-layered tissue architecture that is similar to native blood vessels (Schmidt et al. 2006b). Specific cell phenotypes of each layer could be explained by transdifferentiation of WJ-MSCs due to the applied flow and shear stress or growth factors in the medium.

Bailey et al. reported that WJ-MSCs' expansion on polyglycolic acid (PGA) scaffolds and secretion levels of collagen and glycosaminoglycan (GAG) is higher than temporomandibular joint (TMJ) condylar chondrocytes in vitro (Bailey et al. 2007). Collagen and GAG secretions are markers of chondrogenic differentiation. Likewise, demonstrating differentiation of WJ-MSCs into chondrogenic/osteogenic lineages on PGA scaffolds represents a new cell source for bone tissue engineering (Wang et al. 2009, 2010).

4 Preclinical Studies in Animal Models

WJ-MSCs have been tested in many disease models both in vitro and in vivo due to their proliferation, differentiation capacity, anti-inflammatory, antiapoptotic, and immunomodulatory properties.

Liver diseases are one of the potential uses for stem cell therapy. Limited liver donors and complications of liver transplantations make stem cells a better option for curing these diseases. A previous study has shown that xeno-transplantation of WJ-MSCs could reduce liver fibrosis in rat models. According to the results, WJ-MSCs engrafted in hepatic connective tissue did not differentiate into hepatocyte-like cells. Stem cells restored the liver function by releasing bioactive cytokines (Tsai et al. 2009). In another in vivo study, transplanted WJ-MSCs differentiated into hepatocyte growth factor (HGF), metalloproteinase expressing hepatocyte-like cells and stimulated regeneration of damaged liver (Lin et al. 2010). The following three mechanisms can be suggested for the regenerative effect of WJ-MSCs on liver damages. First, transplanted WJ-MSCs differentiate into hepatocyte-like cells in the environment of host liver cells. By this way, reduced number of hepatocytes can be improved after hepatic commitment of stem cells. The other suggested mechanism is that WJ-MSCs can induce liver regeneration by secreting cytokines such as HGF, IL-6, IL-10 (Li et al. 2013a, b, c). These cytokines suppress hepatic stellate cells that play a key role in hepatic fibrosis by producing matrix components. Lastly, WJ-MSCs can produce matrix metalloproteinase which can degrade extracellular matrix directly, reducing fibrosis (Tsai et al. 2009).

Considering the limited regenerative potential of human cardiovascular system, MSCs can be an attractive candidate for cardiovascular tissue repair. WJ-MSCs could differentiate into cardiomyocyte-like cells after 5-azacytidine or oxytocin treatment (Hollweck et al. 2011; Kaveh et al. 2013). Kaveh et al. have reported that intramyocardial administration of differentiated WJ-MSCs combined with vascular endothelial growth factor (VEGF) improved cardiac function, enhanced angiogenesis, and reduced fibrosis tissue formation after myocardial infarction (MI). VEGF is an inducer of angiogenesis and promotes the proliferation of cardiomyocytes (Hoeben et al. 2004). The combination of differentiated WJ-MSCs and VEGF duplicated the effect of stem cells on MI-induced rabbit models (Kaveh et al. 2013). Transplantation of either differentiated or undifferentiated WJ-MSCs improved left ventricular function 30 days post-MI (Latifpour et al. 2011). There are three possible explanations for the beneficial influence of stem cells in cardiac regeneration: (a) differentiation of WJ-MSCs into myocyte-like cells due to the interaction with the myocardial microenvironment, (b) fusion of engrafted cells with cardiomyocytes, and (c) secretion of growth factors and cytokines by transplanted cells.

Cell therapy strategies have become an increasingly attractive option to develop new treatment techniques for neurodegenerative disorders since the migration ability of MSCs to brain has been demonstrated (Walczak et al. 2008). Transplantation of WJ-MSCs into the brains of hemiparkinsonian rats decreased apomorphine-induced rotations which represent the hypersensitivity of lesioned striatum in the model rats (Weiss et al. 2006). Agah et al. demonstrated that WJ-MSCs could be induced to differentiate into oligodendrocyte-like cells by a combination of trophic factors in vitro. Transplantation of these oligodendrocytes differentiated from WJ-MSCs into the brain ventricles reduced tissue damage in the central nervous system in the animal model of multiple sclerosis (Agah et al. 2013). WJ-MSC-derived oligodendrocyte precursor cells also promote the regeneration of spinal axons and myelin sheaths in animal model of spinal cord injury (Chen et al. 2013). Transplanted WJ-MSCs without any differentiation commitment could survive for 16 weeks and prevent improvements in locomotion in rat spinal cord injury models (Yang et al. 2008).

The capacity of articular cartilage for growth and repair is slow because cartilage does not contain the blood vessels and chondrocytes are fed by diffusion through the synovial fluid. Autologous chondrocyte implantation is used for the treatment of cartilage defects prevalently and has been shown to have some reparative effect on the damaged tissue (Falah et al. 2010). But autologous chondrocyte implantation is a surgical option for young persons and isolation of appropriate number of autologous chondrocytes has technical challenges (Falah et al. 2010; Dahlin et al. 2014). The use of stem cells instead of autologous chondrocytes is an option that would be easier and effective because of their differentiation abilities toward osteogenic and chondrogenic lineages (Wang et al. 2009; Arufe et al. 2011). WJ-MSCs can be considered as candidates for stem cell therapy, based on their expression profile which is positive for cartilage-specific genes, Sox-9 and type II collagen and they can form a cartilage-like tissue in vitro due to the secretion of glycosaminoglycans (GAGs) and type II collagen (Wang et al. 2009; Liu et al. 2014a).

Large skin defects and nonhealing chronic wounds could be life threatening and it has been demonstrated that MSCs enhance wound healing by transdifferentiation and angiogenesis (Wu et al. 2007). Chronic wounds are commonly seen in patients with diabetes mellitus as a consequence of vascular defects. Diabetic wounds have a huge impact on patients' quality of life with a risk of amputation of lower extremities. WJ-MSC infusion could be a curative option for chronic wounds. Transplantation of WJ-MSCs and their conditioned media to animal models showed beneficial therapeutic effects on diabetic wound healing (Shrestha et al. 2013). Enhanced capillary density, increased keratinocyte growth factor (KGF), and platelet-derived growth factor (PDGF) secretion are detected in stem cell transplanted groups. The skin flap is another technique that is used for treating large skin defects. But skin flap necrosis is a common problem in clinical experiments. Injection of WJ-MSCs increased flap survival rate with higher vascular density and improved fibroblast growth factor (FGF) and VEGF levels in a mouse model (Leng et al. 2012).

5 Clinical Applications of WJ-MSCs

5.1 Acute Graft Versus Host Disease (GVHD)

Allogeneic hematopoietic stem cell transplantation (HSCT) is a curative therapy option for a variety of serious malignant and benign diseases such as leukemia, aplastic anemia, and autoimmune diseases. As an immunologically mediated process, involving activation of host antigen-presenting cells and presentation host antigens to donor T cells, GVHD is one of the complications of this treatment procedure (Amorin et al. 2014). Alloactivation of T cells is followed by inflammatory cytokine release and migration of T cells to target tissues such as skin, liver, gut, etc. Alloreactive T cells cause tissue destruction by cytotoxic activity (Sung and Chao 2013).

The immune attack against the patient's own tissues is a major cause of morbidity and mortality of HSCT patients (Sung and Chao 2013). Glucocorticoids are the primary therapy of grade II–IV acute GVHD (Qian et al. 2013). But 60 % of patients have minimal or no responses to corticosteroids which could be defined as steroid-resistant acute GVHD (Magenau and Reddy 2014). Although there is no standard treatment protocol for steroid-resistant acute GVHD, some agents such as antithymocyte globulin, methotrexate (MTX), and mycophenolatemofetil (MMF) are used clinically with low efficiency. Properties of MSCs suggest their potential use for suppressing GVHD without impairing graft versus leukemia effects. Promising treatments for steroid-resistant acute GVHD involve infusion of third-party mesenchymal stem cells. Third-party bone marrow-derived MSCs (BM-MSCs) have been studied for treating GVHD prevalently since Blanc et al. treated a patient with severe treatment-resistant grade IV acute GVHD by infusion of BM-MSCs for the first time (Blanc et al. 2004). Clinical outcomes suggested that BM-MSCs are safe and effective treatment option for GVHD (Prasad et al. 2011; Kurtzberg et al. 2014). But in comparison with cord matrix, bone marrow has more invasive and painful procedure for aspiration and BM-MSCs as adult-derived stem cells have limited expansion potential (Baksh et al. 2007). WJ-MSCs could be substituted for BM-MSCs for treatment of GVHD based on safety, low cost and noninvasive harvesting procedure, quicker expansion ability, being suitable for ready-to-use cell banking, and similar immunomodulatory properties (Baksh et al. 2007; Yoo et al. 2009). First clinical application of WJ-MSCs treatment for GVHD patients was reported in 2011 (Wu et al. 2011). Third-party WJ-MSCs intravenously infused to two patients with severe steroid-resistant acute GVHD and researchers found that these cells had superior proliferative potential and more suppressive effects on peripheral blood mononuclear cell proliferation when compared with BM-MSCs. Both patients had no adverse effects and acute GVHD improved dramatically after four doses of WJ-MSCs infusion. Since then, WJ-MSCs became an alternative to BM-MSCs for treatment of GVHD.Considering our results (unpublished data), 37 % of the patients had complete response and 40 % of the patients had partial response to WJ-MSC injection (Fig. 7).

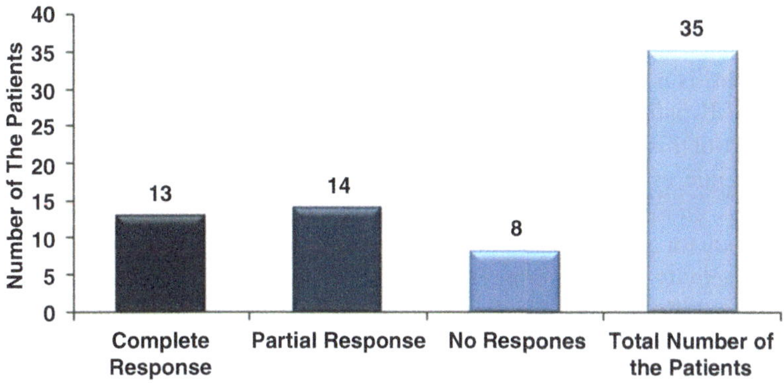

Fig. 7 The response rates to WJ-MSCs infusion of patients with GVHD (Unpublished Data)

5.2 Liver Cirrhosis (LC)

Nearly 350 million people are chronically infected with hepatitis B virus (HBV) and the infection is the tenth leading cause of death (Trépo et al. 2014). The major cause of morbidity and mortality patients with chronic HBV infection is liver fibrosis which is characterized by accumulation of extracellular matrix proteins (Jieanu et al. 2015). If chronic inflammation persists, liver fibrosis results in cirrhosis and liver failure. The most common antifibrotic therapies are treating or removing the underlying stimulus-like infection, metabolic diseases that cause fibrogenesis (Rockey and Friedman 2006). Anti-inflammatory therapies are one of the strategies to abrogate fibrogenesis. Because it is known that persistent inflammation almost always drives the fibrogenic cascade (Lee and Friedman 2011). Liver transplantation is also an option to treat patients with especially end-stage liver cirrhosis. Complications of liver transplantation and limited organ donations forced scientists to focus on alternative therapeutic approaches. MSCs infusion is one of these strategies and it has been shown that BM-MSCs could improve the liver function in patients with LC (Terai et al. 2006; Mohamadnejad et al. 2007). Zhang et al. examined the safety and efficacy of WJ-MSC infusion in patients with LC. According to the study results, WJ-MSC infusion decreased the levels of serum LC markers and improved liver functions (Zhang et al. 2012). Therapeutic effects of WJ-MSCs can be explained as: Stem cells might exert antifibrotic or fibrolytic effects via overexpression of matrix metalloproteinases (MMP) and induction of hepatic stellate cell (HSC) apoptosis.

5.3 Systemic Lupus Erythematosus (SLE)

SLE is a chronic autoimmune disorder characterized by irreversible break in immunological tolerance and the presence of autoreactive immune cells (Ghodke-puranik and Niewold 2015). Combinations of symptoms vary between different patients.

Photosensitivity, arthritis, pleuritis, pericarditis, nephritis, proteinuria, neuropsychiatric disorders, and hematological disorders are common complications of SLE. Lupus nephritis is one of the major complications and cause of morbidity and mortality in SLE patients (Balow 2005). Conventional immunosuppressive therapies cannot control this disease in all of patients with SLE. In 2010, Sun et al. determined the therapeutic effect of WJ-MSCs in refractory SLE patients (Sun et al. 2010). Researchers chose 16 patients with refractory SLE and WJ-MSCs administered intravenously for a single dose. Patients had no adverse effects and had improvement in disease activity for long term. Therapeutic effect could be explained by immunomodulation abilities of MSCs. In this study, it has been observed that WJ-MSCs could up-regulate the percentage of regulatory T cells (Treg) by increasing levels of TGF-β and IL-10. On the other hand, IL-4 levels in the patient's serum were decreased which may cause inhibition of humoral immunity. In another clinical application, a 19-year-old girl who was diagnosed with SLE underwent transplantation of WJ-MSCs for diffuse alveolar hemorrhage (Liang et al 2010). After single dose infusion, this patient showed improvements in her clinical condition, oxygenation level, radiographic and hematological status.

5.4 Spinal Cord Injury

Spinal cord injuries (SCIs) cause disruption of the communication between the brain and parts of the body which are innervated at the lesion area. Two types of SCI are demonstrated:

Complete injuries that patient lose all the ability to feel and move; incomplete injuries that one or more nerve fibers are preserved and patient has some degree of sensation (Crewe and Krause 2002). Corticosteroids and surgeries to stabilize the injury site are common treatment strategies for patients with SCI. Besides, cellular therapy-based clinical trials and researches on the treatment of SCI are still ongoing in recent years. Treatment of spinal cord injuries with MSCs is a new approach promising clinical improvements. Liu et al. transplanted WJ-MSCs intrathecally into 22 patients with SCI (Lv et al. 2013). Referring to experimental results, 81.25 % of patients with incomplete SCI responded to therapy and showed improvements in neurologic functions. But WJ-MSCs transplantation to patients with complete SCI had no effect. Researchers confirmed that intrathecally injection of WJ-MSCs is safe and could improve quality of life of patients with incomplete SCI.

5.5 Cerebellar Ataxia

Spinocerebellar ataxia (SCA) and multiple system atrophy-cerebellar type (MSA-C) are neurodegenerative diseases that are characterized by progressive cerebellarataxia. Astasia, movement disorders, peripheral neuropathy, cognitive dysfunction, and finally losing self-care ability are some of other clinical manifestations. Although preclinical animal studies or clinical trials with sulfamethoxazole and trimethoprim have shown encouraging results, there isn't any effective treatment protocol for controlling the disease (Underwood and Rubinsztein 2008). In 2011, 24 patients with SCA or MSA-C treated with intrathecal injection of WJ-MSC (Dongmei et al. 2011). Based on these results, intrathecal injection of WJ-MSC is safe and 23 patients have shown improvement by International Cooperative Ataxia Rating Scale (ICARS) and Activity of Daily Living Scale (ADL). In another clinical experiment, Improved Berg Balance Scale (BBS) and ICARS were found in majority of 16 patients with SCA who received intravenous and intrathecal injections of WJ-MSCs and these improved scores persisted at least 6 months after transplantation (Jin et al. 2013).

5.6 Stevens–Johnson Syndrome (SJS)

SJS which is an acute skin and mucosal membranes inflammatory syndrome was first described in 1922 (Stevens and Johnson 1922). SJS characterized by keratinocyte death and epidermal necrosis resulting in skin damage, ocular involvement, malaise, cough, rhinorrhea, and anorexia. SJS carries a significant risk of mortality ranging from 1 to 5 %. Some case reports have shown benefits of high-dose intravenous corticosteroids, intravenous immunoglobulin, plasmapheresis, and TNFα inhibitors (Kohanim et al. 2016). Due to immunomodulatory properties of MSCs, the effects of MSC transplantation to SJS patients are being investigated recently. Three female SJS patients ranging in age from 42 to 62 years recovered after WJ-MSC transplantation (Li et al. 2013b) while three patients, who did not respond to treatments with antibiotics, showed improvement in their general condition and in laboratory index due to cell therapy.

5.7 Autism Spectrum Disorders (ASD)

Autism is a neurodevelopmental disorder which is characterized by repetitive behavior patterns, social, and communication deficits. However, although etiology of ASD is not clear, abnormal brain development and rapid head growth could be early signals for diagnosis. Therapeutic approaches involve special educational programming, communication training, and complementary medicine interventions (Myers and Johnson 2007). According to the phase I/II study, infusion of MSCs could be a promising alternative therapy for treating autism (Lv Y-T et al. 2013). Transplantation of the combination of cord blood mononuclear cells (CBMN) and WJ-MSCs is more efficacious comparing the CBMN groups. Synergetic effects between CBMN and WJ-MSCs improved some behavioral symptoms and showed larger therapeutic effects in patients with autism.

5.8 *Diabetes Mellitus (DM)*

DM is defined as a metabolic disorder resulting from the deterioration of glucose and fat metabolism of body. Type-I and type-II diabetes are the most common forms of DM. Type-I diabetes develops by formation of autoimmune response against beta cells which are responsible for insulin secretion in islets of Langerhans. Inadequate insulin secretion by beta cells, in the setting of insulin resistance, leads to the development of type-II diabetes. Although, diet control, physical exercise, and antidiabetic drugs can decrease hyperglycemia in patients with type-II diabetes, beta cells are also degenerated in time, and their number also declines steadily causing elevated blood sugars. Curative therapies for DM involve replacement of the destroyed beta cells and restoring self-tolerance. Pancreas and islet transplantations have been performed to several patients with beneficial results. Limited donors and technical difficulties forced the researchers to explore new treatment strategies. WJ-MSC transplantation could become an alternative therapy for the patients with type-I and type-II diabetes. Hu et al. reported that implantation of WJ-MSCs is safe and restore beta cell function in newly diagnosed patients with type-I diabetes (Hu et al. 2013). Patients were followed up for the 21 months and HbA1c and C-peptide levels which give an opinion about blood levels of insulin were better. WJ-MSC transplantation is also effective for the treatment of type-II diabetes. In 2014, 22 patients with type-II diabetes received a single dose of WJ-MSC transplantation (Liu et al. 2014b). These results show evidence of improved metabolic control and beta cell function with no adverse reactions.

6 Stem Cell-Derived Exosomes/Microvesicles (MV)

Paracrine effect of MSCs plays an important role in efficiency of regenerative therapies. MSCs could modulate the microenvironment by secreting variety of cytokines, chemokines, and growth factors but these small molecules are secreted at low levels. It has been documented that MSCs also release extracellular vesicles (exosomes/microvesicles, ectosomes, membrane particles, etc.) which are taking part to the paracrine activity of the cells (Yeo et al. 2013).

Exosomes/MVs which are derived from intracellular compartments are around 40–100 nm in diameter. The MVs can be purified from conditioned media of stem cells by ultracentrifugation or density gradient centrifugation. According to proteomic studies, MSC-derived MVs include cell adhesion (CD29, CD44, and CD73) and MSC-associated antigens (CD9, CD63, CD81, CD109, CD151, CD248, and CD276), surface receptors, signaling molecules that have a role in differentiation and self-renewal mechanisms (Kim et al. 2012; Yu et al. 2014). It could be considered that MSC-MVs are alternative sources for regenerative therapies through their similar characteristics to MSCs.

A variety of preclinical studies showed that MSC-MVs have beneficial effects on animal models with kidney injury, intervertebral disc (IVD) degeneration, cardiovascular, and liver diseases (Lai et al. 2011; Bruno et al. 2012; Strassburg et al. 2012; Zhou et al. 2013). Transplantation of MV derived from WJ-MSCs reduced hepatic inflammation and collagen deposition in the liver fibrosis (Li et al. 2013c). It has been also reported that MV derived from WJ-MSCs could protect against cisplatin-induced nephrotoxicity (Zhou et al. 2013). Therefore, MVs derived from MSCs can be an opportunity to cell-free regenerative therapy approaches in the future.

7 Future Applications

MSCs' properties such as homing, chemoattraction, survival, multilineage differentiation potential, immunomodulation, antiapoptotic, and anti-inflammatory effects make them an ideal option for possible future cellular therapies. Although BM- and AT-MSCs have been the most commonly used sources of MSCs for clinical applications, WJ-MSCs became an alternative source due to their characteristics mentioned earlier and as a result of their efficient stem cell potency and noninvasive collection procedure.

WJ-MSCs can be used as a vehicle for targeted-cancer therapy due to their tumor tropism. Systematically administration of engineered WJ-MSCs could enable the transport of antitumor agents (anti-VEGF, oncolytic factors, etc.) directly to the tumor area. Therapeutic gene delivery with WJ-MSCs is an option to overcome adverse effects of cytotoxic cytokines and gene therapy with viral vectors which are potentially pathogenic.

WJ-MSCs provide potential for treating autoimmune disorders due to their immunomodulation ability. They are used in an increasing number of clinical trials for the treatment of GVHD and hold the potential for other diseases such as diabetes, SLE. WJ-MSCs offer valuable opportunities for regenerative therapies, as they have minimal ethical concerns and they are suitable for use as third-party stem cell source especially for GVHD treatment. WJ-MSCs are thought to become an important cell source for tissue engineering applications in the future due to their characteristics making them possible to be used in allogeneic applications.

Acknowledgements The authors would like to thank AyberkAkat, M.Sc. for his contributions in the writing of the manuscript.

References

Agah E, Parivar K, Joghataei M (2013) Therapeutic effect of transplanted human Wharton's Jelly stem cell-derived oligodendrocyte progenitor cells (hWJ-MSC-derived OPCs) in an animal model of multiple sclerosis. Mol Neurobiol 49:625–632

Amorin B, Alegretti A, Valim V et al (2014) Mesenchymal stem cell therapy and acute graft-versus-host disease: a review. Hum Cell 27(4):137–150
Antoninus A, Widowati W, Wijaya L et al (2015) Human platelet lysate enhances the proliferation of Wharton's jelly-derived mesenchymal stem cells. Biomark Genom Med 7:87–97
Arufe M, Fuente A, Mateos J et al (2011) Analysis of the chondrogenic potential and secretome of mesenchymal stem cells derived from human umbilical cord stroma. Stem Cells Dev 20:1199–1212
Ayuzawa R, Doi C, Rachakatla R et al (2009) Naïve human umbilical cord matrix derived stem cells significantly attenuate growth of human breast cancer cells in vitro and in vivo. Cancer Lett 280:31–37
Bailey M, Wang L, Bode C et al (2007) A comparison of human umbilical cord matrix stem cells and temporomandibular joint condylar chondrocytes for tissue engineering temporomandibular joint condylar cartilage. Tissue Eng 13:2003–2010
Baksh D, Yao R, Tuan R (2007) Comparison of proliferative and multilineage differentiation potential of human mesenchymal stem cells derived from umbilical cord and bone marrow. Stem Cells 25:1384–1392
Balow J (2005) Clinical presentation and monitoring of lupus nephritis. Lupus 14:25–30
Borhani-Haghighi M, Talaei-Khozani T, Ayatollahi M, Vojdani Z (2015) Wharton's Jelly-derived mesenchymal stem cells can differentiate into hepatocyte-like cells by HepG2 cell line extract. Iran J Med Sci 40:143–151
Broxmeyer H, Douglas G, Hangoc G et al (1989) Human umbilical cord blood as a potential source of transplantable hematopoietic stem/progenitor cells. Proc Natl Acad Sci 86:3828–3832
Bruno S, Grange C, Collino F et al (2012) Microvesicles derived from mesenchymal stem cells enhance survival in a lethal model of acute kidney injury. PLoS One 7(3):e33115
Can A, Karahuseyinoglu S (2007) Concise review: human umbilical cord stroma with regard to the source of fetus-derived stem cells. Stem Cells 25:2886–2895
Carlin R, Davis D, Weiss M et al (2006) Expression of early transcription factors Oct-4, Sox-2 and Nanog by porcine umbilical cord (PUC) matrix cells. Reprod Biol Endocrinol 4:8
Chen H, Zhang Y, Yang Z et al (2013) Human umbilical cord Wharton's jelly-derived oligodendrocyte precursor-like cells for axon and myelin sheath regeneration. Neural Regen Res 8(10):890–899
Conconi MT, Liddo LD, Tommasini M et al (2011) Phenotype and differentiation potential of stromal populations obtained from various zones of human umbilical cord: an overview. Open Tissue Eng Regen Med J 4:6–20
Corotchi M, Popa M, Remes A et al (2013) Isolation method and xeno-free culture conditions influence multipotent differentiation capacity of human Wharton's jelly-derived mesenchymal stem cells. Stem Cell Res Amp Ther 4:81
Crewe N, Krause JS (2002) Spinal cord injuries. In: Brodwin MG, Tellez FA, Brodwin SK (eds) Medical, psychosocial, and vocational aspects of disability, 2nd edn. Athens, Elliott & Fitzpatrick, pp 279–291
Dahlin R, Kinard L, Lam J et al (2014) Articular chondrocytes and mesenchymal stem cells seeded on biodegradable scaffolds for the repair of cartilage in a rat osteochondral defect model. Biomaterials 35:7460–7469
Ding DC, Chou HL, Chang YH et al (2015) Characterization of HLA-G and related immunosuppressive effects in human umbilical cord stroma derived stem cells. Cell Transplant 25:217–228
Doi C, Maurya DK, Pyle MM et al (2010) Cytotherapy with naive rat umbilical cord matrix stem cells significantly attenuates growth of murine pancreatic cancer cells and increases survival in syngeneic mice. Cytotherapy 12(3):408–417
Doitsidou M, Reichman-Fried M, Stebler J et al (2002) Guidance of primordial germ cell migration by the chemokine SDF-1. Cell 111:647–659
Dominici M, Blanc K, Mueller I et al (2006) Minimal criteria for defining multipotent mesenchymal stromal cells. The International Society for Cellular Therapy position statement. Cytotherapy 8:315–317

Dongmei H, Jing L, Mei X et al (2011) Clinical analysis of the treatment of spinocerebellar ataxia and multiple system atrophy-cerebellar type with umbilical cord mesenchymal stromal cells. Cytotherapy 13:913–917

Falah M, Nierenberg G, Soudry M et al (2010) Treatment of articular cartilage lesions of the knee. Int Orthop 34:621–630

Fong C, Richards M, Manasi N et al (2007) Comparative growth behaviour and characterization of stem cells from human Wharton's jelly. Reprod Biomed Online 15:708–718

Fong C-Y, Chak L-L, Biswas A et al (2011) Human Wharton's Jelly stem cells have unique transcriptome profiles compared to human embryonic stem cells and other mesenchymal stem cells. Stem Cell Rev 7:1–16

Gauthaman K, Yee F, Cheyyatraivendran S et al (2012) Human umbilical cord wharton's jelly stem cell (hWJSC) extracts inhibit cancer cell growth in vitro. J Cell Biochem 113:2027–2039

Ghodke-Puranik Y, Niewold T (2015) Immunogenetics of systemic lupus erythematosus: a comprehensive review. J Autoimmun 64:125–136

Gluckman E, Rocha V (2005) History of the clinical use of umbilical cord blood hematopoietic cells. Cytotherapy 7:219–227

Han Y, Chai J, Sun T et al (2011) Differentiation of human umbilical cord mesenchymal stem cells into dermal fibroblasts in vitro. Biochem Bioph Res Co 413:561–565

He H, Nagamura-Inoue T, Takahashi A et al (2015) Immunosuppressive properties of Wharton's jelly-derived mesenchymal stromal cells in vitro. Int J Hematol 102(3):368–378

Hentze H, Soong P, Wang S et al (2009) Teratoma formation by human embryonic stem cells: evaluation of essential parameters for future safety studies. Stem Cell Res 2:198–210

Hoeben A, Landuyt B, Highley MS et al (2004) Vascular endothelial growth factor and angiogenesis. Pharmacol Rev 56(4):549–580

Hoerstrup SP, Kadner A, Breymann C et al (2002) Living, autologous pulmonary artery conduits tissue engineered from umbilical cord cells. Ann Thorac Surg 74(1):46–52

Hollweck T, Hartmann I, Eblenkamp M (2011) Cardiac differentiation of Human Wharton`s jelly stem cells—experimental comparison of protocols. Open Tissue Eng Regen Med J 4:95–102

Horuk R, Peiper S (1996) Chemokines: molecular double agents. Curr Biol 6:1581–1582

Hu J, Yu X, Wang Z et al (2013) Long term effects of the implantation of Wharton's jelly-derived mesenchymal stem cells from the umbilical cord for newly-onset type 1 diabetes mellitus. Endocr J 60(3):347–357

Jacobs SA, Roobrouck VD, Verfaillie CM et al (2013) Immunological characteristics of human mesenchymal stem cells and multipotent adult progenitor cells. Immunol Cell Biol 91(1):32–39

Jieanu CF, Ungureanu BS, Săndulescu DL et al (2015) Quantification of liver fibrosis in chronic hepatitis B virus infection. J Med Life 8(3):285–290

Jin JL, Liu Z, Lu ZJ et al (2013) Safety and efficacy of umbilical cord mesenchymal stem cell therapy in hereditary spinocerebellar ataxia. Curr Neurovasc Res 10(1):11–20

Jonsdottir-Buch S, Lieder R, Sigurjonsson O (2013) Platelet lysates produced from expired platelet concentrates support growth and osteogenic differentiation of mesenchymal stem cells. PLoS One 8(7):e68984

Kalaszczynska I, Ferdyn K (2015) Wharton's jelly derived mesenchymal stem cells: future of regenerative medicine? Recent findings and clinical significance. Biomed Res Int 2015:430847

Kanehira M, Xin H, Hoshino K et al (2007) Targeted delivery of NK4 to multiple lung tumors by bone marrow-derived mesenchymal stem cells. Cancer Gene Ther 14:894–903

Kao SY, Shyu JF, Wang HS et al (2015) Comparisons of differentiation potential in human mesenchymal stem cells from Wharton's jelly, bone marrow, and pancreatic tissues. Stem Cells Int 2015:306158

Kaveh M, Mehdi A, Farid A et al (2013) Restoration of heart function using transplantation of human umbilical cord matrix-derived cardiomyocytes and vascular endothelial growth factor. Open Tissue Eng Regen Med J 6:26–36

Khakoo A, Pati S, Anderson S et al (2006) Human mesenchymal stem cells exert potent antitumorigenic effects in a model of Kaposi's sarcoma. J Exp Med 203:1235–1247

Kholodenko IV, Konieva AA, Kholodenko RV et al (2013) Molecular mechanisms of migration and homing of intravenously transplanted mesenchymal stem cells. J Regen Med 2:4
Killat J, Reimers K, Choi C et al (2013) Cultivation of keratinocytes and fibroblasts in a three-dimensional bovine collagen-elastin matrix (Matriderm®) and application for full thickness wound coverage in vivo. Int J Mol Sci 14(7):14460–14474
Kim H-S, Choi D-Y, Yun S et al (2012) Proteomic analysis of microvesicles derived from human mesenchymal stem cells. J Proteome Res 11:839–849
Klyushnenkova E, Mosca J, Zernetkina V et al (2005) T cell responses to allogeneic human mesenchymal stem cells: immunogenicity, tolerance, and suppression. J Biomed Sci 12:47–57
Knudtzon S (1974) In vitro growth of granulocytic colonies from circulating cells in human cord blood. Blood 43(3):357–361
Kohanim S, Palioura S, Saeed HN (2016) Stevens-Johnson syndrome/toxic epidermal necrolysis - a comprehensive review and guide to therapy. I. Systemic Disease. Ocul Surf 14(1):2–19
Kurtzberg J, Prockop S, Teira P et al (2014) Allogeneic human mesenchymal stem cell therapy (Remestemcel-L, Prochymal) as a rescue agent for severe refractory acute graft-versus-host disease in pediatric patients. Biol Blood Marrow Transplant 20:229–235
Lai R, Chen T, Lim S (2011) Mesenchymal stem cell exosome: a novel stem cell-based therapy for cardiovascular disease. Regen Med 6:481–492
Latifpour M, Nematollahi-Mahani SN, Deilamy M et al (2011) Improvement in cardiac function following transplantation of human umbilical cord matrix-derived mesenchymal cells. Cardiology 120:9–18
Le Blanc K, Rasmusson I, Sundberg B et al (2004) Treatment of severe acute graft-versus-host disease with third party haploidentical mesenchymal stem cells. Lancet 363(9419):1439–1441
Lee U, Friedman S (2011) Mechanisms of hepatic fibrogenesis. Best Pract Res Clin Gastroenterol 25:195–206
Lee H, Lee J, Lee H et al (2010) The therapeutic potential of human umbilical cord blood-derived mesenchymal stem cells in Alzheimer's disease. Neurosci Lett 481:30–35
Leite C, Silva NT, Mendes S et al (2014) Differentiation of human umbilical cord matrix mesenchymal stem cells into neural-like progenitor cells and maturation into an oligodendroglial-like lineage. PLoS One 9(10):e111059
Leng X, Zhang Q, Zhai X, Chen Z (2012) Local transplant of human umbilical cord matrix stem cells improves skin flap survival in a mouse model. Tohoku J Exp Med 227:191–197
Li Z, He C, Xiao J, Chen Z (2013a) Treating end-stage liver diseases with mesenchymal stem cells: an oak is not felled at one stroke. Oa Tissue Eng 1(1):3
Li X, Wang D, Lu Z et al (2013b) Umbilical cord mesenchymal stem cell transplantation in drug-induced Stevens—Johnson syndrome. J Eur Acad Dermatol Venereol 27:659–661
Li T, Yan Y, Wang B et al (2013c) Exosomes Derived from Human Umbilical Cord Mesenchymal Stem Cells Alleviate Liver Fibrosis. Stem Cells Dev 22:845–854
Liang J, Gu F, Wang H et al (2010) Mesenchymal stem cell transplantation for diffuse alveolar hemorrhage in SLE. Nat Rev Rheumatol 6:486–489
Lila N, Carpentier A, Amrein C et al (2000) Implication of HLA-G molecule in heart-graft acceptance. Lancet Lond Engl 355:2138
Lin S-Z, Chang Y-J, Liu J-W et al (2010) Transplantation of human Wharton's jelly-derived stem cells alleviates chemically induced liver fibrosis in rats. Cell Transplant 19:1451–1463
Liu S, Hou K, Yuan M et al (2014a) Characteristics of mesenchymal stem cells derived from Wharton's jelly of human umbilical cord and for fabrication of non-scaffold tissue-engineered cartilage. J Biosci Bioeng 117:229–235
Liu X, Zheng P, Wang X et al (2014b) A preliminary evaluation of efficacy and safety of Wharton's jelly mesenchymal stem cell transplantation in patients with type 2 diabetes mellitus. Stem Cell Res Ther 5:57
Lv Y-T, Zhang Y, Liu M et al (2013) Transplantation of human cord blood mononuclear cells and umbilical cord-derived mesenchymal stem cells in autism. J Transl Med 11:196

Magenau J, Reddy P (2014) Next generation treatment of acute graft-versus-host disease. Leukemia 28:2283–2291

Matsuzuka T, Rachakatla R, Doi C et al (2010) Human umbilical cord matrix-derived stem cells expressing interferon-β gene significantly attenuate bronchioloalveolar carcinoma xenografts in SCID mice. Lung Cancer 70:28–36

McElreavey KD, Irvine AI, Ennis KT et al (1991) Isolation, culture and characterisation of fibroblast-like cells derived from the Wharton's jelly portion of human umbilical cord. Biochem Soc Trans 19(1):29S

Mennan C, Wright K, Bhattacharjee A et al (2013) Isolation and characterisation of mesenchymal stem cells from different regions of the human umbilical cord. Biomed Res Int 2013:916136

Mohamadnejad M, Alimoghaddam K, Mohyeddin-Bonab M et al (2007) Phase 1 trial of autologous bone marrow mesenchymal stem cell transplantation in patients with decompensated liver cirrhosis. Arch Iran Med 10:459–466

Mortezaee K, Minaii B, Sabbaghziarani F et al (2015) Retinoic acid as the stimulating factor for differentiation of Wharton's jelly-mesenchymal stem cells into hepatocyte-like cells. Avicenna J Med Biotechnol 7:106–112

Myers SM, Johnson CP (2007) Management of children with autism spectrum disorders. Pediatrics 120(5):1162–1182

Nagamura-Inoue T, He H (2014) Umbilical cord-derived mesenchymal stem cells: their advantages and potential clinical utility. World J Stem Cells 6:195

Nakamizo A, Marini F, Amano T et al (2005) Human bone marrow-derived mesenchymal stem cells in the treatment of gliomas. Cancer Res 65:3307–3318

Nasef A, Mathieu N, Chapel A et al (2007) Immunosuppressive effects of mesenchymal stem cells: involvement of HLA-G. Transplantation 84:231–237

Nekanti U, Rao V, Bahirvani A et al (2010) Long-term expansion and pluripotent marker array analysis of Wharton's jelly-derived mesenchymal stem cells. Stem Cells Dev 19:117–130

Nicola M, Carlo-Stella C, Magni M et al (2002) Human bone marrow stromal cells suppress T-lymphocyte proliferation induced by cellular or nonspecific mitogenic stimuli. Blood 99:3838–3843

Parry EW (1970) Some electron microscope observations on the mesenchymal structures of full-term umbilical cord. J Anat 107(Pt 3):505–518

Patel S, Meyer J, Greco S et al (2010) Mesenchymal stem cells protect breast cancer cells through regulatory T cells: role of mesenchymal stem cell-derived TGF-β. J Immunol 184:5885–5894

Prasad V, Lucas K, Kleiner G et al (2011) Efficacy and safety of ex vivo cultured adult human mesenchymal stem cells (Prochymal™) in pediatric patients with severe refractory acute graft-versus-host disease in a compassionate use study. Biol Blood Marrow Transplant 17:534–541

Qian L, Wu Z, Shen J (2013) Advances in the treatment of acute graft-versus-host disease. J Cell Mol Med 17:966–975

Qiao L, Xu Z, Zhao T et al (2008) Suppression of tumorigenesis by human mesenchymal stem cells in a hepatoma model. Cell Res 18:500–507

Rachakatla R, Marini F, Weiss M et al (2007) Development of human umbilical cord matrix stem cell-based gene therapy for experimental lung tumors. Cancer Gene Ther 14:828–835

Ramasamy R, Lam E, Soeiro I et al (2006) Mesenchymal stem cells inhibit proliferation and apoptosis of tumor cells: impact on in vivo tumor growth. Leukemia 21:304–310

Ren C, Kumar S, Chanda D et al (2008) Therapeutic potential of mesenchymal stem cells producing interferon-alpha in a mouse melanoma lung metastasis model. Stem Cells 26(9):2332–2338

Rockey D, Friedman SL (2006) Hepatic fibrosis and cirrhosis. In: Section I: Pathophysiology of the liver. Elsevier, Amsterdam, pp 87–109

Salehinejad P, Alitheen N, Ali A et al (2012) Comparison of different methods for the isolation of mesenchymal stem cells from human umbilical cord Wharton's jelly. In Vitro Cell Dev Biol Anim 48:75–83

Sariboyaci AE, Demircan PC, Gacar G et al (2014) Immunomodulatory properties of pancreatic islet-derived stem cells co-cultured with T cells: does it contribute to the pathogenesis of type 1 diabetes? Exp Clin Endocrinol Diabetes 122(3):179–189

Scheers I, Lombard C, Paganelli M et al (2013) Human umbilical cord matrix stem cells maintain multilineage differentiation abilities and do not transform during long-term culture. PLoS One 8(8):e71374
Schmidt D, Mol A, Breymann C et al (2006a) Living autologous heart valves engineered from human prenatally harvested progenitors. Circulation 114:I125–I131
Schmidt D, Asmis L, Odermatt B et al (2006b) Engineered living blood vessels: functional endothelia generated from human TM umbilical cord-derived progenitors. Ann Thorac Surg 82:1465–1471; discussion 1471
Shrestha C, Zhao L, Chen K, et al (2013) Enhanced healing of diabetic wounds by subcutaneous administration of human umbilical cord derived stem cells and their conditioned media. Int J Endocrinol 2013, Article ID 592454, 10 pages
Sohni A, Verfaillie C (2013) Mesenchymal stem cells migration homing and tracking. Stem Cells Int 2013, Article ID 130763, 8 pages
Song C, Li G (2011) CXCR4 and matrix metalloproteinase-2 are involved in mesenchymal stromal cell homing and engraftment to tumors. Cytotherapy 13:549–561
Stevens A, Johnson F (1922) A new eruptive fever associated with stomatitis and ophthalmia: report of two cases in children. Am J Dis Child 24:526–533
Strassburg S, Hodson N, Hill P et al (2012) Bi-directional exchange of membrane components occurs during co-culture of mesenchymal stem cells and nucleus pulposus cells. PLoS One 7(3):e33739
Subramanian A, Fong C-Y, Biswas A, Bongso A (2015) Comparative characterization of cells from the various compartments of the human umbilical cord shows that the Wharton's jelly compartment provides the best source of clinically utilizable mesenchymal stem cells. PLoS One 10(6):e0127992
Sun L, Wang D, Liang J et al (2010) Umbilical cord mesenchymal stem cell transplantation in severe and refractory systemic lupus erythematosus. Arthritis Rheum 62:2467–2475
Sung A, Chao N (2013) Acute graft-versus-host disease: are we close to bringing the bench to the bedside? Best Pract Res Clin Haematol 26:285–292
Terai S, Ishikawa T, Omori K et al (2006) Improved liver function in patients with liver cirrhosis after autologous bone marrow cell infusion therapy. Stem Cells 24:2292–2298
Trépo C, Chan H, Lok A (2014) Hepatitis B virus infection. Lancet 384:2053–2063
Trivanović D, Kocić J, Mojsilović S et al (2013) Mesenchymal stem cells isolated from peripheral blood and umbilical cord Wharton's jelly. Srp Ark Celok Lek 141:178–186
Tsai P, Fu T, Chen Y et al (2009) The therapeutic potential of human umbilical mesenchymal stem cells from Wharton's jelly in the treatment of rat liver fibrosis. Liver Transpl 15:484–495
Underwood B, Rubinsztein D (2008) Spinocerebellar ataxias caused by polyglutamine expansions: a review of therapeutic strategies. Cerebellum 7:215–221
Walczak P, Zhang J, Gilad A et al (2008) Dual-modality monitoring of targeted intraarterial delivery of mesenchymal stem cells after transient ischemia. Stroke 39:1569–1574
Wang H, Hung S, Peng S et al (2004) Mesenchymal stem cells in the Wharton's jelly of the human umbilical cord. Stem Cells 22:1330–1337
Wang L, Tran I, Seshareddy K et al (2009) A comparison of human bone marrow-derived mesenchymal stem cells and human umbilical cord-derived mesenchymal stromal cells for cartilage tissue engineering. Tissue Eng Part A 15:2259–2266
Wang L, Dormer N, Bonewald L, Detamore M (2010) Osteogenic differentiation of human umbilical cord mesenchymal stromal cells in polyglycolic acid scaffolds. Tissue Eng Part A 16:1937–1948
Wang Y, Fan H, Zhou B et al (2011) Fusion of human umbilical cord mesenchymal stem cells with esophageal carcinoma cells inhibits the tumorigenicity of esophageal carcinoma cells. Int J Oncol 40:370–377
Weiss M, Medicetty S, Bledsoe A et al (2006) Human umbilical cord matrix stem cells: preliminary characterization and effect of transplantation in a rodent model of Parkinson's disease. Stem Cells 24:781–792

Weiss M, Anderson C, Medicetty S et al (2008) Immune properties of human umbilical cord Wharton's jelly-derived cells. Stem Cells 26:2865–2874
Wetzig A, Alaiya A, Al-Alwan M et al (2013) Differential marker expression by cultures rich in mesenchymal stem cells. BMC Cell Biol 14:54
Wu Y, Chen L, Scott P, Tredget E (2007) Mesenchymal stem cells enhance wound healing through differentiation and angiogenesis. Stem Cells 25:2648–2659
Wu K-H, Chan C-K, Tsai C et al (2011) Effective treatment of severe steroid-resistant acute graft-versus-host disease with umbilical cord-derived mesenchymal stem cells. Transplantation 91:1412
Xu G, Zhang L, Ren G et al (2007) Immunosuppressive properties of cloned bone marrow mesenchymal stem cells. Cell Res 17:240–248
Yang C-C, Shih Y-H, Ko M-H et al (2008) Transplantation of human umbilical mesenchymal stem cells from Wharton's jelly after complete transection of the rat spinal cord. PLoS One 3:e3336
Yeo R, Chai R, Hian K, Kiang S (2013) Exosome: a novel and safer therapeutic refinement of mesenchymal stem cell. Exosomes Microvesicles 1:1–12
Yoo K, Jang I, Lee M et al (2009) Comparison of immunomodulatory properties of mesenchymal stem cells derived from adult human tissues. Cell Immunol 259:150–156
Yu B, Zhang X, Li X (2014) Exosomes derived from mesenchymal stem cells. Int J Mol Sci 15(3):4142–4157
Zhang Z, Lin H, Shi M et al (2012) Human umbilical cord mesenchymal stem cells improve liver function and ascites in decompensated liver cirrhosis patients. J Gastroen Hepatol 27:112–120
Zheng G, Liu Y, Jing Q, Zhang L (2015) Differentiation of human umbilical cord-derived mesenchymal stem cells into hepatocytes in vitro. Biomed Mater Eng 25(1 Suppl):145–157
Zhou Y, Xu H, Xu W et al (2013) Exosomes released by human umbilical cord mesenchymal stem cells protect against cisplatin-induced renal oxidative stress and apoptosis in vivo and in vitro. Stem Cell Res Amp Ther 4:34
Zhuang H, Zhang R, Zhang S et al (2015) Altered expression of microRNAs in the neuronal differentiation of human Wharton's Jelly mesenchymal stem cells. Neurosci Lett 600:69–74

Perinatal Tissue-Derived Endothelial Progenitor Cells

Abbas Shafiee and Kiarash Khosrotehrani

1 Introduction

Vascularization is an essential physiological process that occurs during tissue development or disease. Traditionally, blood vessels were considered as channels which were essential for growth, feeding the developing tissue with nutrients and removing metabolic waste; but endothelial cells (EC) also stimulate organ morphogenesis in the embryo and maintain tissue homeostasis in adults by providing instructive trophic factors (Red-Horse et al. 2007). This highlights the importance of the vascular system integrity in growth as well as tissue homeostasis. So far, the association of abnormal vascularization with more than 20 diseases including inflammatory disorders such as psoriasis, age-related macular degeneration and cancer has been revealed. Literally, normal blood vessel formation is regulated by signals which induce or inhibit vascularization and any imbalance in the vascular system function results in pathogenesis of various vascular disorders (Carmeliet 2005).

Developmentally, blood vessel formation is initiated by the de novo formation of vessels (vasculogenesis) and followed by expansion of preexisting vessels (angiogenesis). Although, it was previously believed that vasculogenesis was limited

A. Shafiee
UQ Centre for Clinical Research, The University of Queensland, Royal Brisbane and Women's Hospital Building 71/918, Brisbane, QLD 4029, Australia

Institute of Health and Biomedical Innovation, Queensland University of Technology, Brisbane, QLD, Australia

K. Khosrotehrani, M.D., Ph.D., F.A.C.D. (✉)
UQ Centre for Clinical Research, The University of Queensland, Royal Brisbane and omen's Hospital Building 71/918, Brisbane, QLD 4029, Australia

UQ Diamantina Institute, Translational Research Institute, The University of Queensland, Brisbane, QLD, Australia
e-mail: k.khosrotehrani@uq.edu.au

B. Arjmand (ed.), *Perinatal Tissue-Derived Stem Cells*, Stem Cell Biology and Regenerative Medicine, DOI 10.1007/978-3-319-46410-7_4

to embryonic development, in 1997 Asahara et al. changed this paradigm by reporting the differentiation potential of circulating bone marrow (BM) derived cells into EC. Therefore, they showed neovascularization in adults (Asahara et al. 1997) and called this specific population "endothelial progenitor cells" (EPC). Furthermore, participation of EPC in regeneration of ischemic diseases by homing to the damaged site and forming new vessels was demonstrated (Asahara et al. 1999). Since then, EPC were rapidly used as crude preparations in several clinical trials, which demonstrated enhancement of myocardial tissue functions in patients with ischemic heart disease (ClinicalTrials.gov). Levels of engraftment and biological causes and potential activity of the grafted cells and the benefit of EPC therapy remained unexplained (Kalka et al. 2000; Kawamoto et al. 2001, 2003). In this chapter, we will summarize evidence suggesting the existence of fetal EPC in human prenatal tissues, with a special focus on the placenta.

2 Angioblasts Contribute in Early Human Vascular Development

Cardiovascular formation begins very early in the growing embryo. While during embryonic and fetal development, simple diffusion provides oxygen and nutrients for the small developing conceptus, it does not fulfil the needs of the fast growing fetus. Blood vessel formation is therefore a fundamental step in the successful mammalian embryonic development to facilitate this need. Both angiogenesis and vasculogenesis participate in fulfilling this essential demand in the human embryo.

Vasculogenesis, by using unique endothelial precursors called angioblasts, organizes vascularization of the dorsal aorta as well as the tissues of endodermal origin including the spleen, liver, pancreas (Red-Horse et al. 2007). On the other hand, in organs with smaller vascular networks such as the brain and kidneys, angiogenesis is the main process of vascularization (Baldwin 1996; Beck and D'Amore 1997) (Fig. 1).

Developmentally, the first sign of vessel formation occurs at Day 17 of human embryonic development. Cells of extraembryonic splanchnic mesoderm form the yolk sac wall, proliferate and form hemangioblast aggregates or blood islands adjunct to the endoderm. This is the first step where a mesodermal precursor gives rise to cells that will form an endothelium (Fig. 2). In humans, apart from the yolk sac (Yoder et al. 1997), aorta-gonado-mesonephros (AGM) region (Jaffredo et al. 1998), and umbilical arteries, the placenta (Rhodes et al. 2008) also harbors this hemogenic endothelium (HE). These aggregates are the common precursors for primitive hematopoietic stem cells (HSC) and angioblasts (Eichmann et al. 2002). As demonstrated in Fig. 2, progenitors which are situated in the center of islands are the primitive HSC and the surrounding cells are the angioblasts representing the endothelial precursors that, through this process, vascularize the chorionic villi, yolk sac and connecting stalk within 3 weeks. Angioblasts migrate into adjunct locations and organize small vascular structures, configuring the primitive embryonic vascular system and spreading to the entire developing embryo body.

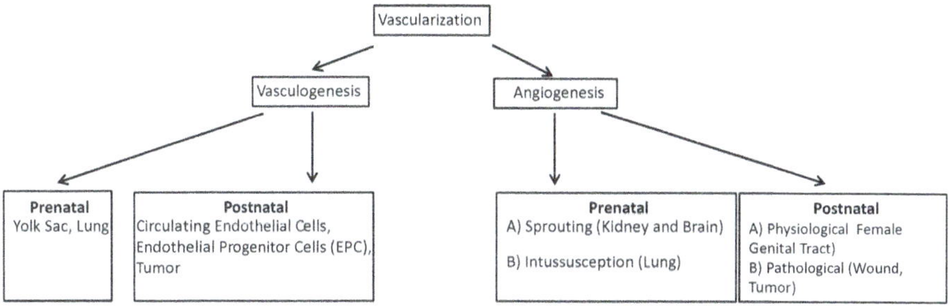

Fig. 1 Vasculogenesis (de novo formation of vessels) and angiogenesis (expansion of preexisting vessels) contribute in vascularization of tissues. Although, it was previously believed that vasculogenesis is limited to embryonic development, recent studies confirmed the existence of vasculogenesis in adult life by using unique endothelial precursor cells. Vasculogenesis organizes vascularization of the dorsal aorta as well as the tissues of endodermal origin including the spleen, liver and pancreas. On the other hand, in organs with smaller vascular networks such as the brain and kidneys, angiogenesis is the main process of vascularization.

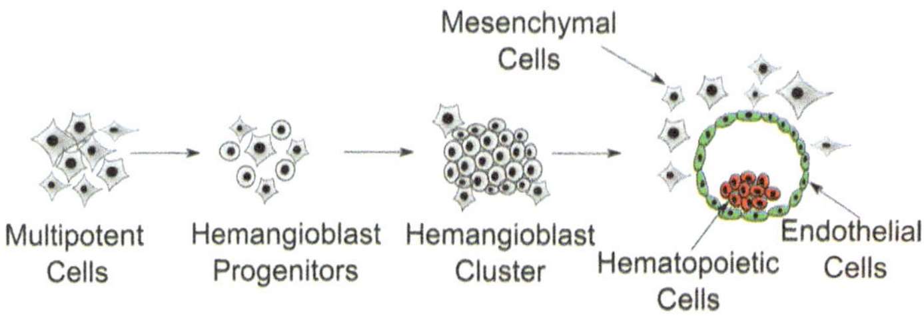

Fig. 2 During human development, mesodermic cells form the yolk sac wall, aorta-gonado-mesonephros region, umbilical arteries and the placenta proliferate and form hemangioblast clusters adjunct to the endoderm. These aggregates are the common precursors for primitive hematopoietic stem cells (HSC) and angioblasts. Progenitors which are situated in the center of islands are the primitive HSC and the surrounding cells are endothelial precursors which migrate into adjunct locations and organize small vascular structures. HSC proliferate and differentiate to hematopoietic cells

Angioblasts therefore work as the leading cells initiating vascular tube and branch formation and are able to migrate to distant secondary sites and participate in the formation of new vessels.

Angiogenesis and vascular intussusceptions, the generation of new vessels by the splitting of existing vessels, contribute to the growth, expansion and remodeling of this primitive vasculature. Perfusion of vessels after the initiation of the heart beat is a key regulator of both vasculogenesis and angiogenesis, and such hemodynamic forces induce vessels' remodeling (Lucitti et al. 2007).

The nascent vessels are immature and the actual vessels require participation of other trailing cells which follow the leading endothelial cells and stabilize the growing vessels to form the tubes entity (Ghabrial and Krasnow 2006).

Overall, distinct areas of the embryo and extraembryonic structures undergo a process of vasculogenesis involving progenitors that develop to form the endothelial system. It was thought that this process was limited to the embryonic and fetal development. In the next chapter, we will provide evidence of its existence in adult age.

3 Postnatal Vasculogenesis in Humans

Initially, it was believed that vasculogenesis was limited to embryonic development, while angiogenesis occurred both in embryogenesis and postnatal life. However, the distinction between these two mechanisms is not quite clear and they overlap, since in both processes EC form new blood vessels by proliferation, differentiation and remodeling. Moreover, the relative contribution of each mechanism in vascularizing the ischemic tissues depends on the existence of angioblasts in the target tissue (Ribatti et al. 2001). In postnatal life, vasculogenesis has been observed in both physiological and pathological situations. In 1997, Asahara et al., for the first time, reported the differentiation potential of circulating BM-derived cells into EC and showed vasculogenesis in adult age (Asahara et al. 1997).

The earlier concept was supported by the coexistence of both mature EC and EPC in postnatal life. Moreover, the recruitment of EPC to ischemic sites and the revascularization of damaged tissues were reported (Asahara et al. 1997). Further studies have verified that EPC has some resemblance to embryonic angioblasts, with a critical role in regeneration. In therapy, several clinical and preclinical studies have shown the formation of new vessels and small enhancement of myocardial tissue functions after EPC delivery (Kalka et al. 2000; Kawamoto et al. 2001, 2003). Moreover, although initially the BM was considered its main source, soon after EPC isolated from other tissues (Ingram et al. 2004; Murohara et al. 2000; Nagano et al. 2007; Patel et al. 2013).

4 Characteristics of Human EPC

In the first report by Asahara et al. they used CD34 magnetic sorting to isolate EPC from BM. But, EPC have been isolated, using several different method, from distinct tissues (Rafii and Lyden 2003; Kalka et al. 2000; Kawamoto et al. 2001; Hristov et al. 2003). EPC's isolation has been performed based on some biological characteristics such as their colony-forming capacity on collagen or fibronectin (known as colony formation assay, CFU) (Hill et al. 2003; Kalka et al. 2000; Kawamoto et al. 2001). It also relies on the detection of endothelial characteristics such as acetylated low-density lipoprotein (Ac-LDL) uptake and lectin binding (Tanaka et al. 2008; Masuda et al. 2011), and finally characterization based on expression of some cell-surface antigens mainly CD133, CD34 and VEGFR-2 (Rafii and Lyden 2003; Hristov et al. 2003).

According to current knowledge, EPC are categorized into two different populations. A growing body of evidence suggested the existence of a shared progenitor for both endothelial and hematopoietic cells during embryonic development. Many attempts have been made to identify similar EPC populations with hemogenic potential using both endothelial (CD31, CD34, VEGFR2 (also called Kinase insert domain receptors KDR or Fetal Liver Kinase 1 Flk-1)) and hematopoietic (CD34, CD133, CD45) markers (Pelosi et al. 2002; Grant et al. 2002; Bailey et al. 2004). But, these methods were proven to result in significant hematopoietic (in particular myeloid) contamination (Case et al. 2007). Additionally, it is believed that these EPC which are expressing hematopoietic markers are able to migrate into periphery areas after vascular injury and secrete certain cytokines in circulation (Grant et al. 2002). More recently, the concept of EPC origin has been shifted, and the exclusion of hematopoietic cells restricts the progenitors to the endothelial lineage. Although briefly debated, human EPC are believed to express CD34, VEGFR-2, CXCR4 and CD105 and to be negative for hematopoietic markers, CD45 in particular (Sukmawati and Tanaka 2015; Timmermans et al. 2009); this is slightly different from mouse EPC expressing c-Kit, Sca-1 and CD34 and VEGFR2 (Sukmawati and Tanaka 2015; Timmermans et al. 2009). Moreover, human EPC are organ or tissue specific (Rafii et al. 2016) and are also called endothelial outgrowth cells (EOC). In vitro studies have confirmed that these cells appear after about 7 days of culture; have a cobblestone-like morphology and express CD144, VEGFR-2, CD31, CD34, CD105 and CD146, but not any of hematopoietic surface markers such as CD45 and CD133 in humans (Timmermans et al. 2009; Yoder et al. 2007). Additionally, EPC are able to form capillary-like structures on Matrigel™ and produce nitric oxide in vitro. EOC showed clinical advantages upon transplantation and enhanced the neovascularization in ischemic sites. EPC also could form de novo vessels upon transplantation (Sukmawati and Tanaka 2015). In 2004, Ingram et al. introduced a new hierarchy of EC in human peripheral and umbilical cord blood (UCB) (Ingram et al. 2004) in accordance with differences in the clonogenic and proliferative potential of the endothelial lineage. They identified a new ex vivo, highly proliferative population (HPP) of "endothelial colony-forming cells (ECFC)" in human cord blood (Ingram et al. 2004).

As discussed earlier, the field of EPC is controversial and there are different descriptions for EPC; therefore, we define human EPC as endothelial cells with the characteristics as presented in Table 1.

5 Placenta

5.1 Placental Tissue

The placenta is a fetal organ that grows with the developing fetus in the gravid maternal uterus and acts as an interface between the mother and the fetus. The placenta originates from the embryo and the fetus, but the very intricate interconnection with the maternal uterus through the decidua results in a sizable maternal fraction. The healthy term placenta is round or oval in shape and dark reddish-blue in color and normally weighs about 500 g. The main roles of the placenta are to

Table 1 Characteristics of human endothelial progenitor cells

Context	Characteristics	Reference
In vivo	EPC express endothelial markers including CD144, VEGFR-2, CD31, CD34, CD105 and CD146, but not any of hematopoietic surface markers such as CD45 and CD133	Timmermans et al. (2009)
	EPC could form de novo vessels upon transplantation	Yoder et al. (2007)
	EPC have clinical advantage and improve neovascularization in ischemic sites	Reinisch et al. (2009)
	EPC derived vessels may be perfused with host blood cells, anastomose with the surrounding host vasculature and form donor–host chimeric vessels	Yoder et al. (2007); Reinisch et al. (2009)
	EPC are different from "endothelial cell colony-forming units" (CFU-ECs) which are derived from the hematopoietic progenitors	Yoder et al. (2007)
In vitro	EPC appear about 7 days after cell seeding	Yoder et al. (2007)
	Cobblestone shaped with differentiation potential into mature EC	Hur et al. (2004)
	Clonogenic with potential to give rise to more than 10,000 progeny in 2-week culture period	Yoder et al. (2007)
	Highly proliferative cells with high levels of telomerase activity and more than 30 population doublings before entering senescence phase	Yoder et al. (2007); Reinisch et al. (2009); Hofmann et al. (2009)
	Expression of only endothelial and not hematopoietic cell surface markers such as CD14, CD45	Yoder et al. (2007)
	EPC could be cryopreserved and maintain their phenotype, proliferation potential, differentiation and functionality upon defrosting	Reinisch et al. (2009)

provide oxygen and nutrients and to remove the waste material from the fetus (Joe et al. 2010). But, placenta also acts as an endocrine organ and produces some important growth factors and hormones such as Insulin Growth Factors 1 and 2 (IGF1and -2), Placental Growth Factor (PlGF), Human Chorionic Gonadotropin (hCG) and Human Placental Lactogen (hPL [Human Chorionic Somatomammotropin]). From developmental point of view, placental formation starts at the implantation of the blastocyst: the outer layer of blastocyst forms a two-layer cover, the cytotrophoblast (inner layer and actively proliferating) and the syncytiotrophoblast (outer layer which erodes uterine tissue). Cytotrophoblast cells start mitotic division and invade the outer layer and form multinucleated cells (Huppertz and Peeters 2005). At Days 11–12 of development in humans, the syncytiotrophoblast invades the maternal capillaries and forms "sinusoids" that consequently become continuous and form lacunae that receive the maternal blood (Schoenwolf 2009).

In subsequent days, trophoblast cells proliferate locally and penetrate the maternal tissues and uteroplacental circulation starts by the end of the second week

(Huppertz and Peeters 2005). At this time, the blastocyst has completely penetrated the uterine wall and cytotrophoblasts form primary villi which are surrounded by the syncytium (Schoenwolf 2009). Next, mesodermal cells come through the villi and form the secondary villi which start differentiation into blood and endothelial cells at the center of the villus and form the tertiary villus ready to provide nutrients for the fetus. These mesodermal cells probably constitute a progenitor population able to give rise to both endothelial progenitors and mesenchymal/stromal stem cells (MSC).

As the fetus develops the need for nutrients increases and causes a substantial adjustment to the placental size that, at term, covers about 15–30 % of the uterus. During the second trimester, maternal decidual cells migrate into the chorionic tissue and are covered by a layer of syncytial cells formed septa that divide the placenta into sections named “cotyledons”. In the following weeks the placenta’s size significantly increases, predominantly during the last third of the pregnancy (Huppertz and Peeters 2005).

5.2 *Placental Vascular Development*

During pregnancy, the gravid uterus is highly vascular, receiving a large blood flow. Similarly, the placenta, formed essentially by fetal cells, is the major vascular link between the mother and the fetus. A third of the placental cells have been reported to be endothelial, making this an accessible model of fetal vascular biology (Robin et al. 2009).

The implantation of the embryo in the uterus is associated with massive angiogenesis, which involves the growth and remodeling of the surrounding maternal blood vessels. Impaired or insufficient vascularization causes altered fetal growth or early embryonic abortion. Fetal and maternal vascular systems individually contribute to the placental vascularization and there is a close relationship between the uterine and umbilical blood flows and placental size in human normal pregnancies (Reynolds and Redmer 2001; Reynolds et al. 2002). The cellular apparatus of the maternal vasculature is beyond the scope of the current study. In regards to fetal vascular system, the first trimester is associated with massive vascularization in placenta. Although no significant vascularization occurs in the second trimester, but the third trimester is accompanied by considerable increase in fetal growth, which comes with substantial vascularization in the placenta (Reynolds and Redmer 2001).

At 12–15 days of gestation, the villous trophoblast forms a small cavity and envelops the mesenchymal tissue containing the pluripotent mesenchymal cells that are the main component of the placental vasculogenesis. At 21 days postgestation, these mesenchymal cells, surrounded by the trophoblast layer, differentiate into hemangioblastic precursors. The hemangiogenic cells maintain a close connection with the trophoblastic peripheral cells and form hemangioblastic cell cords that connect surrounding villi and form the first vessels (Charnock-Jones et al. 2004; Huppertz and Peeters 2005). Hemangiogenic stem cells express distinct proteins, and in the following phase are stimulated to become angioblastic and hematopoietic cells (Demir et al. 2007; Charnock-Jones et al. 2004). At the following stages, vasculogenesis continues and fetoplacental angiogenesis starts to establish mature villous structure. The villi maturation is associated with progress in the vasculogenesis and angiogenesis which establishes the blood flow in the placental villi and the intervillous lumen.

As our understanding of placental tissue formation and development increases, the functional capacities of placental progenitors and their potential to be used as alternative cell sources in clinical cell-based therapy trials appear more feasible (Shafiee et al. 2015). Human term placenta harbors several kinds of progenitors with translational potential. Although HSC and MSC obtained from prenatal tissues are routinely used in transplantation, the benefit of EPC population is less investigated.

5.3 *Placenta as a Robust Source of EPC*

Microchimerism studies support the notion that the placenta harbors populations of fetal stem cells that in the situation of natural transfer, migrate to the mother's tissues, these cells have the capacity to integrate and contribute to many tissues, and most importantly to the endothelium (Kara et al. 2012; Beck et al. 1995). Derivation of fetal or maternal stem cells from term placental villi has been established using several methods including: the explant method (Abumaree et al. 2013; Igura et al. 2004), expression of specific cell surface markers on fetal cells and cultures in specific media (Patel et al. 2014), and selective adhesion of stem cells (Mathews et al. 2015). Beyond trophoblasts and their stem cells, most efforts have focused on MSC (Golos 2011; Castrechini et al. 2010; Ulrich et al. 2013) and hematopoietic progenitors (Robin et al. 2009), but also on endothelial progenitors (Rapp et al. 2012; Patel et al. 2014), given their translational potential. Here, we review EPC as the most common stem cell/progenitor derived from placenta.

Rapp et al., recently isolated resident high proliferative endothelial progenitors from placental culture based on their CD146 expression in vitro (Rapp et al. 2012), and compared their potential to circulating ECFC from cord blood (CB-ECFC); both of them were positive for CD31, CD105, CD144 and KDR and negative for CD14 and CD45, in line with previously reported CD markers for ECFC (Ingram et al. 2004). Placental-ECFCs (P-ECFC) had reduced cell proliferation capacity before senescence, but the number of high proliferative potential (HPP) colonies in P-ECFC was similar to control. To assess the in vivo vessel formation capacity, collagen/fibronectin plugs loaded with ECFC were implanted into Nude mice and the number of vessels/mm^2 in the P-ECFC group was about threefold that of CB-ECFC. In vivo experiments have also confirmed de novo vessel formation upon P-ECFC transplantation and enhanced blood perfusion in the defect areas (Rapp et al. 2012; Patel et al. 2013). This suggests that while both populations have similar phenotypes and kinetics, P-ECFC have better vasculogenic potential (Rapp et al. 2012).

Recently, we introduced an innovative sorting-based strategy and could successfully isolate P-ECFC in clinical relevant quantities (Patel et al. 2013, 2016). Term placental tissue was enzymatically digested and sorted based on expression of hematopoietic (CD45) and endothelial (CD34 and CD31) key surface markers. Fluorescence in situ hybridization (FISH) analysis showed the CD45–CD34+CD31+ population is fetal in origin. Since the placenta is a highly vascular tissue and almost 30 % of placental cells are CD31+, so placenta can be considered as a major source of EPC. Moreover, our results showed that not only the amount of ECFC residing in the placenta was significantly higher than in donor-matched UCB, but in terms of functionality and phenotypical aspects, including gene expression, CB and placental ECFC were similar, and P-ECFC could be used as a potent alternative for EPC and cardiovascular cell-based therapies (Patel et al. 2013).

Moreover, application of placental cells obtained at the term of pregnancy is less challenged compared to other fetal sources which often come from abortion products (Shafiee et al. 2015). Therefore, the human term placenta could be used as a promising cell source for ameliorating disease and promoting tissue regeneration.

6 EPC Application in Regenerative Medicine and Clinical Setting

Vascularization is an emerging field which could revolutionize tissue regeneration approaches. To date, a variety of strategies have been conducted to improve tissue vascularization. In this regards, coadministration of mesenchymal and endothelial cells to improve vessel growth has been proposed (Choong et al. 2006; Santos et al. 2009; Grellier et al. 2009). Cotransplantation of osteoprogenitors with EC resulted in enhanced bone tissue formation (Choong et al. 2006; Santos et al. 2009; Grellier et al. 2009). Similar results have been reported by coadministration of EC for skin (Shepherd et al. 2006), lung (Mondrinos et al. 2008), skeletal (Levenberg et al. 2005) and cardiac muscle (Caspi et al. 2007) (reviewed in Rouwkema et al. (2008)). Cotransplanted EC usually contribute in vascularization of implanted tissues and boundaries and results in proper tissue regeneration (Tremblay et al. 2005). However, current researches have been shifted from using differentiated EC (Hofmann et al. 2008; Choong et al. 2006) to application of EPC (Duttenhoefer et al. 2013; Liu et al. 2012). In comparison to differentiated EC, EPC can be harvested from the patient's own blood and BM, and even from cord blood and placenta bio-banks with greater regenerative potential. To date, application of EPC in enhancement of vascularization of ischemic sites and tissue perfusion improvement has been established in a wide variety of animal models (Kalka et al. 2000; Hirata et al. 2003; He et al. 2004). In situ delivery of a BM CD34+ population to hind limb ischemia (HLI) in a diabetic mice model improved wound healing and enhanced vascular growth (Awad et al. 2006; Yang et al. 2011), indicating their potential in acute clinical settings. Mechanistically this positive impact was explained by increased sensitivity to VEGF and possibly vascularization modulation through angiopoietins (Awad et al. 2006). In addition, the CD34+ population showed higher recruitment toward injured sites upon implantation of fresh BM-derived CD34+ (Yang et al. 2011). Local delivery of fresh CD34+ also improved vessel function through enhancement of blood perfusion in damaged areas (Schatteman et al. 2000).

Promising results in animal models have attracted much interest in clinical applications of EPC. EPC administration has demonstrated promising results in patients with ischemia diseases who do not respond to drugs and where surgical therapy is not applicable (Kawamoto et al. 2009; Matoba et al. 2008; Tanaka et al. 2014). To date, more than 200 interventional studies have been registered at ClinicalTrials.gov. In overall, EPC clinical application can be categorized into two subsets: (1) isolation and injection of EPC from BM, and CB into ischemic tissues; (2) application of capture stents which selectively attach to EPC in circulation and induce endothelialization. EPC local injection in periphery artery injuries resulted in improved leg pain scale, increased vascular perfusion and free walking distance (Tanaka et al. 2014). Of note, clinical improvements lasted for at least 2 years (Tanaka et al. 2014). A similar outcome was reported in patients with chronic limb ischemia (Matoba et al. 2008). In addition EPC therapy also has been used for acute myocardial infarctions (MI) (Tateishi-Yuyama et al. 2002; Dimmeler and Zeiher 2009). Leistner et al. demonstrated the reduced functional infarct size and positive effects of intracoronary infusion of circulating or BM-derived EPC in patients with acute MI (Leistner et al. 2011). However, long-term efficacy of EPC for MI remains to be determined.

Apart from direct injection, several clinical trials were conducted to capture stents in patients with coronary artery disease (Sethi and LEE 2012; Aoki et al. 2005). HEALING-II registry manufactured a bioengineered stent with potential to attract circulating EPC using coating of stainless steel stents with antihuman CD34 antibodies (Aoki et al. 2005). Only at 1 h after implantation the coated stent showed more than 90 % cell coverage with minimal inflammation. This newly formed endothelial layer suggested reducing the risk of restenosis and stent thrombosis (Sethi and LEE 2012). In addition, platelets have less adhesion to capture stents and there is no need for administration of anticoagulative medicines (Sethi and LEE 2012). However, CD34 antibody is not specific for EPC and only a small portion of the CD34-positive cells are true EPC, so other antibodies against KDR or CD144 also applied with increased efficiency of capture and accelerated endothelialization over stents in vitro (Markway et al. 2008) and in experimental animal models (Lee et al. 2012). In contrast to CD34, CD144 supposed to be expressed exclusively on the late EPC and CD144 coated stents capture circulating late EPC and not the early or myeloid EPC (Lim et al. 2011) however the clinical application is still uncertain.

7 Remaining Challenges in EPC Application for Cardiovascular Repair

Due to significant increase in vascular repair using EPC, EPC therapy is rapidly moving toward clinical applications, but further studies are needed to define the mechanism behind the advantageous application of EPC. There is much variability upon EPC treatment which might be due to our incomplete knowledge on EPC preparations. This can potentially affect the quality and quantity of tissue repair. However, the essential limitation of EPC therapy is the low quantity of cells recovered from circulating blood, which is even lower in patients with cardiovascular disease (CVD) (Vasa et al. 2001; Hill et al. 2003), certain health conditions, including diabetes (Tepper et al. 2010), and advanced age (Scheubel et al. 2003). In addition, patients with CVD are reported to have impaired EPC function (Fadini et al. 2012). Moreover, EPC proved to have the dose-dependent effects for vasculogenesis and cardiomyogenesis (Iwasaki et al. 2006). Accordingly, application of EPC may result in few or even no clinical benefits. Therefore, autologous EPC therapy is challenged by low cell quantity and impaired function. Lately, enhancement of cell survival as well as cell proliferation and inducement of EPC migration and differentiation toward tissue of interest have been introduced as new methods to overcome current limitations. For instance, allogeneic EPC application could improve the cell number limitations by proving clinical cell dosage. In this regard, early allogeneic cell therapy has demonstrated benefits in patients with ischemic or cardiomyopathic diseases (Leeper et al. 2010). Allogeneic cell application also comes with some limitations. The long-term cell expansion step is necessary to provide enough cell number, which is a burden for emergency patients (Leeper et al. 2010). Some marker-based EPC isolation strategies are associated with isolation of heterogeneous population and since there is not a definite marker for EPC, so application of heterogeneous EPC is challenging and EPC isolation based on markers which are not overlapped with other cell populations is recommended.

In regards to clinical application, the in vitro expansion step for bulking up the EPC is inevitable and makes these cells more challenging to use for clinical application. BM- and peripheral blood-derived EPC are being exclusively used for EPC therapy in CVD and despite of significant enthusiasm on ECFC, their application is limited to experimental animals. This limitation might be due to lack of thorough ECFC characterization, in vivo biology and limited survival upon ECFC transplantation which have not being addressed yet.

Although EPC application for CVD is promising, it is associated with some limitations which need to be improved before worldwide clinical usage. Limitations include: (1) low cell numbers; (2) impaired cell function, particularly in the elderly and people affected with CVD or diabetes; (3) nonunique EPC isolation strategy.

8 Perspectives

The limited or absence of blood flow to the ischemic sites is the most common reason for CVD death. It is well established that EPC participate in regeneration of ischemic disease by homing to the damaged site and forming new vessels. The human term placenta is a highly vascularized tissue and contains great numbers of EPC. Fetal EPC at clinically relevant quantities were isolated from placenta, and donor-matched comparison with UCB has demonstrated that placenta harbors 27 times more EPC than UCB. In vivo functional assay and gene expression analysis confirmed the similarity between placenta- and UCB-derived EPC. Altogether, placental EPC can be considered as a significant source for future clinical EPC therapy.

Acknowledgements This study was supported by the National Health and Medical Research Council (Project Grant 1023368). K.K. was supported by the National Health and Medical Research Council Career Development Fellowship (Grant 1023371).

Conflict of Interest. The authors report no potential conflicts of interest.

References

Abumaree M, Al Jumah M, Kalionis B, Jawdat D, Al Khaldi A, AlTalabani A, Knawy B (2013) Phenotypic and functional characterization of mesenchymal stem cells from chorionic villi of human term placenta. Stem Cell Rev Rep 9(1):16–31

Aoki J, Serruys PW, van Beusekom H, Ong AT, McFadden EP, Sianos G, van der Giessen WJ, Regar E, de Feyter PJ, Davis HR (2005) Endothelial progenitor cell capture by stents coated with antibody against CD34: the HEALING-FIM (Healthy Endothelial Accelerated Lining Inhibits Neointimal Growth-First In Man) registry. J Am Coll Cardiol 45(10):1574–1579

Asahara T, Murohara T, Sullivan A, Silver M, van der Zee R, Li T, Witzenbichler B, Schatteman G, Isner JM (1997) Isolation of putative progenitor endothelial cells for angiogenesis. Science 275(5302):964–966

Asahara T, Masuda H, Takahashi T, Kalka C, Pastore C, Silver M, Kearne M, Magner M, Isner JM (1999) Bone marrow origin of endothelial progenitor cells responsible for postnatal vasculogenesis in physiological and pathological neovascularization. Circ Res 85(3):221–228

Awad O, Dedkov EI, Jiao C, Bloomer S, Tomanek RJ, Schatteman GC (2006) Differential healing activities of CD34+ and CD14+ endothelial cell progenitors. Arterioscler Thromb Vasc Biol 26(4):758–764

Bailey AS, Jiang S, Afentoulis M, Baumann CI, Schroeder DA, Olson SB, Wong MH, Fleming WH (2004) Transplanted adult hematopoietic stems cells differentiate into functional endothelial cells. Blood 103(1):13–19

Baldwin HS (1996) Early embryonic vascular development. Cardiovasc Res 31Spec No:E34–45

Beck L Jr, D'Amore PA (1997) Vascular development: cellular and molecular regulation. FASEB J 11(5):365–373

Beck F, Erler T, Russell A, James R (1995) Expression of Cdx-2 in the mouse embryo and placenta: possible role in patterning of the extra-embryonic membranes. Dev Dyn 204(3):219–227

Carmeliet P (2005) Angiogenesis in life, disease and medicine. Nature 438(7070):932–936

Case J, Mead LE, Bessler WK, Prater D, White HA, Saadatzadeh MR, Bhavsar JR, Yoder MC, Haneline LS, Ingram DA (2007) Human CD34< sup>+ AC133< sup>+ VEGFR-2< sup>+ cells are not endothelial progenitor cells but distinct, primitive hematopoietic progenitors. Exp Hematol 35(7):1109–1118

Caspi O, Lesman A, Basevitch Y, Gepstein A, Arbel G, Habib IHM, Gepstein L, Levenberg S (2007) Tissue engineering of vascularized cardiac muscle from human embryonic stem cells. Circ Res 100(2):263–272

Castrechini N, Murthi P, Gude NM, Erwich JJ, Gronthos S, Zannettino A, Brennecke SP, Kalionis B (2010) Mesenchymal stem cells in human placental chorionic villi reside in a vascular Niche. Placenta 31(3):203–212

Charnock-Jones D, Kaufmann P, Mayhew T (2004) Aspects of human fetoplacental vasculogenesis and angiogenesis. I. Molecular regulation. Placenta 25(2):103–113

Choong CS, Hutmacher DW, Triffitt JT (2006) Co-culture of bone marrow fibroblasts and endothelial cells on modified polycaprolactone substrates for enhanced potentials in bone tissue engineering. Tissue Eng 12(9):2521–2531

Demir R, Seval Y, Huppertz B (2007) Vasculogenesis and angiogenesis in the early human placenta. Acta Histochem 109(4):257–265

Dimmeler S, Zeiher AM (2009) Cell therapy of acute myocardial infarction: open questions. Cardiology 113(3):155–160. doi:10.1159/000187652

Duttenhoefer F, Lara de Freitas R, Meury T, Loibl M, Benneker LM, Richards RG, Alini M, Verrier S (2013) 3D scaffolds co-seeded with human endothelial progenitor and mesenchymal stem cells: evidence of prevascularisation within 7 days. Eur Cells Mater 26:49–65

Eichmann A, Pardanaud L, Yuan L, Moyon D (2002) Vasculogenesis and the search for the hemangioblast. J Hematother Stem Cell Res 11(2):207–214

Fadini GP, Losordo D, Dimmeler S (2012) Critical reevaluation of endothelial progenitor cell phenotypes for therapeutic and diagnostic use. Circ Res 110(4):624–637

Ghabrial AS, Krasnow MA (2006) Social interactions among epithelial cells during tracheal branching morphogenesis. Nature 441(7094):746–749

Golos TG (2011) Stem cells from the placenta. In: Kay HH, Michael Nelson D, Wang Y (eds) The placenta: from development to disease, from development to disease. Wiley, New York, pp 327–333

Grant MB, May WS, Caballero S, Brown GA, Guthrie SM, Mames RN, Byrne BJ, Vaught T, Spoerri PE, Peck AB (2002) Adult hematopoietic stem cells provide functional hemangioblast activity during retinal neovascularization. Nat Med 8(6):607–612

Grellier M, Bordenave L, Amedee J (2009) Cell-to-cell communication between osteogenic and endothelial lineages: implications for tissue engineering. Trends Biotechnol 27(10):562–571

He T, Smith LA, Harrington S, Nath KA, Caplice NM, Katusic ZS (2004) Transplantation of circulating endothelial progenitor cells restores endothelial function of denuded rabbit carotid arteries. Stroke 35(10):2378–2384

Hill JM, Zalos G, Halcox JP, Schenke WH, Waclawiw MA, Quyyumi AA, Finkel T (2003) Circulating endothelial progenitor cells, vascular function, and cardiovascular risk. N Engl J Med 348(7):593–600. doi:10.1056/NEJMoa022287

Hirata K, Li T-S, Nishida M, Ito H, Matsuzaki M, Kasaoka S, Hamano K (2003) Autologous bone marrow cell implantation as therapeutic angiogenesis for ischemic hindlimb in diabetic rat model. Am J Phys Heart Circ Phys 284(1):H66–H70

Hofmann A, Ritz U, Verrier S, Eglin D, Alini M, Fuchs S, Kirkpatrick CJ, Rommens PM (2008) The effect of human osteoblasts on proliferation and neo-vessel formation of human umbilical vein endothelial cells in a long-term 3D co-culture on polyurethane scaffolds. Biomaterials 29(31):4217–4226. doi:10.1016/j.biomaterials.2008.07.024

Hofmann NA, Reinisch A, Strunk D (2009) Isolation and large scale expansion of adult human endothelial colony forming progenitor cells. Journal of visualized experiments: JoVE 32:1524

Hristov M, Erl W, Weber PC (2003) Endothelial progenitor cells: mobilization, differentiation, and homing. Arterioscler Thromb Vasc Biol 23(7):1185–1189. doi:10.1161/01.ATV.0000073832.49290.B5

Huppertz B, Peeters LL (2005) Vascular biology in implantation and placentation. Angiogenesis 8(2):157–167

Hur J, Yoon C-H, Kim H-S, Choi J-H, Kang H-J, Hwang K-K, Oh B-H, Lee M-M, Park Y-B (2004) Characterization of two types of endothelial progenitor cells and their different contributions to neovasculogenesis. Arteriosclerosis, thrombosis, and vascular biology 24 (2):288–293

Igura K, Zhang X, Takahashi K, Mitsuru A, Yamaguchi S, Takahashi T (2004) Isolation and characterization of mesenchymal progenitor cells from chorionic villi of human placenta. Cytotherapy 6(6):543–553

Ingram DA, Mead LE, Tanaka H, Meade V, Fenoglio A, Mortell K, Pollok K, Ferkowicz MJ, Gilley D, Yoder MC (2004) Identification of a novel hierarchy of endothelial progenitor cells using human peripheral and umbilical cord blood. Blood 104(9):2752–2760

Iwasaki H, Kawamoto A, Ishikawa M, Oyamada A, Nakamori S, Nishimura H, Sadamoto K, Horii M, Matsumoto T, Murasawa S (2006) Dose-dependent contribution of CD34-positive cell transplantation to concurrent vasculogenesis and cardiomyogenesis for functional regenerative recovery after myocardial infarction. Circulation 113(10):1311–1325

Jaffredo T, Gautier R, Eichmann A, Dieterlen-Lievre F (1998) Intraaortic hemopoietic cells are derived from endothelial cells during ontogeny. Development 125(22):4575–4583

Joe AW, Yi L, Natarajan A, Le Grand F, So L, Wang J, Rudnicki MA, Rossi FM (2010) Muscle injury activates resident fibro/adipogenic progenitors that facilitate myogenesis. Nat Cell Biol 12(2):153–163. doi:10.1038/ncb2015

Kalka C, Masuda H, Takahashi T, Kalka-Moll WM, Silver M, Kearney M, Li T, Isner JM, Asahara T (2000) Transplantation of ex vivo expanded endothelial progenitor cells for therapeutic neovascularization. Proc Natl Acad Sci U S A 97(7):3422–3427. doi:10.1073/pnas.070046397

Kara RJ, Bolli P, Karakikes I, Matsunaga I, Tripodi J, Tanweer O, Altman P, Shachter NS, Nakano A, Najfeld V (2012) Fetal cells traffic to injured maternal myocardium and undergo cardiac differentiation. Circ Res 110(1):82–93

Kawamoto A, Gwon HC, Iwaguro H, Yamaguchi JI, Uchida S, Masuda H, Silver M, Ma H, Kearney M, Isner JM, Asahara T (2001) Therapeutic potential of ex vivo expanded endothelial progenitor cells for myocardial ischemia. Circulation 103(5):634–637

Kawamoto A, Tkebuchava T, Yamaguchi J, Nishimura H, Yoon YS, Milliken C, Uchida S, Masuo O, Iwaguro H, Ma H, Hanley A, Silver M, Kearney M, Losordo DW, Isner JM, Asahara T (2003) Intramyocardial transplantation of autologous endothelial progenitor cells for therapeutic neovascularization of myocardial ischemia. Circulation 107(3):461–468

Kawamoto A, Katayama M, Handa N, Kinoshita M, Takano H, Horii M, Sadamoto K, Yokoyama A, Yamanaka T, Onodera R (2009) Intramuscular transplantation of G-CSF-mobilized CD34+ cells in patients with critical limb ischemia: a phase I/IIa, multicenter, single-blinded, dose-escalation clinical trial. Stem Cells 27(11):2857–2864

Lee JM, Choe W, Kim B-K, Seo W-W, Lim W-H, Kang C-K, Kyeong S, Eom KD, Cho H-J, Kim Y-C (2012) Comparison of endothelialization and neointimal formation with stents coated with antibodies against CD34 and vascular endothelial-cadherin. Biomaterials 33(35):8917–8927

Leeper NJ, Hunter AL, Cooke JP (2010) Stem cell therapy for vascular regeneration adult, embryonic, and induced pluripotent stem cells. Circulation 122(5):517–526

Leistner DM, Fischer-Rasokat U, Honold J, Seeger FH, Schächinger V, Lehmann R, Martin H, Burck I, Urbich C, Dimmeler S (2011) Transplantation of progenitor cells and regeneration enhancement in acute myocardial infarction (TOPCARE-AMI): final 5-year results suggest long-term safety and efficacy. Clin Res Cardiol 100(10):925–934

Levenberg S, Rouwkema J, Macdonald M, Garfein ES, Kohane DS, Darland DC, Marini R, van Blitterswijk CA, Mulligan RC, D'Amore PA (2005) Engineering vascularized skeletal muscle tissue. Nat Biotechnol 23(7):879–884

Lim W-H, Seo W-W, Choe W, Kang C-K, Park J, Cho H-J, Kyeong S, Hur J, Yang H-M, Cho H-J (2011) Stent coated with antibody against vascular endothelial-cadherin captures endothelial progenitor cells, accelerates re-endothelialization, and reduces neointimal formation. Arterioscler Thromb Vasc Biol 31(12):2798–2805

Liu Y, Teoh SH, Chong MS, Lee ES, Mattar CN, Randhawa NK, Zhang ZY, Medina RJ, Kamm RD, Fisk NM, Choolani M, Chan JK (2012) Vasculogenic and osteogenesis-enhancing potential of human umbilical cord blood endothelial colony-forming cells. Stem Cells 30(9):1911–1924. doi:10.1002/stem.1164

Lucitti JL, Jones EA, Huang C, Chen J, Fraser SE, Dickinson ME (2007) Vascular remodeling of the mouse yolk sac requires hemodynamic force. Development 134(18):3317–3326. doi:10.1242/dev.02883

Markway BD, McCarty OJ, Marzec UM, Courtman DW, Hanson SR, Hinds MT (2008) Capture of flowing endothelial cells using surface-immobilized anti-kinase insert domain receptor antibody. Tissue Eng Part C Methods 14(2):97–105

Masuda H, Alev C, Akimaru H, Ito R, Shizuno T, Kobori M, Horii M, Ishihara T, Isobe K, Isozaki M (2011) Methodological development of a clonogenic assay to determine endothelial progenitor cell potential. Circ Res 109(1):20–37

Mathews S, Rao KL, Prasad KS, Kanakavalli M, Reddy AG, Raj TA, Thangaraj K, Pande G (2015) Propagation of pure fetal and maternal mesenchymal stromal cells from terminal chorionic villi of human term placenta. Sci Rep 5

Matoba S, Tatsumi T, Murohara T, Imaizumi T, Katsuda Y, Ito M, Saito Y, Uemura S, Suzuki H, Fukumoto S (2008) Long-term clinical outcome after intramuscular implantation of bone marrow mononuclear cells (Therapeutic Angiogenesis by Cell Transplantation [TACT] trial) in patients with chronic limb ischemia. Am Heart J 156(5):1010–1018

Mondrinos MJ, Koutzaki SH, Poblete HM, Crisanti MC, Lelkes PI, Finck CM (2008) In vivo pulmonary tissue engineering: contribution of donor-derived endothelial cells to construct vascularization. Tissue Eng A 14(3):361–368. doi:10.1089/tea.2007.0041

Murohara T, Ikeda H, Duan J, Shintani S, K-i S, Eguchi H, Onitsuka I, Matsui K, Imaizumi T (2000) Transplanted cord blood–derived endothelial precursor cells augment postnatal neovascularization. J Clin Investig 105(11):1527–1536

Nagano M, Yamashita T, Hamada H, Ohneda K, K-i K, Nakagawa T, Shibuya M, Yoshikawa H, Ohneda O (2007) Identification of functional endothelial progenitor cells suitable for the treatment of ischemic tissue using human umbilical cord blood. Blood 110(1):151–160

Patel J, Seppanen E, Chong MS, Yeo JS, Teo EY, Chan JK, Fisk NM, Khosrotehrani K (2013) Prospective surface marker-based isolation and expansion of fetal endothelial colony-forming cells from human term placenta. Stem Cells Transl Med 2(11):839–847

Patel J, Shafiee A, Wang W, Fisk N, Khosrotehrani K (2014) Novel isolation strategy to deliver pure fetal-origin and maternal-origin mesenchymal stem cell (MSC) populations from human term placenta. Placenta 35(11):969–971

Patel J, Wong HY, Wang W, Alexis J, Shafiee A, Stevenson AJ, Gabrielli B, Fisk NM, Khosrotehrani K (2016) Self-renewal and high proliferative colony forming capacity of late-outgrowth endothelial progenitors is regulated by cyclin-dependent kinase inhibitors driven by notch signaling. Stem Cells 34(4):902–912

Pelosi E, Valtieri M, Coppola S, Botta R, Gabbianelli M, Lulli V, Marziali G, Masella B, Müller R, Sgadari C (2002) Identification of the hemangioblast in postnatal life. Blood 100(9):3203–3208

Rafii S, Lyden D (2003) Therapeutic stem and progenitor cell transplantation for organ vascularization and regeneration. Nat Med 9(6):702–712. doi:10.1038/nm0603-702

Rafii S, Butler JM, Ding B-S (2016) Angiocrine functions of organ-specific endothelial cells. Nature 529(7586):316–325

Rapp BM, Saadatzedeh MR, Ofstein RH, Bhavsar JR, Tempel ZS, Moreno O, Morone P, Booth DA, Traktuev DO, Dalsing MC (2012) Resident endothelial progenitor cells from human placenta have greater vasculogenic potential than circulating endothelial progenitor cells from umbilical cord blood. Cell Med 2(3):85–96

Red-Horse K, Crawford Y, Shojaei F, Ferrara N (2007) Endothelium-microenvironment interactions in the developing embryo and in the adult. Dev Cell 12(2):181–194

Reinisch A, Hofmann NA, Obenauf AC, Kashofer K, Rohde E, Schallmoser K, Flicker K, Lanzer G, Linkesch W, Speicher MR (2009) Humanized large-scale expanded endothelial colony–forming cells function in vitro and in vivo. Blood 113 (26):6716–6725

Reynolds LP, Redmer DA (2001) Angiogenesis in the placenta. Biol Reprod 64(4):1033–1040

Reynolds LP, Grazul-Bilska AT, Redmer DA (2002) Angiogenesis in the female reproductive organs: pathological implications. Int J Exp Pathol 83(4):151–164

Rhodes KE, Gekas C, Wang Y, Lux CT, Francis CS, Chan DN, Conway S, Orkin SH, Yoder MC, Mikkola HK (2008) The emergence of hematopoietic stem cells is initiated in the placental vasculature in the absence of circulation. Cell Stem Cell 2(3):252–263

Ribatti D, Vacca A, Nico B, Roncali L, Dammacco F (2001) Postnatal vasculogenesis. Mech Dev 100(2):157–163

Robin C, Bollerot K, Mendes S, Haak E, Crisan M, Cerisoli F, Lauw I, Kaimakis P, Jorna R, Vermeulen M (2009) Human placenta is a potent hematopoietic niche containing hematopoietic stem and progenitor cells throughout development. Cell Stem Cell 5(4):385–395

Rouwkema J, Rivron NC, van Blitterswijk CA (2008) Vascularization in tissue engineering. Trends Biotechnol 26(8):434–441. doi:10.1016/j.tibtech.2008.04.009

Santos MI, Unger RE, Sousa RA, Reis RL, Kirkpatrick CJ (2009) Crosstalk between osteoblasts and endothelial cells co-cultured on a polycaprolactone–starch scaffold and the< i> in vitro development of vascularization. Biomaterials 30(26):4407–4415

Schatteman GC, Hanlon HD, Jiao C, Dodds SG, Christy BA (2000) Blood-derived angioblasts accelerate blood-flow restoration in diabetic mice. J Clin Invest 106(4):571–578. doi:10.1172/JCI9087

Scheubel RJ, Zorn H, Silber R-E, Kuss O, Morawietz H, Holtz J, Simm A (2003) Age-dependent depression in circulating endothelial progenitor cells in patients undergoing coronary artery bypass grafting. J Am Coll Cardiol 42(12):2073–2080

Schoenwolf GC (2009) Larsen's human embryology. Churchill Livingstone, Philadelphia

Sethi R, LEE CH (2012) Endothelial progenitor cell capture stent: safety and effectiveness. J Interv Cardiol 25(5):493–500

Shafiee A, Fisk NM, Hutmacher DW, Khosrotehrani K, Patel J (2015) Fetal endothelial and mesenchymal progenitors from the human term placenta: potency and clinical potential. Stem Cells Transl Med 4(5):419–423

Shepherd BR, Enis DR, Wang F, Suarez Y, Pober JS, Schechner JS (2006) Vascularization and engraftment of a human skin substitute using circulating progenitor cell-derived endothelial cells. FASEB J 20(10):1739–1741

Sukmawati D, Tanaka R (2015) Introduction to next generation of endothelial progenitor cell therapy: a promise in vascular medicine. Am J Transl Res 7(3):411–421

Tanaka R, Wada M, Kwon SM, Masuda H, Carr J, Ito R, Miyasaka M, Warren SM, Asahara T, Tepper OM (2008) The effects of flap ischemia on normal and diabetic progenitor cell function. Plast Reconstr Surg 121(6):1929–1942

Tanaka R, Masuda H, Kato S, Imagawa K, Kanabuchi K, Nakashioya C, Yoshiba F, Fukui T, Ito R, Kobori M (2014) Autologous G-CSF-mobilized peripheral blood CD34+ cell therapy for diabetic patients with chronic nonhealing ulcer. Cell Transplant 23(2):167–179

Tateishi-Yuyama E, Matsubara H, Murohara T, Ikeda U, Shintani S, Masaki H, Amano K, Kishimoto Y, Yoshimoto K, Akashi H (2002) Therapeutic angiogenesis for patients with limb ischaemia by autologous transplantation of bone-marrow cells: a pilot study and a randomised controlled trial. Lancet 360(9331):427–435

Tepper OM, Carr J, Allen RJ, Chang CC, Lin CD, Tanaka R, Gupta SM, Levine JP, Saadeh PB, Warren SM (2010) Decreased circulating progenitor cell number and failed mechanisms of stromal cell-derived factor-1α mediated bone marrow mobilization impair diabetic tissue repair. Diabetes 59(8):1974–1983

Timmermans F, Plum J, Yöder MC, Ingram DA, Vandekerckhove B, Case J (2009) Endothelial progenitor cells: identity defined? J Cell Mol Med 13(1):87–102

Tremblay PL, Hudon V, Berthod F, Germain L, Auger FA (2005) Inosculation of tissue-engineered capillaries with the host's vasculature in a reconstructed skin transplanted on mice. Am J Transplant 5(5):1002–1010. doi:10.1111/j.1600-6143.2005.00790.x

Ulrich C, Rolauffs B, Abele H, Bonin M, Nieselt K, Hart ML, Aicher WK (2013) Low osteogenic differentiation potential of placenta-derived mesenchymal stromal cells correlates with low expression of the transcription factors Runx2 and Twist2. Stem Cells Dev 22(21):2859–2872

Vasa M, Fichtlscherer S, Aicher A, Adler K, Urbich C, Martin H, Zeiher AM, Dimmeler S (2001) Number and migratory activity of circulating endothelial progenitor cells inversely correlate with risk factors for coronary artery disease. Circ Res 89(1):e1–e7

Yang J, Ii M, Kamei N, Alev C, Kwon S-M, Kawamoto A, Akimaru H, Masuda H, Sawa Y, Asahara T (2011) CD34+ cells represent highly functional endothelial progenitor cells in murine bone marrow. PLoS One 6(5):e20219

Yoder MC, Hiatt K, Mukherjee P (1997) In vivo repopulating hematopoietic stem cells are present in the murine yolk sac at day 9.0 postcoitus. Proc Natl Acad Sci 94(13):6776–6780

Yoder MC, Mead LE, Prater D, Krier TR, Mroueh KN, Li F, Krasich R, Temm CJ, Prchal JT, Ingram DA (2007) Redefining endothelial progenitor cells via clonal analysis and hematopoietic stem/progenitor cell principals. Blood 109(5):1801–1809

Human Amniotic Membrane as a Biological Source for Regenerative Medicine

Mazaher Gholipourmalekabadi, Narendra Pal Singh Chauhan, Behrouz Farhadihosseinabad, and Ali Samadikuchaksaraei

1 Introduction

The properties of amniotic membrane to promote wound healing have been known for over 70 years. Amniotic membrane was shown to have several beneficial effects: it promotes epithelialization, has antimicrobial effects, and decreases inflammation and fibrosis. The discovery of cell populations in amniotic membrane which are capable of differentiating into a variety of cell types has stimulated the research aimed for characterizing the cells and evaluating their potential utility in

M. Gholipourmalekabadi
Cellular and Molecular Research Center, Iran University of Medical Sciences, Tehran, Iran

Department of Tissue Engineering & Regenerative Medicine, Bhupal Nobles Post Graduate (B.N.P.G.) College, Hemmat Highway, Tehran 144961-4535, Iran
e-mail: mazaher.gholipour@gmail.com

N.P.S. Chauhan
Department of Chemistry, Faculty of Advanced Technologies in Medicine, Iran University of Medical Sciences, Udaipur, Rajasthan, India
e-mail: narendrapalsingh14@gmail.com

B. Farhadihosseinabad
Biotechnology Department, School of Advanced Technologies in Medicine, Shahid Beheshti University of Medical Sciences, Tehran, Iran
e-mail: beh_far1364@yahoo.com

A. Samadikuchaksaraei, M.D., Ph.D., D.I.C., F.R.S.P.H. (✉)
Cellular and Molecular Research Center, Iran University of Medical Sciences, Tehran, Iran

Department of Tissue Engineering & Regenerative Medicine, Faculty of Advanced Technologies in Medicine, Iran University of Medical Sciences, Hemmat Highway, Tehran 144961-4535, Iran

Department of Medical Biotechnology, Faculty of Allied Medicine, Iran University of Medical Sciences, Tehran, Iran
e-mail: samadikuchaksaraei@yahoo.com

B. Arjmand (ed.), *Perinatal Tissue-Derived Stem Cells*, Stem Cell Biology and Regenerative Medicine, DOI 10.1007/978-3-319-46410-7_5

regenerative medicine. While a major focus of research has been the use of amniotic membrane in tissue engineering and cell replacement, also for repair of injured tissues via paracrine actions to treat injury and diseases.

Every year, more and more of the world's population suffer from organ failure and tissue loss due to different pathologies and traumas. Therefore, we expect to be faced with an increased number of patients requiring organ or tissue replacement therapies. Tissue engineering (TE) or regenerative medicine is a multidisciplinary and rapidly developing area of science that has opened a new frontier to solve the shortage of transplantable tissues or organs. The aim of tissue engineering is to design and fabricate a biological replacement and thereby restore the structure and functionality of injured tissues by a triplet of scaffolds, growth factors, and cells, including stem cells. To date, many synthetic (Azami et al. 2012) and natural (Mobini et al. 2013a) scaffolds have been developed for regeneration of various tissues such as bone (Saki et al. 2009; Mobini et al. 2013b), skin (Jafari et al. 2011), etc. As a natural scaffold, human amniotic membrane (HAM) has been found to be a promising biological scaffold for regeneration of damaged soft tissue (Gholipourmalekabadi et al. 2015a). In this chapter, we describe the basic structure and properties of HAM that makes it an excellent source for TE applications.

2 Structure

2.1 The Matrix

The HAM consists of three main layers including an epithelial monolayer, basement membrane, and avascular stroma. The epithelial layer is composed of a single layer of epithelial cells that are uniformly arranged on the basement membrane. Amniotic epithelial cells exhibit quite characteristic morphological features. They have a relatively small number of intracytoplasmic organelles, microvilli on the apical surface, abundant cytoplasmic processes to the lateral and basal sides, and loose intercellular connections between each other. The epithelial basement membrane has positive immunoreactivity for collagen type IV (α1, α2, α5, and α6 chains); laminin (α3, ß1, γ1, and γ2 chains); laminin-1 and -5; perlecan; nidogen-1 and -2; agrin; fibronectin; collagen types VII, XV, XVI, and XVII; matrilin-4; and tenascin-C. On the other hand, it has negative immunoreactivity for collagen type IV (α3 and α4 chains), laminin (α4, ß2, ß3, and γ3 chains), and collagen type V (Dietrich-Ntoukas et al. 2012).

The main fibrous skeleton is formed from the densely compacted collagen fibers that are bounded with each other and also basement layer by fibronectin (glycoprotein). The main collagens type of this layer are interstitial collagens (types I and III) that are formed in parallel bundles in order to maintain the mechanical integrity of HAM. It has been reported that such collagens are secreted by mesenchymal stem cells and fibroblasts present in avascular stroma layer. The filamentous connections between interstitial collagen and epithelial basement membrane are formed by collagens type V and VI. The nearest layer to chorionic membrane is known as intermediate layer. This layer seems spongy in histological observations due to its nonfibrillar meshwork structure and also presence of a large content of glycoproteins and proteoglycans. The connection between spongy layer and chorionic membrane is weak, so that it can be simply disconnected via blunt dissection (Epstein et al. 1998).

It has been well documented that presence of some biological active substances such as different cytokines and signaling molecules including the Tumor Necrosis Factor (TNF), Transforming Growth Factor alpha (TGF-a) and beta (TGF-b), Interferon, basic Fibroblastic Growth Factor (bFGF), Epidermal Growth Factor (EGF), Hepatic Growth Factor (HGF), keratinocyte growth factor (Yu et al. 2009), interleukin-4 (IL-4) (Jones et al. 1995), IL-6, IL-8 (Keelan et al. 1997), natural inhibitors of metalloproteases, β-defensins, and prostaglandins, etc. (Insausti et al. 2010; Koizumi et al. 2000b; Parolini et al. 2008) endow such promising properties to HAM. Some highlighted properties of HAM are described later in some details.

2.2 *Amniotic Membrane-Derived Stem Cells*

Cell therapy is an inseparable part of regenerative medicine. A reliable, available, and safe cell source should be considered to provide the cells needed for cell replacement therapies. Therefore, many investigations have been conducted to find out the most appropriate cell source in the context. Although many types of cells such as pluripotent stem cells (Samadikuchaksaraei and Bishop 2007; Van Vranken et al. 2007) and bone marrow-derived mesenchymal stem cells (Eftekharzadeh et al. 2015) have been considered for cell therapy and tissue engineering applications (Siti-Ismail et al. 2012; Samadikuchaksaraei and Bishop 2006; Shoae-Hassani et al. 2015), none of them showed the features of an ideal cell source for clinical applications. Therefore, many investigations are ongoing to find a suitable cell source for cell therapy. Amniotic mesenchymal stem cells (AMSCs) and amniotic epithelial cells (AECs), which are derived from AM, have excellent self-renewal and differentiation properties. The AMSCs express the mesenchymal-specific markers including CD44, CD73, CD29, CD105, and CD90 and also are negative for hematopoietic markers and human leukocyte antigen including CD34, CD45, CD11b, CD19, HLA-A, HLA-B, and DR antigens (Kim et al. 2007). Human AECs have some favorable properties that make them a suitable cell source for cell therapy. For example, these cells suppress local immune response, show low immunogenicity, and thereby minimize transplant rejection rate (Bilic et al. 2008).

AFSCs express embryonic stem cell-specific markers and are morphologically similar to fibroblast cells. These cells also express MSC markers such as CD105, CD73, CD44, CD166, CD29, CD58, CD90, CD117. It has been reported that the AFSCs express MHC I molecular antigen, while do not or weakly express MHC class II antigen (HLA-DR) (Moschidou et al. 2013). Also, these cells do not express CK-19 gene, the gene responsible for teratoma formation in pluripotent stem cells (Bilic et al. 2008).

AMSCs and AEC have been utilized as a therapeutic cell source in different experimental models of diseases such as myocardial infarction (Fang et al. 2012), neuronal regeneration (Sankar and Muthusamy 2003; Roh et al. 2013), kidney disease (Perin et al. 2010; Chang et al. 2011; Baulier et al. 2014), liver disease (Zheng et al. 2008, 2012; Hodge et al. 2014), skin and burn wounds (Yoon et al. 2009; Skardal et al. 2012), and so on.

3 Properties

3.1 *Angiogenic Property*

Several studies have been conducted to survey the angiogenic properties of HAM, although their results were controversial in the context. Accumulating evidence has revealed that the HAM possesses both angiogenesis and antiangiogenesis properties. The angiogenic factors including Vascular Endothelial Growth Factor (VEGF), Interleukin-8 (IL-8), angiogenin, interferon-γ, Interleukin-6 (IL-6), basic Fibroblast Growth Factor (bFGF), Epidermal Growth Factor (EGF) and Platelet-Derived Growth Factor (PDGF) (Burgos 1986; Wolbank et al. 2009) and antiangiogenic factors such as IL-1 receptor antagonist, TIMP3 and TIMP4 are secreted by amniotic epithelial cells (Hao et al. 2000).

3.2 *Antiscarring Effects*

The HAM bio-scaffold modulates the wound healing process by promoting tissue reconstruction. This membrane also prevents scar formation through down-regulation of TGF-b and its receptor in the remaining living cells of damaged skin tissue (Lee et al. 2000; Tseng et al. 1998). Late or low reepithelialization as well as prolonged inflammatory response cause chronic wound, resulting in scar formation. Therefore, prevention of infection and stimulation of epithelialization may profoundly minimize formation of scar. The anti-inflammation potential of HAM has been well documented in several studies. The HAM down-regulate the expression of proinflammatory cytokines such as IL-1α and IL-1β in the injured tissue (Solomon et al. 2001).

3.3 *Cell Adhesion Property*

The inflammatory cells such as lymphocytes can be attached to hyaluronic acid (a ligand for CD44) which exists largely in HAM via their CD44. In addition, fibronectin, laminin, and various types of collagens and proteoglycans that exist in basement

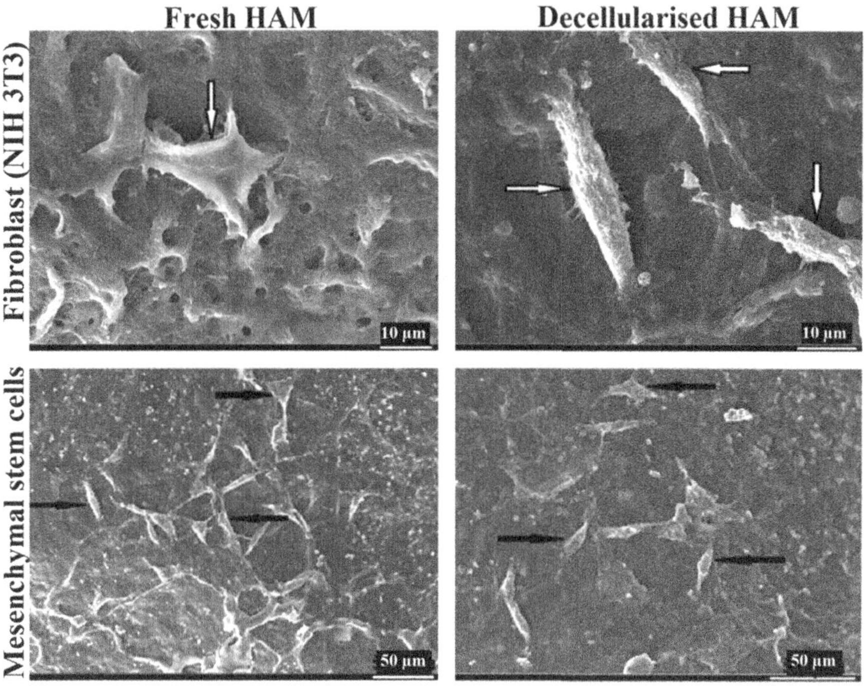

Fig. 1 SEM micrographs of the fibroblast NIH 3T3 and bone marrow-derived mesenchymal stem cells on both fresh and decellularized HAM after 72 h incubation time. *White and black arrows* indicate fibroblast and mesenchymal stem cells, respectively, grown on the membrane (Gholipourmalekabadi et al. 2015a, b)

layer of HAM can act as a ligand for integrin receptors. Such cell adhesion properties in HAM make it a promising biological membrane and create a favorable condition for overlying cell growth, expansion, and eventually tissue regeneration (Higa et al. 2005). In our previously published study (Gholipourmalekabadi et al. 2015c, 2016), favorable cell adhesion and growth support of both fresh and decellularized HAM for mesenchymal stem cells and fibroblasts have been shown by SEM micrographs (Fig. 1).

3.4 Antibacterial Activity

As mentioned earlier in this chapter, HAM has a broad-spectrum antibacterial activity. It has been shown that existence of a low-molecular-mass group of proteins including defensins, secretory leukocyte proteinase inhibitor (SLPI), and elafin within the HAM matrix are responsible for its antimicrobial properties. The majority of defensins which are expressed by amniotic epithelial cells are β3-defensin. This protein is a member of β-defensins family that is secreted by epithelial cells of mucosal surface and also considered as an important part of the innate immune system (Harder et al. 2000; Higa et al. 2005). The innate immune system is the first line of defense against the entrance of microorganism and infections into the body (King et al. 2003). Several studies have investigated the antibacterial activity of HAM against various strains of bacteria (Inge et al. 1991; Lo and Pope 2009; Tehrani et al. 2013). For example, Mohammadi and coworkers (Mohammadi et al. 2013) showed that the fresh human amniotic membrane grafted in patients with chronic infected burn wounds profoundly decreased the rate of infections. This is an interesting property that helps to decrease the dosage of routinely used antibiotics in the condition that leads to application of amniotic membrane. Some of these antibiotics negatively affect the behavior of stem cells (Cohen et al. 2006).

4 Decellularization of Human Amniotic Membrane

It has been reported that there are some complications such as graft rejection regarding the implantation of HAM as an allograph. At least three types of cells, epithelial, fibroblast, and mesenchymal stem cells, exist within the HAM that may evoke recipient body immune responses. Denudation of HAM profoundly decreases its graft rejection rate (Zhang et al. 2013; Riau et al. 2010). On the other hand, it is generally accepted that the collection and preservation conditions of tissue engineered constructs before surgery remarkably affect the success in any implantation (Riau et al. 2010; Sutherland et al. 2015; Thibault et al. 2013). Therefore, several attempts have been made to develop an efficient method for decellularization and preservation of HAM. It has been well defined that denudation of HAM will make it a better supporter and less immunogenic (Zhang et al. 2013; Riau et al. 2010). Several studies have been conducted for development of an efficient method for decellularization and preservation of HAM with various rates of success. The majority of these methods are based on the detergent and enzyme-based techniques (He et al. 2002; Mligiliche et al. 2002; Wilshaw et al. 2006).

In a study, Wilshaw et al. (2006) used various detergent-based materials such as protease inhibitors, sodium dodecyl sulfate (SDS), tris-buffered saline (TBS), aprotinin, DNase, and RNase to present an efficient protocol for HAM denudation. Although the protocol was time consuming and expensive, the results obtained from their study confirmed the full removal of the cells from HAM matrix without any detectable effects on biomechanical and cytotoxicity behaviors of the membrane. In another study, Luo et al. (2004) prepared a decellularized HAM by chemical detergent-enzymatic agents. Such protocols have been widely used for many years for the decellularization of HAM worldwide. Recently, we have developed a simple, reproducible, and cost-effective method for decellularization and preservation of HAM without using any enzymatic agent. After confirmation of full removal of the cells, we showed that the basement membrane proteins of HAM such as collagen types I, III, and IV remained intact after decellularization process (Fig. 2) (Gholipourmalekabadi et al. 2015c). We also found that the number of lymphocytes

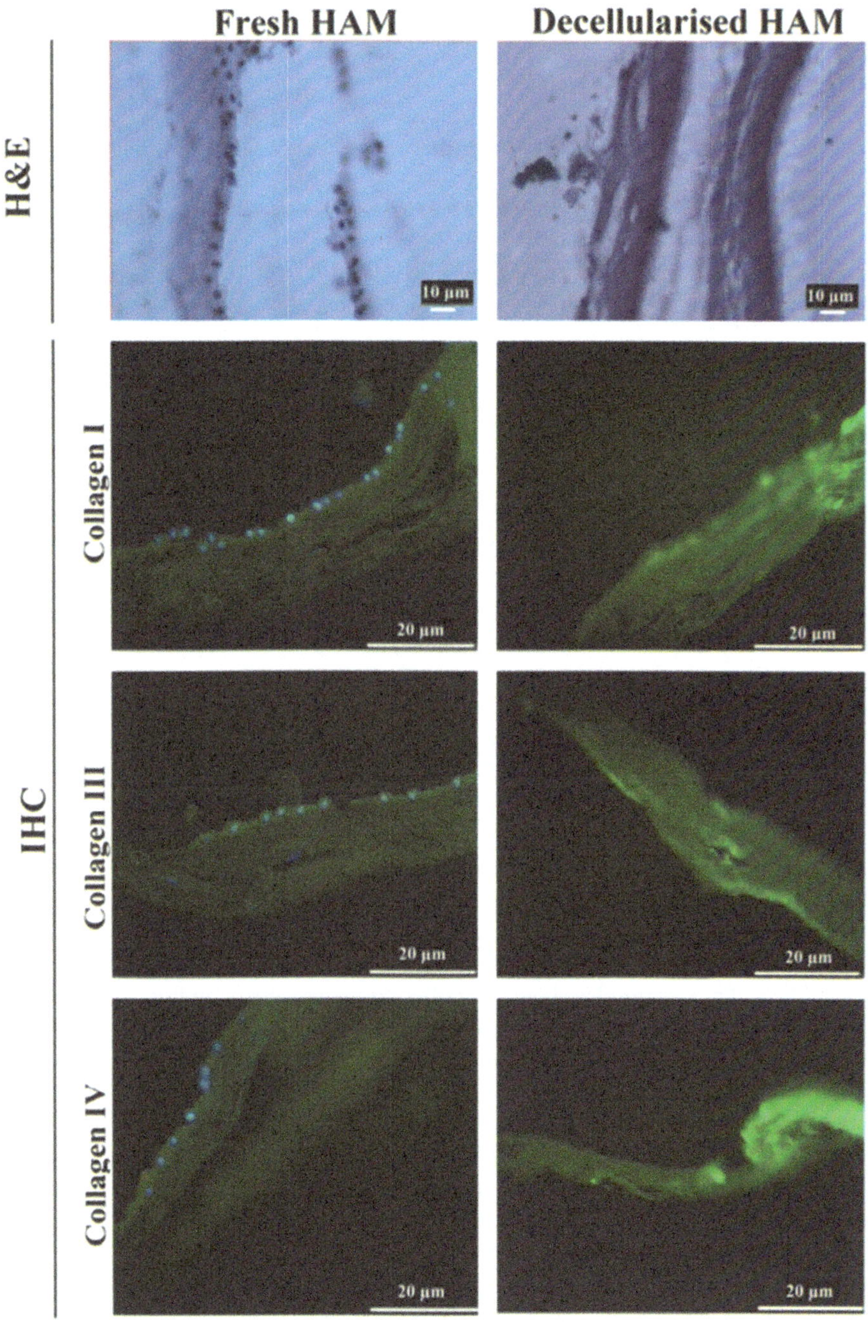

Fig. 2 H&E and IHC stained samples for both fresh and decellularized HAM. IHC was performed for collagen types I, III, and IV (Gholipourmalekabadi et al. 2015c)

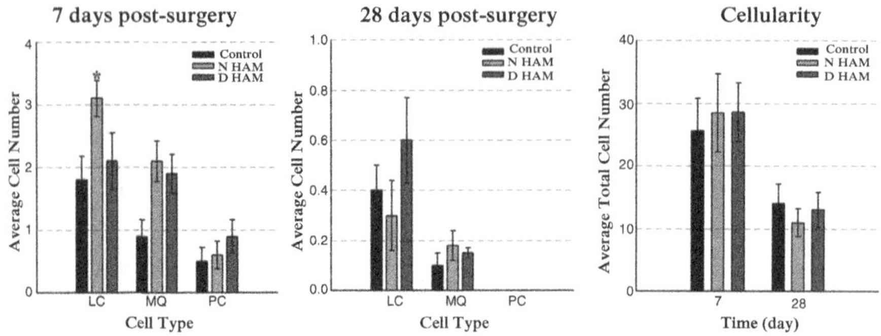

Fig. 3 Average cell number of lymphocytes (LC), macrophages (MQ), and plasma cells (PC) after 7 and 28 days postimplantation as well as cellularity in the implanted site. An *asterisk* indicates significant differences (Gholipourmalekabadi et al. 2015c)

(LC) recruited to the subcutaneously implanted fresh HAM area was higher than those in decellularized HAM implanted group by day 7, the difference was insignificant after 28 days postsurgery. Unlike LC, the number of macrophage (MQ), plasma cell (PC), and even cellularity in the implanted site did not change between the experimental groups (Fig. 3) (Gholipourmalekabadi et al. 2015c).

5 Collection and Preservation of HAM

As generally accepted, the collection and preservation conditions of tissue before surgery can profoundly affect the effectiveness of grafting and success in transplantation. Therefore, many attempts have been made to optimize the best condition for collection and preservation of HAM (Riau et al. 2010; Sutherland et al. 2015; Thibault et al. 2013). In using AM, donors should be tested for infecting organisms. In addition, the AM samples should be collected under sterile conditions and stored in frozen state (John and Oommen 2010; Loeffelbein et al. 2014). On the other hand, fresh HAM may not always be available when needed due to its short shelf life. Therefore, various preservation techniques have been developed to collect and preserve the HAM, among which cryopreservation of HAM in liquid nitrogen (Bravo et al. 2000; Bujang-Safawi et al. 2010), silver nitrate (Bujang-Safawi et al. 2010; Haberal et al. 1987), preservation in antibiotics solution, glycerol-preserved sheets, gamma-irradiated sheets, and dried sheets (Bujang-Safawi et al. 2010) are notable.

The dried irradiated AM (Bujang-Safawi et al. 2010) seems to be more comfortable than glycerol-preserved HAM (Bravo et al. 2000; Kesting et al. 2008) and its application is quick, easy, and also greatly relives pain compared with the membrane prepared through other preservation methods (Bujang-Safawi et al. 2010). Additionally, the dried irradiated AM sheets are free of microbial contaminations and also can be stored at room temperature for 5 years (Bujang-Safawi et al. 2010).

6 Applications in Tissue Engineering

In view of the unique characteristics of HAM, it has been widely used in surgical interventions and tissue engineering such as wound dressing (Stern 1913; Davis 1910), neurosurgery (Trelford and Trelford-Sauder 1979), ophthalmic surgery (Shimazaki et al. 1997), and vagina surgery (Bleggi-Torres et al. 1997). In addition, this membrane is easily available and cost effective. Here, some applications of HAM in tissue regeneration are described.

6.1 *Ocular Surface*

The outermost layer of the eye is composed of the cornea and conjunctive parts. The cornea may have been influenced and damaged by different environmental factors and diseases such as chemical or thermal burns, Stevens–Johnson syndrome, severe keratitis, etc. These can lead to limbal stem cell deficiency (LSCD) and eventually cause corneal opacification and blindness (Dua and Forrester 1990; Shapiro et al. 1981). Despite of the efforts in the context of corneal transplantation, the techniques which are commonly used often fail and even cause persistent epithelial defects and secondary infection. However, the conventional therapies show no effect on the restoration of the limbal stem cell population (Kuckelkorn et al. 2001). In addition to corneal surface reconstruction, there is a major therapeutic challenge in conjunctival reconstruction and LSCD. In the case of LSCD treatment, despite of the limbal epithelial stem cells (LESCs) transplantation method that is used routinely for treating the patients with severe LSCD, a limited success has been achieved. Using of autologous LESC transplantation is associated with increased risk of inducing OS disease in the donor eye due to a large requirement of limbal graft (Chen and Tseng 1991; Dua and Azuara-Blanco 2000; Holland 1996). In another hand, in allogeneic LESCs transplantation, long-term systemic immunosuppressant treatment may lead to irreversible damages (Daya et al. 2000; Tchah et al. 2003).

The first use of fetal membrane for ocular treatment was published in the 1940. In that study, a full-thickness fetal membrane including both amnion and chorion was implanted for the reconstruction of the conjunctiva defects and repair of symblepharon (Trelford and Trelford-Sauder 1979). The results were not promising, presumably because of presence of the chorion. Further studies date back to the 1946 and 1947 when a chemically modified amniotic membrane called amnioplastin was used for ocular burns as a temporary patch (Sorsby et al. 1947; Sorsby and Symons 1946). Although some favorable results such as reduced scar formation, improved comfort, and shorted hospitalization stays (Sorsby et al. 1947; Sorsby and Symons 1946) were achieved in that study, further studies in the context were not pursued for a long period of time. The resumption of studies on use of preserved human amniotic membrane for ocular surface reconstruction was carried out by Kim and Tseng in 1995 in a rabbit model (Kim and Tseng 1995). A model of limbal stem cell deficiency was created where the lack of corneal epithelial progenitor cells resulted in the corneal conjunctivalization. The notable success in that investigation for reconstruction of the cornea encouraged the other researchers for utilizing the amniotic membrane as a therapeutic biomaterial for various ophthalmic problems such as corneal burns, corneal ulcerations, and also as a support surface for growth and expansion of epithelial cells (Chen et al. 2000; Choi et al. 1998; Heiligenhaus et al. 2008; Koizumi et al. 2000a; Kruse et al. 1999; Lee and Tseng 1997; Meller and Tseng 1999; Shimazaki et al. 1997; Tsai et al. 2000; Tseng et al. 1998). Structural similarity between amniotic basement membrane and conjunctiva that is composed of a tensile reticular fibers network and also some other special characteristics such as transparency, elasticity, structural integrity, and its positive effects on epithelial cell migration and proliferation have made HAM an ideal substitution for ocular reconstruction (Shimazaki et al. 1997; Niknejad et al. 2008; Fukuda et al. 1999; Dua et al. 2004).

The structural similarity of HAM with corneal and conjunctival layers such as existing type V, IV, VII collagen, fibronectin, laminin I and V provide a suitable surface for attachment and anchorage of corneal epithelial cells (Heiligenhaus et al. 2008; Sangwan et al. 2007).

Indeed, HAM can play the role of a feeder layer for growth and proliferation of epithelial cells and their progenitors and also keep their normal morphology. In addition to the aforementioned effects of amniotic membrane on epithelial cells, it has been found to have a great impact on differentiation and clonogenicity of progenitor cells (Niknejad et al. 2008). The HAM has also been used for remediation of limbal stem cell deficiencies in ophthalmology (Azuara-Blanco et al. 1999; Heiligenhaus et al. 2008; Kim and Tseng 1995; KIM et al. 2000).

In a study conducted by Lee and Tseng (Lee and Tseng 1997), amniotic membrane was applied in treatment of patients with persistent epithelial defects. In another study, Augusto Azuara-Blanco et al. (Azuara-Blanco et al. 1999) evaluated the efficacy of the amniotic membrane transplantation (AMT) in ocular surface reconstruction. It has been showed that the AMT was helpful in patients with persistent epithelial defect and also had a great inhibitory effect on corneo conjunctival adhesion after surgery.

The positive effectiveness of amniotic membrane transplantation in decrease of stromal scarring and ocular surface inflammation has been well documented (Azuara-Blanco et al. 1999; Heiligenhaus et al. 2008; Niknejad et al. 2008; Sangwan et al. 2007).

Shimmura and colleagues implanted HAM to 20 patients with persistent corneal epithelial defects to study its anti-inflammatory effects in vivo. The local body immune response to the membrane was evaluated by histopathological examinations at 1 week postsurgery. According to their results, the majority of the infiltrated inflammatory cells were stained positively with anti-CD14 antibodies, indicating that these cells were originated from monocyte/macrophage lineage. Also, various subsets of T cells, including CD4(+) and CD8(+) cells and also CD20(+) cells were observed sporadically. Anti-inflammatory properties of the amniotic membrane can be explained by its high adhesion property for inflammatory cells and also the pro-apoptotic agents exist within it. TUNEL assays showed that the inflammatory cells cultured on the HAM surface represented morphological and molecular characteristics of the cells undergoing apoptosis (Shimmura et al. 2001). Although such evidences clearly explain anti-inflammatory properties of the HAM, some other factors may also be involved in the context. For instance, the HAM can reduce the secretion of some cytokines involved in inflammatory process from ocular epithelial cells (Sangwan et al. 2007). In a study conducted by Solomon et al. (2001), it has been shown that the HAM stromal matrix suppressed IL-1α and IL-1β, two critical pro-inflammatory cytokines, at both protein and mRNA levels, while caused an up-regulated expression of the anti-inflammatory cytokines such as interleukin-1 receptor antagonist (IL 1 RA).

On the other hand, it is important to be noted that the conventional methods for isolation and expansion of corneal limbal epithelial cells have been failed. Therefore, there is an urgent need in development of a reliable technique for ex vivo expansion of corneal limbal epithelial cells as a therapeutic method for patients with severe ocular surface diseases (OSD) (Pellegrini et al. 1997). In 2000, the possible application of amniotic membrane as a carrier for autologous corneal limbal epithelial cells was investigated by researchers. They showed that the implantation of the HAM seeded with autologous corneal limbal epithelial cells had a positive effect on reconstruction of corneal surface of patients with unilateral OSD (Schwab et al. 2000).

Limbal stem cells (LSC) deficiency is a visually disabling condition in which the corneal surface epithelium is not able to heal spontaneously. Therefore, autologous transplantation of limbal epithelial cells (HLECs) has been considered as an efficient method for treatment of such patients. Grueterich et al. (2003) reported that the HAM can be considered as a niche for the limbal epithelial stem cells. These cells can be easily isolated from limbal biopsy using enzyme treatment and seeded onto HAM (Nakamura et al. 2006).

In a study conducted by Esther et al. (Rendal-Vázquez et al. 2012), a thin layer of (1–2 cell layer) of epithelium was formed over the HAM. According to their results, the epithelial cells cultured on basal layer (the epithelial side) of HAM showed a remarkable expression of LSC markers such as p63 and ABCG2, while similar results were not found on the chorionic side.

Implantation of the HAM loaded with ex vivo cultured LSCs has opened a new window of hope to treatment of patients with the deficiencies or destruction of LSC population. The successes achieved in this treatment strategy have caused a growing interest in utilizing of ex vivo LSC-loaded HAM for patients with LSC.

6.2 Skin

Skin is largest organ in body and plays a key role in protecting body against damaging sunlight, harmful chemicals, pathogens, and water loss. Skin functions also as an insulation, temperature regulation, sensation, and protection of D folates. Therefore, damage of skin tissue can cause many health problems. Mammalian skin consists of two major layers: (1) Epidermis provides waterproofing and also protect body from microorganisms and, (2) dermis containing blood vessels and skin appendices. Skin is damaged in various conditions such as diabetic, cancer, accident, microbial infections, burning, etc. (Alibardi 2003; Madison 2003; Proksch et al. 2008). The nature of wound healing process includes a delicate balance of inflammatory, vascular, connective tissue, and epithelial cells activities (Baskovich et al. 2008) that proceed via three overlapping phases: inflammation, proliferation, and remodeling.

Upon skin injury, platelet cells activate clot formation and then secrete some cytokines and growth factor such as PDGF to promote wound healing. The inflammatory cells migrate to the defect site for phagocytosis of dead cells as well as microorganisms. Epithelial cells proliferate and migrate from margins to the middle of wound to avoid water evaporation (reepithelialization). Inflammatory cells such as macrophage secrete some cytokines to mediate various aspects of wound healing. The fibroblasts proliferate, migrate, and synthesize collagens to form ECM. Finally, the newly formed tissue in the defect site is remodeled through some proteases to restore the ECM contents and structure of natural skin (Yildirimer et al. 2012; Kondo and Ishida 2010). In large-surface and full-thickness wounds, reepithelialization cannot completely proceed (Baskovich et al. 2008; Bello and Phillips 2000). There is an increased risk of scar formation in the wounds that not heal by 10 days (DEITCH et al. 1983).

Success in the wound healing process depends on depth, size, and location of damaged area. Superficial wounds can heal spontaneously, while several studies have shown that full-thickness wounds require to be covered with skin substitutes.

According to the reports in literatures, an ideal skin substitute should improve healing, reduce water loss and infections, minimize scar formation, relieve pain and be flexible in thickness, readily available, easily applicable, and cost effective (Bujang-Safawi et al. 2010; Eaglstein 1985; Halim et al. 2010; Shores et al. 2007).

Autologous skin grafting is considered as a standard strategy in severe skin wounds tissue engineering. However, there are some complications regarding the use of such grafts including long time hospitalization and donor site morbidity (Bello and Phillips 2000). Xenotransplant has also been reported to be effective in skin wound regeneration. Despite of favorable results in using allotransplant and xenotransplant, some limitations have been reported in such strategies, especially if tissue banks do not exist. For instance, there is an urgent need for skin graft, about 6000 square centimeters, in 50 % of patients with full thickness burn. It is obvious that preparation of donor area in such cases is a serious problem. Therefore, the importance of temporary skin substitute, at least until donor area could be used again, is undeniable (Robson and Krizek 1973). There is still no ideal skin scaffold available that fulfills all the features mentioned earlier. During the past two decades, various commercial products such as Integra™, Biobrane™, Alloderm™, Trans Cyte™/ dermagraft-TC™, Apligraft™, and Epicel™ have been applied for treatment of skin wound in clinic with various rates of advantages and disadvantages (Meyer et al. 2009).

Tissue engineering scaffolds provide a suitable environment for normal cellular growth, differentiation, and angiogenesis through mimicking the target tissue that guarantee long-term viability of implanted graft (Epstein et al. 1999; Bello and Phillips 2000; Atiyeh et al. 2005; Bujang-Safawi et al. 2010; Halim et al. 2010).

Skin tissue engineering scaffolds have often made from two major biomaterials: synthetic biodegradable polymers such as hyaluronic acid-based polymers, poly (glycolic) acid and Poly Tetra Fluoro Ethylene (PTFE), and naturally derived from mammalian tissue sources, well known as naturally occurring biopolymers (Bromberg et al. 1965; Ghalambor et al. 2000; Bujang-Safawi et al. 2010; Halim et al. 2010; Gholipourmalekabadi et al. 2015c). Synthetic biodegradable polymers used in wound healing have some disadvantages. For example, such scaffolds are unable to fully restore the normal structure and function of injured vascular tissues (Chen et al. 1997; Hodde 2002), lead to tissue deposition that is less than optimal (Cao et al. 1998; Hodde 2002), and result in the formation of excessive scar tissue or infection (Mendelsohn and Dunlop 1998; Ghalambor et al. 2000; Hodde 2002).

Naturally occurring biopolymers include cadaveric fascia, small intestinal submucosa, acellular dermis (e.g., Pig skin containing Collagen types: I, IV, VII (Hodde 2002; Gholipourmalekabadi et al. 2015a, 2015c), proteins: elastin, glycosaminoglycans, and growth factors), bladder acellular matrix graft, and amniotic membrane (containing Collagen types: I, III, IV (Aplin et al. 1985; Lei et al. 1999; Koizumi et al. 2000b; Meinert et al. 2001; Hodde 2002), proteins: decorin, glycosaminoglycans: hyaluronic acid and Growth factors: EGF, TGF-b, TGF-a, FGF-2, KGF, HGF/SF).

Nevertheless, some complications and variable success rates (Ghalambor et al. 2000) have been reported regarding the use of naturally occurring biopolymers and none of these scaffolds have served as an ideal artificial skin substitute (Ghalambor et al. 2000). Taken together, amniotic membrane (AM) seems preferable over other skin dressing due to its promising characteristics, as mentioned earlier in this paper. Several studies have reported various advantages of amnion membrane skin wound healing (Modesti et al. 1989; Stern 1989; Wolf et al. 1991; Barret et al. 2000; Ghalambor et al. 2000; Mermet et al. 2007; Kim et al. 2008; Halim et al. 2010). Amniotic membrane has been widely used as a biological skin dressing for decades, dated back to 1910 (Davis 1910). In addition to many promising properties of HAM as a skin substitute, no human leukocyte antigen (HLA-A, -B, or -DR) has been detected on AM. This unique feature of AM eliminates the possibility of graft rejection (Shimazaki et al. 1998). Furthermore, AM is a cost-effective and easy access wound dressing (Bose 1979; Lynch and Blocker 1979).

In a study conducted by Loeffelbein et al. (2012), 40 experimental full-thickness skin wounds were implanted with an autologous split-thickness skin graft (STSG) solely or in combination with a monolayer or multilayer of human amniotic membrane. Subsequently, on days 5, 7, 10, 20, 40, and 60 postsurgery, the biopsy samples were obtained for both clinical examinations and immunohistochemical staining for smooth muscle actin (aSMA), laminin, von-Willebrand-factor (vWF), and Ki-67. According to their results, multilayered HAM improved reepithelialization and minimized scar formation. In another study, Fijan et al. (2014) implanted the HAM to 30 patients with full-thickness fingertip injuries and have found promising results when compared to its treatment efficacy with commercial skin graft dressing.

Poor vascularization of skin grafts has also been an unsolved problem in wound healing (Eppler et al. 2002). Neovascularization is a critical step in healing process of partial-thickness thermal injury (Pan et al. 2010). Angiogenesis or neovascularization phenomenon is identified by capillary formation from preexisting microvessels

(Carmeliet 2003). In the recent years, the potency of therapeutic angiogenesis in treatment of some special diseases such as ischemic heart disease, cerebrovascular disease, and delayed wound healing has been well cleared. A variety of biological agents and bioactive materials have been investigated for promoting the angiogenesis within damaged area (Höckel et al. 1993; Thompson et al. 2000).

As described earlier in this chapter, angiogenic or antiangiogenic characteristics of amniotic membrane are controversial and results have been different in the related studies. Niknejad et al. (2013) showed that two sides of HAM, epithelial and mesenchymal sides, had different angiogenic properties. According to their findings, the angiogenic and antiangiogenic properties were attributed with the mesenchymal and epithelial sides of the HAM, respectively. Therefore, such unique feature of HAM gives an excellent opportunity to be applied in various diseases. As wound dressing, the mesenchymal side of amniotic membrane can be utilized for improving the angiogenesis process and decreasing the healing time of full-thickness wounds.

6.3 Burn Wounds

Thermal trauma is one of the most common causes of severe skin defect. The keratin layers of skin and its lipid content play a critical role in keeping the water content of body at a normal level. In fact, skin provides a natural barrier to prevent excessive evaporation of body fluids. The lipids are thermosensitive and easily destroyed by heat (Jelenko 1967). It has been reported that the effective vapor pressure gradient increases 15–20 times (normal 1.5+0.08 mm above atmospheric pressure) when this natural barrier is destroyed by heat.

Complications in treatment of deep burn wounds have remained challenging, so that threatens the health of burn patients, especially those with low socioeconomic class (Mohammadi et al. 2015). In the burn wound area, the body loses a large amount of water and electrolytes, which is lethal for burn patients, and provides a suitable culture medium for growth of infectious agents. Therefore, replacing the body fluids should be considered as a first step during 24 h after burning. Also, appropriate measures should be taken to avoid high water loss through lost epidermal layer (Moncrief and Mason 1964).

The main goals of treatment in the burn patients are as follows: promoting the healing, control of pain, prevention or treatment of postburn infections, and decreasing the repeated trauma (Ghalambor et al. 2000; Halim et al. 2010). Healing of skin contains a complicated event that begins in the moment the skin is injured (Diegelmann and Evans 2004; Midwood et al. 2004; Groeber et al. 2011). It has been shown that superficial partial-thickness burns (second A degree) often heal spontaneously in 3 weeks, whereas healing of deep partial-thickness burns (second B degree) and full-thickness burns (third degree) takes several weeks (Barret et al. 2000). According to the literatures, early dermabrasion, or escharectomy, followed by coverage with skin grafts is the most effective way to treat second and third degree burn wounds (Barret et al. 2000). As reported, wound healing in patients with third degree burn has remained challenging. Full-thickness burn wounds with more than 1 cm in diameter require skin grafting to prevent cosmic deformities, scar formation, and resulting in an impaired mobility (Herndon et al. 1989; Papini 2004; Shevchenko et al. 2010; Groeber et al. 2011).

1 % silver sulfadiazine (SSD) and topical silver cream have widely been used as a treatment method for patients with burns and chronic wounds. Although silver has a broad spectrum of antimicrobial activity and low development of bacterial resistance, such strategies require frequent application, are care intensive to apply and remove, and is sometimes painful (Barret et al. 2000). On the other hand, toxicity of silver for human cells is controversial. Indeed, cytotoxicity effect of silver is silver dose dependent and could have an irreversible toxicity effect on human cells in high concentrations (Gholipourmalekabadi et al. 2015b, d; Nezafati et al. 2012).

Unfortunately, nowadays many patients with deep partial-thickness burns (second degree) are treated by daily washing and SSD dressing in many burns centers, especially in developing countries. Although potential of biological dressings in treatment of severe skin wounds has been confirmed by many studies, their applications in developing and Islamic countries are still not common. Many skin wound dressing have been developed during last decades, among which the HAM was the most promising skin substitute for treatment of burn wounds due to its favorable characteristics in wound healing and preventing bacterial infections. As mentioned before, there are some growth factors and cytokines such as TGF-a, TGF-b, bFGF, EGF within the matrix of HAM. These factors are also secreted by the cells around the damaged skin and mediate process of wound healing in normal situation. For example, epidermal growth factor plays a key role in reconstruction of epithermal layer of damaged skin. This growth factor exists abundantly in wound fluid and its mitogenic activity on epithelial, endothelial, and mesothelial cells has been well documented (Cribbs et al. 1998; Jahovic et al. 2004; Werner and Grose 2003). EGF has been found to improve reepithelialization of damaged skin and also accelerate proliferation and tensile strength of dermis (Alemdaroğlu et al. 2006, 2008; Brown et al. 1989; Greenhalgh 1996). TGF-α is another example that affect proliferation of keratinocytes (Cribbs et al. 2002; McCarthy et al. 1996; Werner and Grose 2003). In this direction, Houng and colleagues (Gu et al. 2011) showed a significantly increased level of EGF expression in rat alkali-burned corneas 2 weeks after implantation with HAM.

Infection, especially caused by bacterial sepsis, is believed to be the primary cause of mortality among burn patients. It has been revealed that two gram negative strains of bacteria, *Pseudomonas aeruginosa* (*P. aeruginosa*) and *Escherichia coli* (*E. coli*), and a gram positive strain of bacteria, *Staphylococcus aureus* (*S. aureus*) are the most common infectious agents after burn injuries (Ghalambor et al. 2000; Gholipourmalekabadi et al. 2015a, d, e; Nezafati et al. 2012). Infections also negatively affect the healing process of wounds, especially burn wound. Therefore, control of infection during the treatment period without doubt would have a great influence on the healing efficiency (Atiyeh et al. 2007). In a study conducted by Robson and Krizek (1973), burn wound created in rat skin was first inoculated with *Pseudomonas aeruginosa*. The wounds were then separately implanted with human skin and HAM. According to their reports, HAM grafting was more effective in decreasing bacterial population than human skin.

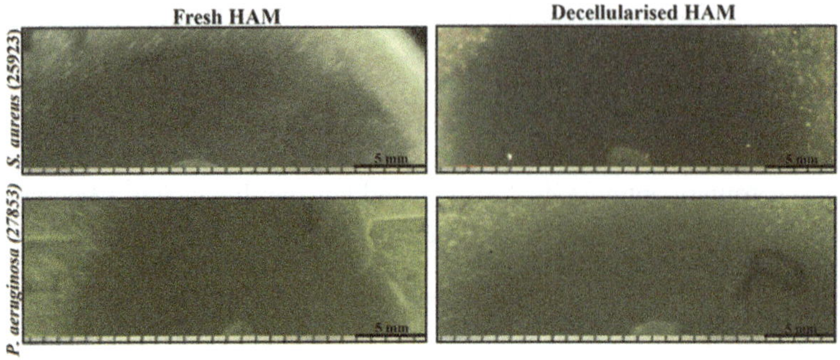

Fig. 4 Growth inhibition zone of standard strains of *Staphylococcus aureus* and Pseudomonas around both fresh and decellularized HAM (Gholipourmalekabadi et al. 2015d)

On the other hand, overuse of antibiotics caused an increased prevalence of antibiotic-resistant bacteria. Development of multidrug-resistant (MDR) strains has remained as a major concern in the healthcare community (Abbasi-Montazeri et al. 2013; Kardas et al. 2005; Toomey et al. 2009). In our previously published study, MDR strains of bacteria were isolated from exudate of patients with burn. The antibacterial activity of HAM against both standard and MDR strains of bacteria was tested. According to the results, neither fresh nor decellularized HAM showed a strong antibacterial activity against MDR, which had a broad-spectrum antibacterial activity for bacterial standard strains (ATCC) (Fig. 4). In that study, it has been concluded that the HAM cannot protect burn wounds against the infections caused by MDR and an alternative strategy is required to fully treat such infections (Gholipourmalekabadi et al. 2015a).

6.4 *Diabetic Ulcers*

Diabetes type 2 (diabetes mellitus) is a common clinical disorder which globally affect 5–10 % of the general population. Diabetes mellitus is associated with cardiovascular diseases, peripheral neuropathy, deformity, and trauma which more often result in diabetic foot ulcers (DFU) (Huang et al. 2009). Approximately 15–25 % of diabetic patients are in high risk of developing foot ulcers. Due to prolonged healing time in diabetic patients, there is an increased risk of infection, severe morbidity, and amputation; about 15 % of DFU cases lead to lower extremity amputation. In fact, it has been reported that healing rates in such cases were 24.2 % and 30.9 % after 12 and 20 weeks, respectively. Therefore, rapid wound healing is the main purpose of DFU management (Boulton et al. 2004; Margolis et al. 1999; Snyder et al. 2010). Although control of infection, off-loading, and hyperbaric oxygen therapy are not effective and should be replaced with a new therapeutic method, such strategies are still offered for treatment of diabetic ulcers. Recently, researchers showed that the HAM is a promising dressing material for treatment of DFU. In a clinical study conducted by Alap P et al. (Shah 2014), amniotic membrane patches were applied as a novel therapeutic method for patients with nonhealing DFU. Their findings strongly suggested the potential of HAM as an excellent dressing and an alternative treatment strategy for patients with DFU.

7 Future Perspective

The regenerative properties of amniotic membrane make it a good biological source of cells and matrices for regenerative medicine applications. More studies on preparation and preservation of this membrane are needed to optimize its commercial use.

Author Contributions Designed: Samadikuchaksaraei. Wrote the manuscript: Gholipourmalekabadi, Chauhan, Farhadihosseinabad. Commented and finalized: Samadikuchaksaraei.

References

Abbasi-Montazeri E, Khosravi AD, Feizabadi MM, Goodarzi H, Khoramrooz SS, Mirzaii M, Kalantar E, Darban-Sarokhalil D (2013) The prevalence of methicillin resistant Staphylococcus aureus (MRSA) isolates with high-level mupirocin resistance from patients and personnel in a burn center. Burns 39:650–654

Alemdaroğlu C, Değim Z, Çelebi N, Zor F, Özturk S, Erdoğan D (2006) An investigation on burn wound healing in rats with chitosan gel formulation containing epidermal growth factor. Burns 32:319–327

Alemdaroğlu C, Degim Z, Celebi N, Şengezer M, Alömeroglu M, Nacar A (2008) Investigation of epidermal growth factor containing liposome formulation effects on burn wound healing. J Biomed Mater Res A 85:271–283

Alibardi L (2003) Adaptation to the land: the skin of reptiles in comparison to that of amphibians and endotherm amniotes. J Exp Zool B Mol Dev Evol 298:12–41

Aplin J, Campbell S, Allen TD (1985) The extracellular matrix of human amniotic epithelium: ultrastructure, composition and deposition. J Cell Sci 79:119–136

Atiyeh BS, Hayek SN, Gunn SW (2005) New technologies for burn wound closure and healing—review of the literature. Burns 31:944–956

Atiyeh BS, Costagliola M, Hayek SN, Dibo SA (2007) Effect of silver on burn wound infection control and healing: review of the literature. Burns 33:139–148

Azami M, Tavakol S, Samadikuchaksaraei A, Hashjin MS, Baheiraei N, Kamali M, Nourani MR (2012) A porous hydroxyapatite/gelatin nanocomposite scaffold for bone tissue repair: in vitro and in vivo evaluation. J Biomater Sci Polym Ed 23:2353–2368

Azuara-Blanco A, Pillai C, Dua HS (1999) Amniotic membrane transplantation for ocular surface reconstruction. Br J Ophthalmol 83:399–402

Barret JP, Dziewulski P, Ramzy PI, Wolf SE, Desai MH, Herndon DN (2000) Biobrane versus 1% silver sulfadiazine in second-degree pediatric burns. Plast Reconstr Surg 105:62–65

Baskovich B, Sampson EM, Schultz GS, Parnell LK (2008) Wound dressing components degrade proteins detrimental to wound healing. Int Wound J 5:543–551

Baulier E, Favreau F, LE Corf A, Jayle C, Schneider F, Goujon J-M, Feraud O, Bennaceur-Griscelli A, Hauet T, Turhan AG (2014) Amniotic fluid-derived mesenchymal stem cells prevent fibrosis and preserve renal function in a preclinical porcine model of kidney transplantation. Stem Cells Transl Med 3(7):809–820

Bello YM, Phillips TJ (2000) Recent advances in wound healing. JAMA 283:716–718

Bilic G, Zeisberger SM, Mallik AS, Zimmermann R, Zisch AH (2008) Comparative characterization of cultured human term amnion epithelial and mesenchymal stromal cells for application in cell therapy. Cell Transplant 17:955–968

Bleggi-Torres L, Werner B, Piazza M (1997) Ultrastructural study of the neovagina following the utilization of human amniotic membrane for treatment of congenital absence of the vagina. Braz J Med Biol Res 30:861–864

Bose B (1979) Burn wound dressing with human amniotic membrane. Ann R Coll Surg Engl 61:444

Boulton AJ, Kirsner RS, Vileikyte L (2004) Neuropathic diabetic foot ulcers. N Engl J Med 351:48–55

Bravo D, Rigley TH, Gibran N, Strong DM, Newman-Gage H (2000) Effect of storage and preservation methods on viability in transplantable human skin allografts. Burns 26:367–378

Bromberg BE, Song IC, Mohn MP (1965) The use of pig skin as a temporary biological dressing. Plast Reconstr Surg 36:80–90

Brown GL, Nanney LB, Griffen J, Cramer AB, Yancey JM, Curtsinger LJ III, Holtzin L, Schultz GS, Jurkiewicz MJ, Lynch JB (1989) Enhancement of wound healing by topical treatment with epidermal growth factor. N Engl J Med 321:76–79

Bujang-Safawi E, Halim A, Khoo T, Dorai A (2010) Dried irradiated human amniotic membrane as a biological dressing for facial burns—A 7-year case series. Burns 36:876–882

Burgos H (1986) Angiogenic factor from human term placenta. Purification and partial characterization. Eur J Clin Investig 16:486–493

Cao Y, Rodriguez A, Vacanti M, Ibarra C, Arevalo C, Vacanti CA (1998) Comparative study of the use of poly (glycolic acid), calcium alginate and pluronics in the engineering of autologous porcine cartilage. J Biomater Sci Polym Ed 9:475–487

Carmeliet P (2003) Angiogenesis in health and disease. Nat Med 9:653–660

Chang J-W, Hung S-P, Wu H-H, Wu W-M, Yang A-H, Tsai H-L, Yang L-Y, Lee OK (2011) Therapeutic effects of umbilical cord blood-derived mesenchymal stem cell transplantation in experimental lupus nephritis. Cell Transplant 20:245–257

Chen J, Tseng S (1991) Abnormal corneal epithelial wound healing in partial-thickness removal of limbal epithelium. Invest Ophthalmol Vis Sci 32:2219–2233

Chen C, Lumsden AB, Ofenloch JC, Noe B, Campbell EJ, Stratford PW, Yianni YP, Taylor AS, Hanson SR (1997) Phosphorylcholine coating of ePTFE grafts reduces neointimal hyperplasia in canine model. Ann Vasc Surg 11:74–79

Chen H-J, Pires RT, Tseng SC (2000) Amniotic membrane transplantation for severe neurotrophic corneal ulcers. Br J Ophthalmol 84:826–833

Choi YS, Kim JY, Wee WR, Lee JH (1998) Effect of the application of human amniotic membrane on rabbit corneal wound healing after excimer laser photorefractive keratectomy. Cornea 17:389–395

Cohen S, Samadikuchaksaraei A, Polak JM, Bishop AE (2006) Antibiotics reduce the growth rate and differentiation of embryonic stem cell cultures. Tissue Eng 12:2025–2030

Cribbs R, Luquette M, Besner G (1998) Acceleration of partial-thickness burn wound healing with topical application of heparin-binding EGF-like growth factor (HB-EGF). J Burn Care Rehabil 19:95–101

Cribbs RK, Harding PA, Luquette MH, Besner GE (2002) Endogenous production of heparin-binding EGF-like growth factor during murine partial-thickness burn wound healing. J Burn Care Rehabil 23:116–125

Davis JS (1910) Skin transplantation. Johns Hopkins Hosp Rep 15:307–396

Daya SM, Bell RD, Habib NE, Powell-Richards A, Dua HS (2000) Clinical and pathologic findings in human keratolimbal allograft rejection. Cornea 19:443–450

Deitch EA, Wheelahan TM, Rose MP, Clothier J, Cotter J (1983) Hypertrophic burn scars: analysis of variables. J Trauma Acute Care Surg 23:895–898

Diegelmann RF, Evans MC (2004) Wound healing: an overview of acute, fibrotic and delayed healing. Front Biosci 9:283–289

Dietrich-Ntoukas T, Hofmann-Rummelt C, Kruse FE, Schlotzer-Schrehardt U (2012) Comparative analysis of the basement membrane composition of the human limbus epithelium and amniotic membrane epithelium. Cornea 31:564–569

Dua HS, Azuara-Blanco A (2000) Autologous limbal transplantation in patients with unilateral corneal stem cell deficiency. Br J Ophthalmol 84:273–278

Dua HS, Forrester JV (1990) The corneoscleral limbus in human corneal epithelial wound healing. Am J Ophthalmol 110:646–656

Dua HS, Gomes JA, King AJ, Maharajan VS (2004) The amniotic membrane in ophthalmology. Surv Ophthalmol 49:51–77

Eaglstein WH (1985) Experiences with biosynthetic dressings. J Am Acad Dermatol 12:434–440

Eftekharzadeh M, Nobakht M, Alizadeh A, Soleimani M, Hajghasem M, Kordestani Shargh B, Karkuki Osguei N, Behnam B, Samadikuchaksaraei A (2015) The effect of intrathecal delivery of bone marrow stromal cells on hippocampal neurons in rat model of Alzheimer's disease. Iran J Basic Med Sci 18:520–525

Eppler SM, Combs DL, Henry TD, Lopez JJ, Ellis SG, Yi JH, Annex BH, McCluskey ER, Zioncheck TF (2002) A target-mediated model to describe the pharmacokinetics and hemodynamic effects of recombinant human vascular endothelial growth factor in humans. Clin Pharmacol Ther 72:20–32

Epstein FH, Parry S, Strauss JF (1998) Premature rupture of the fetal membranes. N Engl J Med 338:663–670

Epstein FH, Singer AJ, Clark RA (1999) Cutaneous wound healing. N Engl J Med 341:738–746

Fang C-H, Jin J, Joe J-H, Song Y-S, So B-I, Lim SM, Cheon GJ, Woo S-K, Ra J-C, Lee Y-Y (2012) In vivo differentiation of human amniotic epithelial cells into cardiomyocyte-like cells and cell transplantation effect on myocardial infarction in rats: comparison with cord blood and adipose tissue-derived mesenchymal stem cells. Cell Transplant 21:1687–1696

Fijan A, Hashemi A, Namazi H (2014) A novel use of amniotic membrane for fingertip injuries. J Wound Care 23:255–258

Fukuda K, Chikama T-I, Nakamura M, Nishida T (1999) Differential distribution of subchains of the basement membrane components type IV collagen and laminin among the amniotic membrane, cornea, and conjunctiva. Cornea 18:73–79

Ghalambor A, Pipelzadeh MH, Khodadadi A (2000) The amniotic membrane: a suitable biological dressing to prevent infection in thermal burns. Med J Islamic Acad Sci 13:115–118

Gholipourmalekabadi M, Nezafati N, Hajibaki L, Mozafari M, Moztarzadeh F, Hesaraki S, Samadikuchaksaraei A (2015a) Detection and qualification of optimum antibacterial and cytotoxic activities of silver-doped bioactive glasses. IET Nanobiotechnol 9(4):209–214

Gholipourmalekabadi M, Bandehpour M, Mozafari M, Hashemi A, Ghanbarian H, Sameni M, Salimi M, Gholami M, Samadikuchaksaraei A (2015b) Decellularized human amniotic membrane: more is needed for an efficient dressing for protection of burns against antibiotic-resistant bacteria isolated from burn patients. Burns 41:1488–1497

Gholipourmalekabadi M, Mozafari M, Bandehpour M, Salehi M, Sameni M, Caicedo HH, Mehdipour A, Hamidabadi HG, Samadikuchaksaraei A, Ghanbarian H (2015c) Optimization of nanofibrous silk fibroin scaffold as a delivery system for bone marrow adherent cells: in vitro and in vivo studies. Biotechnol Appl Biochem 62(6):785–794

Gholipourmalekabadi M, Mozafari M, Salehi M, Seifalian A, Bandehpour M, Ghanbarian H, Urbanska AM, Sameni M, Samadikuchaksaraei A, Seifalian AM (2015d) Development of a cost-effective and simple protocol for decellularization and preservation of human amniotic membrane as a soft tissue replacement and delivery system for bone marrow stromal cells. Adv Healthc Mater 4:918–926

Gholipourmalekabadi M, Sameni M, Hashemi A, Zamani F, Rostami A, Mozafari M (2015e) Silver-and fluoride-containing mesoporous bioactive glasses versus commonly used antibiotics: activity against multidrug-resistant bacterial strains isolated from patients with burns. Burns 42(1):131–140

Gholipourmalekabadi M, Sameni M, Radenkovic D, Mozafari M, Mossahebi-Mohammadi M, Seifalian A (2016) Decellularized human amniotic membrane: how viable is it as a delivery system for human adipose tissue-derived stromal cells? Cell Prolif 49:115–121

Greenhalgh DG (1996) The role of growth factors in wound healing. J Trauma Acute Care Surg 41:159–167

Groeber F, Holeiter M, Hampel M, Hinderer S, Schenke-Layland K (2011) Skin tissue engineering—in vivo and in vitro applications. Adv Drug Deliv Rev 63:352–366

Grueterich M, Espana EM, Tseng SC (2003) Ex vivo expansion of limbal epithelial stem cells: amniotic membrane serving as a stem cell niche. Surv Ophthalmol 48:631–646
Gu H-W, Bian D-M, Hu N, Zhang J-F (2011) Effects of amniotic membrane transplantation on cytokines expression in chemically burned rat corneas. Int J Ophthalmol 4:33
Haberal M, Oner Z, Bayraktar U, Bilgin N (1987) The use of silver nitrate-incorporated amniotic membrane as a temporary dressing. Burns 13:159–163
Halim AS, Khoo TL, Yussof SJM (2010) Biologic and synthetic skin substitutes: An overview. Indian journal of plastic surgery 43:S23
Hao Y, Ma DH-K, Hwang DG, Kim W-S, Zhang F (2000) Identification of antiangiogenic and antiinflammatory proteins in human amniotic membrane. Cornea 19:348–352
Harder J, Meyer-Hoffert U, Teran LM, Schwichtenberg L, Bartels J, Maune S, Schroder J-M (2000) Mucoid Pseudomonas aeruginosa, TNF-α, and IL-1 β, but Not IL-6, induce human β-defensin-2 in respiratory epithelia. Am J Respir Cell Mol Biol 22:714–721
He Q, Chen B, Wang Z, Li Q (2002) [The experimental study of culture in vitro of fibroblasts seeded onto human amnion extracellular matrix (HA-ECM)]. Zhonghua zheng xing wai ke za zhi 18:229–231
Heiligenhaus A, Heinz C, Schmitz K, Tappeiner C, Bauer D, Meller D (2008) Amniotic membrane transplantation for the treatment of corneal ulceration in infectious keratitis. In: Cornea and external eye disease. Springer, New York
Herndon DN, Barrow RE, Rutan RL, Rutan TC, Desai MH, Abston S (1989) A comparison of conservative versus early excision. Therapies in severely burned patients. Ann Surg 209:547
Higa K, Shimmura S, Shimazaki J, Tsubota K (2005) Hyaluronic acid-CD44 interaction mediates the adhesion of lymphocytes by amniotic membrane stroma. Cornea 24:206–212
Höckel M, Schlenger K, Doctrow S, Kissel T, Vaupel P (1993) Therapeutic angiogenesis. Arch Surg 128:423–429
Hodde J (2002) Naturally occurring scaffolds for soft tissue repair and regeneration. Tissue Eng 8:295–308
Hodge A, Lourensz D, Vaghjiani V, Nguyen H, Tchongue J, Wang B, Murthi P, Sievert W, Manuelpillai U (2014) Soluble factors derived from human amniotic epithelial cells suppress collagen production in human hepatic stellate cells. Cytotherapy 16:1132–1144
Holland EJ (1996) Epithelial transplantation for the management of severe ocular surface disease. Trans Am Ophthalmol Soc 94:677
Huang ES, Basu A, O'Grady M, Capretta JC (2009) Projecting the future diabetes population size and related costs for the US. Diabetes Care 32:2225–2229
Inge E, Talmi YP, Sigler L, Finkelstein Y, Zohar Y (1991) Antibacterial properties of human amniotic membranes. Placenta 12:285–288
Insausti CL, Blanquer M, Bleda P, Iniesta P, Majado Martínez M, Castellanos G, Moraleda Jimenez JM (2010) The amniotic membrane as a source of stem cells. Histol Histopathol 25(1):91–98
Jafari J, Emami SH, Samadikuchaksaraei A, Bahar MA, Gorjipour F (2011) Electrospun chitosan-gelatin nanofiberous scaffold: fabrication and in vitro evaluation. Biomed Mater Eng 21:99–112
Jahovic N, Guzel E, Arbak S, Yeğen BÇ (2004) The healing-promoting effect of saliva on skin burn is mediated by epidermal growth factor (EGF): role of the neutrophils. Burns 30:531–538
Jelenko C III (1967) Studies in burns. I. Water loss from the body surface. Ann Surg 165:83
John A, Oommen J (2010) Use of amniotic membrane in dermatology. Indian Journal of Dermatology, Venereology, and Leprology 76:196
Jones CA, Williams KA, Finlay-Jones JJ, Hart PH (1995) Interleukin 4 production by human amnion epithelial cells and regulation of its activity by glycosaminoglycan binding. Biol Reprod 52:839–847
Kardas P, Devine S, Golembesky A, Roberts C (2005) A systematic review and meta-analysis of misuse of antibiotic therapies in the community. Int J Antimicrob Agents 26:106–113
Keelan JA, Sato T, Mitchell MD (1997) Interleukin (IL)-6 and IL-8 production by human amnion: regulation by cytokines, growth factors, glucocorticoids, phorbol esters, and bacterial lipopolysaccharide. Biol Reprod 57:1438–1444

Kesting MR, Wolff K-D, Hohlweg-Majert B, Steinstraesser L (2008) The role of allogenic amniotic membrane in burn treatment. Journal of burn care & research 29:907–916

Kim JC, Tseng SC (1995) Transplantation of preserved human amniotic membrane for surface reconstruction in severely damaged rabbit corneas. Cornea 14:473–484

Kim JS, Kim JC, Na BK, Jeong JM, Song CY (2000) Amniotic membrane patching promotes healing and inhibits proteinase activity on wound healing following acute corneal alkali burn. Exp Eye Res 70:329–337

Kim J, Kang HM, Kim H, Kim MR, Kwon HC, Gye MC, Kang SG, Yang HS, You J (2007) Ex vivo characteristics of human amniotic membrane-derived stem cells. Cloning and stem cells 9:581–594

Kim CH, Kim SS, Shon SK, Kim DH, Song CG, Kim HJ (2008) The effect of human amniotic membrane, epidermal cells and marrow mesenchymal stem cells in healing a skin defect. Journal of the Korean Orthopaedic Association 43:276–286

King AE, Critchley HO, Sallenave J-M, Kelly RW (2003) Elafin in human endometrium: an antiprotease and antimicrobial molecule expressed during menstruation. The Journal of Clinical Endocrinology & Metabolism 88:4426–4431

Koizumi N, Inatomi T, Quantock AJ, Fullwood NJ, Dota A, Kinoshita S (2000a) Amniotic membrane as a substrate for cultivating limbal corneal epithelial cells for autologous transplantation in rabbits. Cornea 19:65–71

Koizumi N, Inatomi T, Sotozono C, Fullwood NJ, Quantock AJ, Kinoshita S (2000b) Growth factor mRNA and protein in preserved human amniotic membrane. Curr Eye Res 20:173–177

Kondo T, Ishida Y (2010) Molecular pathology of wound healing. Forensic Sci Int 203:93–98

Kruse FE, Rohrschneider K, Völcker HE (1999) Multilayer amniotic membrane transplantation for reconstruction of deep corneal ulcers. Ophthalmology 106:1504–1511

Kuckelkorn R, Keller G, Redbrake C (2001) Long-term results of large diameter keratoplasties in the treatment of severe chemical and thermal eye burns. Klin Monatsbl Augenheilkd 218:542–552

Lee S-H, Tseng SC (1997) Amniotic membrane transplantation for persistent epithelial defects with ulceration. Am J Ophthalmol 123:303–312

Lee S-B, Li D-Q, Tan DT, Meller D, Tseng SC (2000) Suppression of TGF-ß signaling in both normal conjunctival fibroblasts and pterygial body fibroblasts by amniotic membrane. Curr Eye Res 20:325–334

Lei H, Kalluri R, Furth EE, Baker AH, Strauss JF (1999) Rat amnion type IV collagen composition and metabolism: implications for membrane breakdown. Biol Reprod 60:176–182

Lo V, Pope E (2009) Amniotic membrane use in dermatology. Int J Dermatol 48:935–940

Loeffelbein DJ, Baumann C, Stoeckelhuber M, Hasler R, Mucke T, Steinsträßer L, Drecoll E, Wolff KD, Kesting MR (2012) Amniotic membrane as part of a skin substitute for full-thickness wounds: an experimental evaluation in a porcine model. J Biomed Mater Res B Appl Biomater 100:1245–1256

Loeffelbein DJ, Rohleder NH, Eddicks M, Baumann CM, Stoeckelhuber M, Wolff KD, Drecoll E, Steinstraesser L, Hennerbichler S, Kesting MR (2014). Evaluation of human amniotic membrane as a wound dressing for split-thickness skin-graft donor sites. Biomed Res Int 2014 Article ID 572183, 12 pages

Luo J, Li X, Yang Z (2004) Preparation of human acellular amniotic membrane and its cytocompatibility and biocompatibility. Zhongguo xiu fu chong jian wai ke za zhi 18:108–111

Lynch J, Blocker T (1979) Thermal burns. Plastic Surgery 2:611–620

Madison KC (2003) Barrier function of the skin: "la raison d'etre" of the epidermis. J Investig Dermatol 121:231–241

Margolis DJ, Kantor J, Berlin JA (1999) Healing of diabetic neuropathic foot ulcers receiving standard treatment. A meta-analysis. Diabetes Care 22:692–695

McCarthy DW, Downing MT, Brigstock DR, Luquette MH, Brown KD, Abad MS, Besner GE (1996) Production of heparin-binding epidermal growth factor-like growth factor (HB-EGF) at sites of thermal injury in pediatric patients. J Investig Dermatol 106:49–56

Meinert M, Eriksen GV, Petersen AC, Helmig RB, Laurent C, Uldbjerg N, Malmström A (2001) Proteoglycans and hyaluronan in human fetal membranes. Am J Obstet Gynecol 184:679–685

Meller D, Tseng SC (1999) Conjunctival epithelial cell differentiation on amniotic membrane. Investig Ophthalmol Vis Sci 40:878–886
Mendelsohn M, Dunlop G (1998) Gore-tex augmentation grafting in rhinoplasty--Is it safe? Journal of Otolaryngology-Head & Neck Surgery 27:337
Mermet I, Pottier N, Sainthillier JM, Malugani C, Cairey-Remonnay S, Maddens S, Riethmuller D, Tiberghien P, Humbert P, Aubin F (2007) Use of amniotic membrane transplantation in the treatment of venous leg ulcers. Wound Repair Regen 15:459–464
Meyer U, Handschel J, Meyer T, Wiesmann HP (2009) Fundamentals of tissue engineering and regenerative medicine. Springer, New York
Midwood KS, Williams LV, Schwarzbauer JE (2004) Tissue repair and the dynamics of the extracellular matrix. Int J Biochem Cell Biol 36:1031–1037
Mligiliche N, Endo K, Okamoto K, Fujimoto E, Ide C (2002) Extracellular matrix of human amnion manufactured into tubes as conduits for peripheral nerve regeneration. J Biomed Mater Res 63:591–600
Mobini S, Hoyer B, Solati-Hashjin M, Lode A, Nosoudi N, Samadikuchaksaraei A, Gelinsky M (2013a) Fabrication and characterization of regenerated silk scaffolds reinforced with natural silk fibers for bone tissue engineering. J Biomed Mater Res A 101:2392–2404
Mobini S, Solati-Hashjin M, Peirovi H, Osman NAA, Gholipourmalekabadi M, Barati M, Samadikuchaksaraei A (2013b) Bioactivity and biocompatibility studies on silk-based scaffold for bone tissue engineering. J Med Biol Eng 33:207–214
Modesti A, Scarpa S, D'Orazi G, Simonelli L, Caramia FG (1989) Localization of type IV and V collagens in the stroma of human amnion. Prog Clin Biol Res 296:459
Mohammadi AA, Jafari SMS, Kiasat M, Tavakkolian AR, Imani MT, Ayaz M, Tolide-Ie HR (2013) Effect of fresh human amniotic membrane dressing on graft take in patients with chronic burn wounds compared with conventional methods. Burns 39:349–353
Mohammadi AA, Sabet B, Riazi H, Tavakko-Lian AR, Mohammadi MK, Iranpak S (2015) Human amniotic membrane dressing: an excellent method for outpatient management of burn wounds. Iranian Journal of Medical Sciences 34:61–64
Moncrief JA, Mason AD Jr (1964) Evaporative water loss in the burned patient. J Trauma Acute Care Surg 4:180–185
Moschidou D, Drews K, Eddaoudi A, Adjaye J, DE Coppi P, Guillot PV (2013) Molecular signature of human amniotic fluid stem cells during fetal development. Current stem cell research & therapy 8:73–81
Nakamura T, Inatomi T, Sotozono C, Ang LP, Koizumi N, Yokoi N, Kinoshita S (2006) Transplantation of autologous serum-derived cultivated corneal epithelial equivalents for the treatment of severe ocular surface disease. Ophthalmology 113:1765–1772
Nezafati N, Moztarzadeh F, Hesaraki S, Mozafari M, Samadikuchaksaraei A, Hajibaki L, Gholipour M (2012) Effect of silver concentration on bioactivity and antibacterial properties of SiO2-CaO-P2O5 sol-gel derived bioactive glass. Key Eng Mater 493–494:74–79
Niknejad H, Peirovi H, Jorjani M, Ahmadiani A, Ghanavi J, Seifalian AM (2008) Properties of the amniotic membrane for potential use in tissue engineering. Eur Cells Mater 15:88–99
Niknejad H, Paeini-Vayghan G, Tehrani F, Khayat-Khoei M, Peirovi H (2013) Side dependent effects of the human amnion on angiogenesis. Placenta 34:340–345
Pan SC, Wu LW, Chen CL, Shieh SJ, Chiu HY (2010) Deep partial thickness burn blister fluid promotes neovascularization in the early stage of burn wound healing. Wound Repair Regen 18:311–318
Papini R (2004) ABC of burns: management of burn injuries of various depths. BMJ: British Medical Journal 329:158
Parolini O, Alviano F, Bagnara GP, Bilic G, Buhring HJ, Evangelista M, Hennerbichler S, Liu B, Magatti M, Mao N (2008) Concise review: isolation and characterization of cells from human term placenta: outcome of the first international Workshop on Placenta Derived Stem Cells. Stem Cells 26:300–311
Pellegrini G, Traverso CE, Franzi AT, Zingirian M, Cancedda R, DE Luca M (1997) Long-term restoration of damaged corneal surfaces with autologous cultivated corneal epithelium. Lancet 349:990–993

Perin L, Sedrakyan S, Giuliani S, DA Sacco S, Carraro G, Shiri L, Lemley KV, Rosol M, Wu S, Atala A (2010) Protective effect of human amniotic fluid stem cells in an immunodeficient mouse model of acute tubular necrosis. PLoS One 5:e9357

Proksch E, Brandner JM, Jensen JM (2008) The skin: an indispensable barrier. Exp Dermatol 17:1063–1072

Rendal-Vázquez ME, San-Luis-Verdes A, Yebra-Pimentel-Vilar MT, López-Rodríguez I, Domenech-García N, Andión-Núñez C, Blanco-García F (2012) Culture of limbal stem cells on human amniotic membrane. Cell Tissue Bank 13:513–519

Riau AK, Beuerman RW, Lim LS, Mehta JS (2010) Preservation, sterilization and de-epithelialization of human amniotic membrane for use in ocular surface reconstruction. Biomaterials 31:216–225

Robson MC, Krizek TJ (1973) The effect of human amniotic membranes on the bacteria population of infected rat burns. Ann Surg 177:144

Roh D-H, Seo M-S, Choi H-S, Park S-B, Han H-J, Beitz AJ, Kang K-S, Lee J-H (2013) Transplantation of human umbilical cord blood or amniotic epithelial stem cells alleviates mechanical allodynia after spinal cord injury in rats. Cell Transplant 22:1577–1590

Saki M, Narbat MK, Samadikuchaksaraei A, Ghafouri HB, Gorjipour F (2009) Biocompatibility study of a hydroxyapatite-alumina and silicon carbide composite scaffold for bone tissue engineering. Yakhteh 11:55–60

Samadikuchaksaraei A, Bishop AE (2006) Derivation and characterization of alveolar epithelial cells from murine embryonic stem cells in vitro. Methods Mol Biol 330:233–248

Samadikuchaksaraei A, Bishop AE (2007) Effects of growth factors on the differentiation of murine ESC into type II pneumocytes. Cloning Stem Cells 9:407–416

Sangwan VS, Burman S, Tejwani S, Mahesh SP, Murthy R (2007) Amniotic membrane transplantation: a review of current indications in the management of ophthalmic disorders. Indian J Ophthalmol 55:251

Sankar V, Muthusamy R (2003) Role of human amniotic epithelial cell transplantation in spinal cord injury repair research. Neuroscience 118:11–17

Schwab IR, Reyes M, Isseroff RR (2000) Successful transplantation of bioengineered tissue replacements in patients with ocular surface disease. Cornea 19:421–426

Shah AP (2014) Using amniotic membrane allografts in the treatment of neuropathic foot ulcers. J Am Podiatr Med Assoc 104:198–202

Shapiro M, Friend J, Thoft R (1981) Corneal re-epithelialization from the conjunctiva. Invest Ophthalmol Vis Sci 21:135–142

Shevchenko RV, James SL, James SE (2010) A review of tissue-engineered skin bioconstructs available for skin reconstruction. J R Soc Interface 7:229–258

Shimazaki J, Yang H-Y, Tsubota K (1997) Amniotic membrane transplantation for ocular surface reconstruction in patients with chemical and thermal burns. Ophthalmology 104:2068–2076

Shimazaki J, Shinozaki N, Tsubota K (1998) Transplantation of amniotic membrane and limbal autograft for patients with recurrent pterygium associated with symblepharon. Br J Ophthalmol 82:235–240

Shimmura S, Shimazaki J, Ohashi Y, Tsubota K (2001) Antiinflammatory effects of amniotic membrane transplantation in ocular surface disorders. Cornea 20:408–413

Shoae-Hassani A, Mortazavi-Tabatabaei SA, Sharif S, Seifalian AM, Azimi A, Samadikuchaksaraei A, Verdi J (2015) Differentiation of human endometrial stem cells into urothelial cells on a three-dimensional nanofibrous silk-collagen scaffold: an autologous cell resource for reconstruction of the urinary bladder wall. J Tissue Eng Regen Med 9:1268–1276

Shores JT, Gabriel A, Gupta S (2007) Skin substitutes and alternatives: a review. Adv Skin Wound Care 20:493–508

Siti-Ismail N, Samadikuchaksaraei A, Bishop AE, Polak JM, Mantalaris A (2012) Development of a novel three-dimensional, automatable and integrated bioprocess for the differentiation of embryonic stem cells into pulmonary alveolar cells in a rotating vessel bioreactor system. Tissue Eng Part C Methods 18:263–272

Skardal A, Mack D, Kapetanovic E, Atala A, Jackson JD, Yoo J, Soker S (2012) Bioprinted amniotic fluid-derived stem cells accelerate healing of large skin wounds. Stem cells translational medicine 1:792

Snyder RJ, Kirsner RS, Warriner R III, Lavery LA, Hanft JR, Sheehan P (2010) Consensus recommendations on advancing the standard of care for treating neuropathic foot ulcers in patients with diabetes. Ostomy Wound Manage 56:S1–S24

Solomon A, Rosenblatt M, Monroy D, Ji Z, Pflugfelder SC, Tseng SC (2001) Suppression of interleukin 1α and interleukin 1β in human limbal epithelial cells cultured on the amniotic membrane stromal matrix. Br J Ophthalmol 85:444–449

Sorsby A, Symons H (1946) Amniotic membrane grafts in caustic burns of the eye:(Burns of the second degree). Br J Ophthalmol 30:337

Sorsby A, Haythorne J, Reed H (1947) Further experience with amniotic membrane grafts in caustic burns of the eye. Br J Ophthalmol 31:409

Stern M (1913) The grafting of preserved amniotic membrane to burned and ulcerated surfaces, substituing skin grafts: a preliminary report. J Am Med Assoc 60:973–974

Stern HS (1989) Silver sulphadiazine and the healing of partial thickness burns: a prospective clinical trial. Br J Plast Surg 42:581–585

Sutherland AJ, Converse GL, Hopkins RA, Detamore MS (2015) The bioactivity of cartilage extracellular matrix in articular cartilage regeneration. Advanced healthcare materials 4:29–39

Tchah H, Lee S-A, Sung K, Cho BJ, Kook MS (2003) Apoptosis in keratocytes caused by mitomycin C. Invest Ophthalmol Vis Sci 44:1912–1917

Tehrani FA, Ahmadiani A, Niknejad H (2013) The effects of preservation procedures on antibacterial property of amniotic membrane. Cryobiology 67:293–298

Thibault RA, Mikos AG, Kasper FK (2013) Scaffold/extracellular matrix hybrid constructs for bone-tissue engineering. Advanced healthcare materials 2:13–24

Thompson WD, Li WW, Maragoudakis M (2000) The clinical manipulation of angiogenesis: pathology, side-effects, surprises, and opportunities with novel human therapies. J Pathol 190:330–337

Toomey N, Monaghan Á, Fanning S, Bolton D (2009) Transfer of antibiotic resistance marker genes between lactic acid bacteria in model rumen and plant environments. Appl Environ Microbiol 75:3146–3152

Trelford JD, Trelford-Sauder M (1979) The amnion in surgery, past and present. Am J Obstet Gynecol 134:833–845

Tsai RJ-F, Li L-M, Chen J-K (2000) Reconstruction of damaged corneas by transplantation of autologous limbal epithelial cells. N Engl J Med 343:86–93

Tseng SC, Prabhasawat P, Barton K, Gray T, Meller D (1998) Amniotic membrane transplantation with or without limbal allografts for corneal surface reconstruction in patients with limbal stem cell deficiency. Arch Ophthalmol 116:431–441

Van Vranken BE, Rippon HJ, Samadikuchaksaraei A, Trounson AO, Bishop AE (2007) The differentiation of distal lung epithelium from embryonic stem cells. Curr Protoc Stem Cell Biol Chapter 1, Unit 1G 1

Werner S, Grose R (2003) Regulation of wound healing by growth factors and cytokines. Physiol Rev 83:835–870

Wilshaw S-P, Kearney JN, Fisher J, Ingham E (2006) Production of an acellular amniotic membrane matrix for use in tissue engineering. Tissue Eng 12:2117–2129

Wolbank S, Hildner F, Redl H, VAN Griensven M, Gabriel C, Hennerbichler S (2009) Impact of human amniotic membrane preparation on release of angiogenic factors. J Tissue Eng Regen Med 3:651–654

Wolf HJ, Schmidt W, Drenckhahn D (1991) Immunocytochemical analysis of the cytoskeleton of the human amniotic epithelium. Cell Tissue Res 266:385–389

Yildirimer L, Thanh NT, Seifalian AM (2012) Skin regeneration scaffolds: a multimodal bottom-up approach. Trends Biotechnol 30:638–648

Yoon BS, Moon J-H, Jun EK, Kim J, Maeng I, Kim JS, Lee JH, Baik CS, Kim A, Cho KS (2009) Secretory profiles and wound healing effects of human amniotic fluid–derived mesenchymal stem cells. Stem Cells Dev 19:887–902

Yu SJ, Soncini M, Kaneko Y, Hess DC, Parolini O, Borlongan CV (2009) Amnion: a potent graft source for cell therapy in stroke. Cell Transplant 18:111–118

Zhang T, Yam GH-F, Riau AK, Poh R, Allen JC, Peh GS, Beuerman RW, Tan DT, Mehta JS (2013) The effect of amniotic membrane de-epithelialization method on its biological properties and ability to promote limbal epithelial cell culture the effect of amniotic membrane denudation methods. Invest Ophthalmol Vis Sci 54:3072–3081

Zheng Y-B, Gao Z-L, Xie C, Zhu H-P, Peng L, Chen J-H, Chong YT (2008) Characterization and hepatogenic differentiation of mesenchymal stem cells from human amniotic fluid and human bone marrow: a comparative study. Cell Biol Int 32:1439–1448

Zheng Y-B, Zhang X-H, Huang Z-L, Lin C-S, Lai J, Gu Y-R, Lin B-L, Xie D-Y, Xie S-B, Peng L (2012) Amniotic-fluid–derived mesenchymal stem cells overexpressing interleukin-1 receptor antagonist improve fulminant hepatic failure. PLoS One 7(7):e41392

Current Understanding Realities of Umbilical Cord Stem Cells Biology and Future Perspectives in Clinical Application

Somayeh Ebrahimi-Barough, Reza Rahbarghazi, Zohreh Bagher, Jafar Ai, and Elham Hoveizi

1 Introduction

Stem cells are found in all multicellular organisms that can renew themselves and differentiate into a range of specialized cell types, making them interesting to research and cell therapy (Weissman et al. 2001). There are generally three major types of stem cells based on their time of isolation during development: embryonic stem cells (ESCs), fetal stem cells (FSCs), and adult (somatic) stem cells (Weissman et al. 2001). Embryonic stem cells (ESCs) can be derived from the blastocyst stage of early embryonic period. ESCs from the inner cell mass are pluripotent cells that are distinguished by their ability to differentiate into multiple cell types and by their ability to propagate (Fritsch and Singer 2008). The first mouse and human embryonic stem cells were isolated in 1953 at the Jackson Laboratory in Bar Harbor,

The original version of this chapter was revised. An erratum to this chapter can be found at DOI 10.1007/978-3-319-46410-7_12

S. Ebrahimi-Barough, Ph.D. (✉) • J. Ai
Department of Tissue Engineering and Applied Cell Sciences, Faculty of Advanced Technologies in Medicine, Tehran University of Medical Sciences, Tehran, Iran
e-mail: ebrahimi_s@sina.tums.ac.ir

R. Rahbarghazi
Stem Cell Research Center, Tabriz University of Medical Sciences, Tabriz, Iran

Department of Applied Cell Sciences, Faculty of Advanced Medical Sciences, Tabriz University of Medical Sciences, Tabriz, Iran

Z. Bagher
ENT and Head & Neck Research Center and Department, Hazrat Rasoul Akram Hospital, Iran University of Medical Sciences, Tehran, Iran

E. Hoveizi
Department of Biology, Faculty of Sciences, Shahid Chamran University of Ahvaz, Ahvaz, Iran

B. Arjmand (ed.), *Perinatal Tissue-Derived Stem Cells*, Stem Cell Biology and Regenerative Medicine, DOI 10.1007/978-3-319-46410-7_6

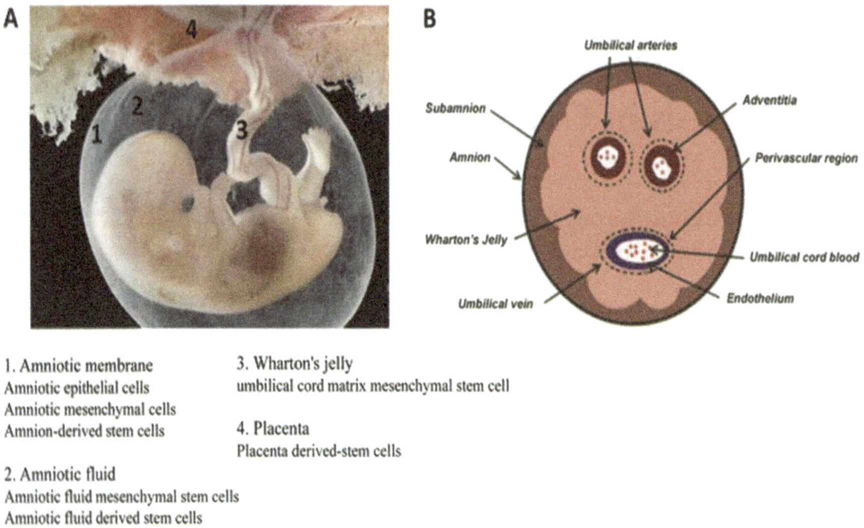

Fig. 1 (**a**) Extra-embryonic stem cell sources. (**b**) Cross-sectional diagram of human umbilical cord showing the compartments from which stem cells have been isolated (amnion, subamnion, Whartons jelly, perivascular, adventiatia, endothelium, and umbilical cord blood)

Maine that lead to start the many researches in this field (Bongso et al. 1994; Thomson et al. 1998). Embryonic stem cells have several distinctive properties include having a normal karyotype, maintaining high telomerase activity, and high proliferation rate. The pluripotent property of ESC presents a significant potential in clinical applications, but the use of these cells in clinic is limited by ethical, political, biological, and regulatory hurdles (Henningson et al. 2003). Other stem cells can be found in several tissues, such as bone marrow (BM), skin, ovary, sperm, adipose tissue, endometrium and pregnant products of umbilical cord blood (UCB), amniotic fluid, and placenta (Pappa and Anagnou 2009; Ebrahimi-Barough et al. 2013, 2015). FSCs can be derived from fetus proper, such as blood, liver, bone marrow, pancreas, spleen, and kidney, and supportive extraembryonic tissues, such as chorionic villus, amniotic fluid, placenta, or umbilical cord (Pappa and Anagnou 2009; Niknamasl et al., 2014). The extraembryonic stem cells are valuable due to their potential clinical utility. Extraembryonic stem cells are obtained from Umbilical cord Wharton's jelly, amniotic fluid (AF), amnion, and placenta (Ballen 2010) (Fig. 1). The first population of stem cells was identified in adult mouse bone marrow by McCulloch and Till in the early 1960s (Friedenstein et al. 1966). After 20 years, the first report of a successful cord blood transplant was reported in a child with Fanconi's anemia (Gluckman et al. 1989). The first isolated fetal stem cells were hematopoietic, derived from human umbilical cord blood (Broxmeyer et al. 1989). The isolated cells were capable of long-term self-renewal and differentiation to multiple hematopoietic lineages. In some countries banking of the cord blood routinely was done for newborns against the advent of childhood hematological maladies (Broxmeyer et al. 1989). The utility of UCB-derived stem cells in the medical field is expanded owing to the facts that UCB is easy to obtain from discard tissues without risk to the donor, and more tolerant to human leukocyte antigen (HLA) mismatches for lowering the risk of graft-versus-host disease (GVHD) and these source of cells are younger than adult BM. Due to the earlier properties, UCB are considered as a new treatment option and therapeutic agent in regenerative medicine. Today, there are over 400,000 cord blood units donated with dramatic growth and stored worldwide for unrelated use (Ballen 2010; Welte et al. 2010).

2 Umbilical Cord Mesenchymal Stem Cell

In placental mammals, the embryo or fetus and placenta is connected by the umbilical cord (also called the navel string, birth cord, or funiculus umbilicalis) which avoid umbilical vessels from collapse, compression, and torsion and provide a good blood circulation between the maternal circulation and the fetal circulation (Kalaszczynska and Ferdyn 2015). Anatomically and histologically, the umbilical cord (UC) consists of the amniotic membrane—made up of one or several layers of cuboidal and squamous epithelia and known as the umbilical epithelium, which is thought to derive from the amniotic epithelium; the umbilical blood vessels, namely, two arteries and one vein which are surrounded by mucoid connective tissue rich in proteoglycans and mucopolysaccharides, known as Wharton's jelly (WJ), that can be subdivided into the subamniotic zone, the intervascular zone, and the perivascular zone (Batsali et al. 2013). The umbilical vein supplies the fetus with oxygenated, nutrient-rich blood from the placenta. Contrasting, the fetal heart pumps deoxygenated, nutrient-depleted blood through the umbilical arteries back to the placenta (Wang et al. 2004). WJ is a mucous connective tissue which contains specialized cells such as multipotent fibroblast-like MSC population and also some mast cells, imbedded in an amorphous substance rich in collagen (type III collagen) and in glycosaminoglycans, especially hyaluronic acid (Batsali et al. 2013). More than 10 years ago, fibroblast-like MSC population were first obtained and these cells have been isolated from three compartments of umbilical cord: MSCs isolated from UC blood (UCB-MSCs), the subendothelium of the umbilical vein, and MSCs in WJ were named as umbilical cord matrix stem cells (UCMSCs) originating from extraembryonic mesoderm at day 13 of embryonic development (Wang et al. 2004). This population of MSCs which derives from the Wharton's jelly cells displays MSC characteristics as defined by the ISCT (International Society for Cellular Therapy) that they adhere to plastic surfaces when proliferating and multiplying; they have self-renewal capabilities just like mesenchymal cells; they display the surface markers of mesenchymal stem cells, such as CD44, CD73, CD90, CD105, CD117; and they have the capacity to differentiate into different cell lines when exposed to suitable growth media, giving rise to bone tissue cells, cartilage cells, adipose tissue cells, muscle tissue cells, and neuronal cells (Batsali et al. 2013; Wang et al. 2004; Kim et al. 2013). Therefore, this source of MSCs can be used as a potential tool for allogeneic transplants, especially in regenerative medicine. There are two possible theories on how stem cells arrived and exist in the WJ. According to Wang et al. studies first, in early human development, there were two waves of migration of fetal MSCs. In the first wave, MSCs migrated via the

UC to the placenta from the yolk sac and aorta-gonadal mesonephros (AGM) and in a second migration MSCs reverse migrated from the placenta via the UC to home in the fetal liver and bone marrow. During this migration some of these MSCs got trapped and stayed in gelatinous WJ of the UC (Kim et al. 2013; Bongso and Fong 2013). Due to their new environment, their stemness characteristics appear to get modified and make them different from bone marrow MSCs (hBMMSCs). Second hypothesis is that the cells in the WJ are really primitive mesenchymal stromal cells originating from mesenchyme that was already there within the UC matrix. The role of these cells may be to produce the various glycoproteins, mucopolysaccharides, glycosaminoglycans, and extracellular matrix proteins to form a gelatinous ground substance (Prasanna and Jahnavi 2011). Stem cells in WJ can be different from other UC compartments stem cells because the WJ is rich in mucopolysaccharides and possess a network of glycoprotein and collagen microfibrils and environment can influence stemness characteristics of cells. There are several bioactive molecules such as interferons, growth factors, interleukins, GAGs, cell adhesion molecules in the secretions released by hWJSCs (Kim et al. 2013) and these features seem to be the building blocks for immunomodulatory mechanisms and tissue repair (Weiss et al. 2008a). As previously mentioned, stem cells have been identified in the amniotic compartment (outer epithelial layer and inner subamniotic mesenchymal layer), the perivascular compartment surrounding the vessels, the WJ compartment, the media and adventitia compartment of the walls of UC blood vessels, the endothelial compartment (inner lining of the vein), and the vascular compartment (blood lying within the UC blood vessels) (Can and Karahuseyinoglu 2007). In vitro experimental analysis shows differences in the number and nature of cells among these regions and it has been demonstrated that they have different properties (Kim et al. 2013) and these led to this hypothesis that these regions might be originating from different preexisting structures (Can and Karahuseyinoglu 2007). Umbilical cord blood (UCB) is a rich source of hematopoietic stem cells and studies showed that hematopoietic stem cells (HSCs), such as erythroid (BFU-E), multipotential (CFU-GEMM), and granulocyte-macrophage (CFU-GM) progenitor cells exist in human cord blood (Marcus and Woodbury 2008a). UCB stem cells can be obtained easily and noninvasively without harm to the mother or infant, and enough stem cells can be easily cryopreserved and stored without significant loss of the features. Unlike adult bone morrow, UCB stem cells are not exposed to the environment toxins or radiation (Kim et al. 2013; Marcus and Woodbury 2008a). Some scientists also succeed for isolation of MSCs from UCB that were able to self-renew with a high proliferative capacity and possessed differentiation potential to many lineages such as osteoblasts, muscle cells, chondrocytes, adipocytes, insulin-producing cells, vascular endothelial cells, cardiomyocytes, hepatocyte-like cells, and hematopoietic cell-supporting stroma (Batsali et al. 2013). Plasticity, homing, and engraftment are three important functions of UCB stem cells. There are three indistinct regions of the WJ for the stem cells derived from the Wharton's jelly of human umbilical cords (hWJSCs) including the perivascular, the intervascular, and the subamnion zones (Fig. 1) (Kim et al. 2013). Moreover, these cells have a fibroblast-like morphology with high proliferative rate and multipotent differentiation capacity and retain their stemness properties for a long time in vitro (9–10 passages). Recently, it has thus been demonstrated that at least two apparently distinct progenitor cell populations exist in the umbilical cord matrix. Based on morphological feature, these type 1 and type 2 cells can be further identified by their differential expression of vimentin and cytokeratins (Marcus and Woodbury 2008a; Karahuseyinoglu et al. 2007). Type 1 cells obtained from perivascular region stain strongly for pancytokeratin and give rise in vitro to flat, wide cytoplasmic cells. Intervascular and subamniotic regions generate type 2 cells with fibroblast-like morphology persisting throughout culture. Based on this finding, it has shown that UC-MSCs are actually type 2 cells and after exposing to inductive media, these cells have adipogenic, chondrogenic, osteogenic, and neuronal differentiation capacity (Batsali et al. 2013; Karahuseyinoglu et al. 2007). There is an important difference between adult bone marrow stromal cells (MSCs) and UCMSCs in differentiation capacity to adipocyte and chondrocyte. In vitro studies showed that adult MSCs more differentiate to adipocytes, whereas UCMSCs were more capable of chondrogenic differentiation than adult MSCs and because of these findings we can suggest that UCMSCs may be used as a good candidate for cartilage repair in future clinical applications (Baksh et al. 2007). Also, it is shown type 1 and type 2 UCMSCs had different ability to generate nonmesodermal derivatives. When these two types of cells were exposed to neural induction media, type 2 differentiated to putative neurons and expressed beta-III-tubulin, neurofilament-M (NF-M), and NeuN, while type 1 cells remained unaltered, with no significant morphological changes apparent and did not express neural cell markers (Batsali et al. 2013; Karahuseyinoglu et al. 2007; Baksh et al. 2007). UC perivascular zone-derived MSCs named as Human Umbilical Cord Perivascular Cells (HUCPVCs) differ in their growth characteristics from UC-MSCs and have a colony forming unit (CFU) frequency of 1:333 (Batsali et al. 2013; Kim et al. 2013) and exhibit higher proliferative rate and osteogenic potential as compared to BM-MSCs (Batsali et al. 2013). The phenotype of HUCPVCs is similar with WJ-derived MSCs but these cells express the high level of CD146 that absent or weakly expressed by MSCs derived from other regions of WJ (Batsali et al. 2013).

3 The Immunomodulatory Properties of UC-MSCs

Low immunogenicity of cells is a more important factor for allogeneic transplantation. MSCs possess the immune properties because of low MHC-I level and absence of MHC-II expression that protect them from Natural killer cell-mediated lysis. Also, MSCs downregulate the interleukin (IL2) and CD38 on phytohemagglutinin-activated lymphocytes (Wang et al. 2009; Zhou et al. 2011). MSCs escape from the immune system by modulation of host dendritic and T-cell function and enhancement of regulatory T-cell induction (Weiss et al. 2008a). Low immunogenicity in MSCs is suggested to be due to secretion of the soluble factors such as transforming growth factor -1, hepatocyte growth factor, interleukin-6, prostaglandin E2, indoleamine2,3-dioxygenase-mediated tryptophan depletion, or nitric oxide (Weiss et al. 2008a). UC-MSCs similar to other sources of mesenchymal stem cells express

MHC class I (HLA-ABC) at low levels but not class II (HLA-DR); costimulatory antigens such as CD40, CD80, CD86 implicated in activation of both T and B cell responses; and high levels of inhibitors of immune response including indoleamine-2,3-dioxygenase (IDO) and prostaglandin E2 (PGE2) (Batsali et al. 2013; Kim et al. 2013). The most important fact related to UC-MSCs is that these cells express high levels of leukocyte antigen G6 (HLA-G6) which is produced by trophoblast and plays important role in the immune tolerance during pregnancy by inducing the expansion of regulatory T cells and protects the embryo from maternal immune response (Kim et al. 2013; Weiss et al. 2008a; Barry et al. 2005). There are six or seven splice variants of HLA-G that HLA-G1, HLA-G2, HLA-G3, and HLA-G4 are membrane-bound isoforms; HLA-G5 and HLA-G6 are soluble forms (Weiss et al. 2008a). The soluble forms of HLA-G inhibit T-cell activation and have immunoregulatory functions (Riteau et al. 2001). The UCB only needs to be matched at four of six HLA class I and II molecules between the donor and patient on other hand, bone marrow generally requires a high degree of HLA match (Weiss et al. 2008a). This provides advantages for graft-versus-host disease (Wang et al. 2009). In comparison to BM-MSCs derived from aged donors, lower levels of HLA-I expressed in UC-MSCs and immunorejection of WJ-MSC seems not to pose a threat and HLA matching may not be required before MSC transplantation (Riteau et al. 2001). Therefore, administration of immunosuppressive drugs is not required; thereby the patient is protected against drug side effects. However, another study suggests that UC-MSCs may not be as immune-privileged as thought, as they can trigger weak allogeneic immune activation in vitro and will eventually undergo rejection following xenogeneic transplantation (Wang et al. 2009; Zhou et al. 2011; Barry et al. 2005). Several studies have been published on the immunosuppressive potential of UC-MSCs on T lymphocytes. One group reported that UC-MSCs target CD4+ and CD8+ T subpopulation equally and are able to inhibit T-cell proliferation regardless of the stimulus used to activate T cells (Barry et al. 2005). Some studies reported that immune characteristics of UC-MSCs can be changed by inflammatory cytokines treatment. For example, it has been shown that UC-MSCs immunogenicity increases when these cells expose to low doses of INF-γ (below 50 ng/mL) because this dose leads to upregulation of HLA-DR and HLA-I expression and following this event the immunogenicity of UC-MSC increases (Barry et al. 2005). The studies on immunomodulatory effects of UC-MSCs show that these cells can inhibit B cell proliferation, differentiation by soluble factors with downregulation of Blimp-1 and upregulation of PAX-5 and inhibition of Akt and p38 MAPK phosphorylation as essential master regulators and signal transduction pathways involved in the regulation of B cell proliferation and differentiation (Che et al. 2012). In addition to, recently has been shown WJ-MSCs express IL-6 and VEGF, which is more important in the immunosuppressive capability of MSCs (Weiss et al. 2008a). WJ-MSCs are less immunogenic than other sources of MSCs but under certain circumstances, UCMSCs can be immunogenic. It happens when cells are injected in an inflamed region or stimulated with IFN-γ prior to injection (Cho et al. 2008). Therefore, after cell therapy care must be taken to avoid sensitization against the injected cells, especially if these cells are used for repairing damaged, inflamed tissue that needs repeated injection into the same location (Kim et al. 2013).

The main outstanding issue relating to cell-based therapy remains whether immune-privilege of allogeneic WJ-MSC transplantation upon differentiation is maintained or subsequently faded out. Despite to secure use of allogeneic MSCs in clinical approaches from different source of tissues, there have been however some reports of significant reduction in survival and long-term engraftment, peculiarly during endothelial and myogenic differentiation of MSCs in vivo milieu (Kim et al. 2013). Noticeably, a shift of expression pattern of immune antigens MHC-I and MHC-II made BMSCs vulnerable to immune rejection in a rat model of myocardial infarction or osteogenic differentiation of hydroxyapatite composite containing allogeneic BMSC while application of immunosuppressant agents, FK506 for example, improved dramatically the level of transplantation efficiency (Carlin et al. 2006). In contrary, the past and present research trends unveiled some encouraging results for WJ-MSCs with preserving its immune-modulatory capabilities upon multidirectional differentiation. Interestingly, a minor amount of gene expression increase in the level of MHC class I was reported for chondrogenic differentiation of human WJ-MSC that coincide with lack of expression of MHC costimulatory molecules, although some authorities proved a different potent inhibitor of immune response factors including IDO, HLA-G, and PGE2 through WJ-MSC committed differentiation into tissues (Fong et al. 2011). In other work, tyrosine hydroxylase-positive catecholaminergic cells derived from porcine WJ-MSC were emerged in rat model of brain injury with of necessity of immunosuppressive treatment (Cho et al. 2008). Collectively, it seems that immune-modulatory effects of WJ-MSCs exert by either upregulation of negative costimulatory ligands, downregulation of immunosuppressive agents, and establishment of different anergy and tolerance activities (Kim et al. 2013).

4 Advantages of UC-MSCs over Embryonic and Adult Stem Cells

Research on ESCs and ASCs has shown that UC-MSCs have also attracted great interest because of their advantages over embryonic and adult counterparts. It is already well known that UC-MSCs show a phenotype closely resembling that of embryonic stem cells (ESC) and have a broad spectrum of differentiation potential beyond mesodermal origin (Hoynowski et al. 2007). Biochemical and immunohistochemical studies show that UC-MSCs express low levels of some transcriptional factors belong to embryonic stem cells such as the members of the OCT family, cell-surface markers for ESCs (SSEA-1 (stagespecific embryonic antigen-1), SSEA-4, Tra-1-60 and Tra-1-81), alkaline phosphatase (ALP), DNMT3B and GABRB3 and the genomic markers (SOX2, NANOG, REX2) (Hoynowski et al. 2007; Carlin et al. 2006).

A low expression of the aforementioned pluripotency markers would suggest, although UC-MSCs are not as pluripotent as ESCs, however, they are highly multipotent. UC-MSCs retain to express many of these pluripotent stem cell markers at least nine passages during ex vivo expansion (Batsali et al. 2013). Unlike hESCs, UC-MSCs do not produce teratomas when transplanted as undifferentiated cells (Fong et al. 2007). Modest expression of pluripotency genes and the high expression level of several tumor suppressor genes may explain the reason (Rachakatla et al. 2007). In addition, since the UC-MSCs do not produce teratomas, it is possible suggesting perhaps that UC-derived stem cells unlike embryonic stem cells and tumorigenic cells have a certain limit of telomerase activity (TA) (Greider 1998). In fact, UC-MSCs display several features of ESC, while the use of them does not raise ethical or legal issues and they do not produce teratomas upon transplantation (Marcus and Woodbury 2008b). Moreover, the UC-MSCs are immune privileged cells, which make them ideal for both autologous and allogeneic use in regenerative medicine applications (Gotherstrom et al. 2003; Le 2003; Hoogduijn et al. 2010). All these features set UC-MSCs apart from ESCs and make them as a promising stem cell source for treating various diseases in the clinics.

Among the adult stem cells, BMSCs are still considered as the gold standard in most research and clinical applications but the number of mesenchymal stem cell derived from bone marrow is a very rare population (0.001–0.01 % of mononuclear cells) (Castro-Malaspina et al. 1980). To obtain sufficient numbers of the cells for therapeutic purposes an extensive in vitro expansion of cells is usually required, thereby enhancing the risk of loss of stemness properties and contaminations (Bongso and Fong 2013; Pittenger et al. 1999). In comparison, 1 cm of umbilical cord yields approximately 5×10^4 stem cells, which is 5000-fold greater than the number of MSCs (Weiss et al. 2006). So, it is quickly and easily to get a substantial amount of cells after several passages compared to BMSCs.

Comparative gene expression profiling between UC-MSCs and BM-MSCs demonstrated that the production of Nanog, Dnmt3b, and Gabrb3 and expression of the pluripotent stem cell markers Brix, CD9, Gal, Kit, and Rex1 were significantly higher in UC-MSCs compared with BM-MSCs (Nekanti et al. 2010a). UC-MSCs also released significantly higher levels of genes implicated in the phosphoinositide 3-kinase (PI3K)-AKT survival/proliferation pathway (Hsieh et al. 2010). Their expression of UC-MSCs reflects that they are more primitive and have a shorter doubling time and a broader pluripotency than BMMSCs (Karahuseyinoglu et al. 2007; Troyer and Weiss 2008).

There is growing evidence showing that donor age affects several properties of mesenchymal stem cells. Exposure to environmental stress can lead to DNA damage, cellular senescence, or loss of regenerative function (Stolzing et al. 2008). It has been demonstrated that reactive oxygen species (ROS) levels which maintain health span of mesenchymal stem cells can rise dramatically with age. This can promote MSC aging through significant damage to cell structures (Liang et al. 2014). In addition, during normal aging of an animal or in cell culture, cells divide and telomeres length is commonly shortened. Telomeres length in UC-MSCs is significantly higher compared with adult MSC to maintain the stability of genomes (Batsali et al. 2013). It is also shown that expression of genes related to inflammatory response and also proteins which have beneficial effects on aging-related diseases decreased with aging MSC (Bustos et al. 2014). Further studies demonstrate that the expression levels of genes related to senescence increase while proapoptotic regulators levels decreased in MSC with age (Alt et al. 2012). The potential applications of adult MSC therapies can be also greatly impacted by the donor's variables such as lifestyle and health status at the time of collection. For example, nonsteroidal anti-inflammatory drugs (NSAIDs) which commonly used to treat chronic pain, inflammation, and fever can alter therapeutic potential of MSC (Pountos et al. 2011). Metabolic diseases such as diabetes and obesity may also change MSC microenvironment and reduced effectiveness regeneration properties of MSC (Phadnis et al. 2009). Therefore, the application of autologous stem cell therapy may not be satisfactory for metabolic disorders. In a recent study, the investigators found the impairment of adipose stem cell (derived from diabetics patient) to establish a vascular network in wound healing mouse model (Rennert et al. 2014).

So, an alternative source of stem cells for treatment of aged patients may be required when considering decreased growth and differential capacities of the adult stem cells as well as invasive and painful harvesting procedures (Roobrouck et al. 2008). Thanks to the younger origin, UC-MSCs exhibit relatively high levels of telomerase activity, short population doubling times, and long times to senescence, without loss of stem cell potency compared to adult MSC which makes them more unique and useful for therapeutic applications of aged patients (Troyer and Weiss 2008; Nekanti et al. 2010b).

Side-by-side comparison of adult MSC with UC-MSCs demonstrated that MSC from umbilical cord has unique properties for clinical implication. They have a broader multipotent plasticity, and proliferate faster than adult MSCs with the fewer ethical concerns and the fact that they are from healthy, young donors make them more valuable therapeutic cell for the treatment of various diseases or tissue damage (Taghizadeh et al. 2011). This factor, coupled with the ease of collection with great expansion capabilities and immunomodulatory ability represents UC-MSCs as a unique source of stem cells to be employed for both autologous and allogeneic cellular therapies and regenerative medicine (Stefano et al. 2015). It seems reasonable to conclude that UC-MSCs have advantages over ESCs and adult MSCs and explain the rapidly growing interest of these cells for therapeutic use.

5 Isolation and Characteristics Features of UC-MSCs

According to the various research groups, different stem cell populations with varied stemness properties can be detected from the various parts of cord. MSCs are present in the Wharton's jelly (WJ), perivascular (PV), subendothelium (SE), umbilical cord lining (UCL), and whole umbilical cord (wUC) (Conconi et al. 2011; Subramanian et al. 2015). It should be noted that MSCs have also been isolated in small numbers and with very low proliferation rates from the umbilical cord blood (Wexler et al. 2003; Perdikogianni et al. 2008). Indeed, comparing MSCs obtained from the various parts is difficult as the heterogeneity of extraction, culture media, and analysis procedures is high. Overall, MSCs from all of these regions fit the classical criteria for MSCs. They all share a fibroblast-like morphology with multipotent differentiation capacity and high proliferation rate. In addition, they express human leukocyte antigen (HLA) class I but not HLA class II and also do not express hematopoietic markers (Dominici et al. 2006). In this section, the phenotypic characterization of different parts of the UC with respect to extraction methods and differentiation potential has been described.

5.1 MSCs from the Wharton's Jelly

The conventional method to obtain MSC from the umbilical cord involves explant or enzymatic digestion methods, or a combination of both (Seshareddy et al. 2008). Although, the enzyme method produces more homogenous cell populations and large cell number in shorter period than the explant method the enzymatic digestion followed by centrifugation is a time consuming and stressful process and may induce cellular damage. More simply, instead of enzymatic digestion, the cord mechanically dissociated into very small pieces and the segments which are used as explants transferred in tissue culture plates until the cells will migrate to the plastic bottom. The explant method is simple and cost-effective method for isolating and culturing umbilical cord-derived mesenchymal stem cells and as it does not involve enzymatic treatment the cell damage is minimized (Salehinejad et al. 2012; Yoon et al. 2013).

There is several in vitro and in vivo evidence that reveals Wharton's jelly as the best compartment of umbilical cord to obtained MSCs (Can and Karahuseyinoglu 2007; Ding et al. 2015). To obtain MSCs from Wharton's jelly, vessels should be removed and then the tissue can be treated by enzymatic or explant or both methods (Ishige et al. 2009; Tong et al. 2011). In addition to method, types of collagenase, enzyme concentration, incubation times, and culture media haven't been standardized thus far.

The mesenchymal features of cell populations derived from all regions of the UC were positive for the MSC signature markers such as CD44, CD73, CD90, and HLA-I and negative for CD31 and HLA-DR (Wetzig et al. 2013). The mesenchymal features of Wharton's jelly cells have also been confirmed by the expression of CD13. They did not express B lymphocyte antigen (CD19), hematopoietic markers (CD34) and contradictory results have been obtained on the expression of CD105 (SH2 or endoglin) and CD45 (leukocyte common antigen) (Conconi et al. 2011; Bakhshi et al. 2008; Kadam et al. 2009; Hamad et al. 2015). Furthermore, embryonic stem cell markers, such as Oct-4, SSEA4, nucleostemin, SOX-2, and Nanog have also been expressed (Le 2003). WJMSCs are negative for the expression of CD80 and CD86, which are costimulatory molecules in T-cell activation and are positive for HLA-G, which has been related to immune tolerance in pregnancy or allograft transplantation, suggesting that these cells could be used in clinic without the risk of acute rejection (Weiss et al. 2008b; Friedman et al. 2007). MSCs from Wharton's jelly seem to have a great differentiation potential. They differentiated not only into cells of the mesodermal lineage but also able to differentiate into cells of ecto- and endodermic origin which makes them an attractive tool for the use in cell therapy (Bagher et al. 2015, 2016a; Borhani-Haghighi et al. 2015).

5.2 MSCs from the Perivascular Zone

To collect UC perivascular stem cells (UCPVCs), the vessels are extracted from UC and the ends of the vessels are tied together with a suture creating loops and then placed the vessel into enzymatic solution for isolating cells from the perivascular tissue (Sarugaser et al. 2005). Along with the other common marker between the umbilical cord compartment, UCPVCs expressed CD105 and did not express CD45 and CD34 (Conconi et al. 2011). Contradictory results have been obtained on the expression of CD14, CD106, and CD117 (Sarugaser et al. 2009; Martin-Rendon et al. 2008a). Furthermore, a high expression of CD10 has been reported in the perivascular region of umbilical cord (Farias et al. 2011). Moreover, UCPV cells were negative to the presence of embryonic stem cell markers such as Oct-4 and SSEA-4 suggesting that UCPV cells are more differentiated and mature than the other region of umbilical cord and this may explain why perivascular cells are not able to differentiate into neurons. Furthermore, the ability to differentiate in vitro into osteoblastic, adipogenic, chondrogenic, myogenic, and fibroblastic lineages without expression of both HLA class I and II, representing a wide range of clinical applications for these cells (Sarugaser et al. 2005).

5.3 MSCs from the Subendothelial Layer

To obtain the subendothelial layer cells, umbilical vein removed and then passing through an enzymatic solution to digest and remove the MSC from endothelial layer (Covas et al. 2003). This cell population was found to be positive for the CD29 (integrin β-1), CD13, CD44, CD49e, CD54, CD166, CD73, CD90, CD105, CD166 and HLA-class I markers and negative for CD45, CD31, CD14, CD34, CD117(c-kit), CD133, HLA-DR, vWf (Covas et al. 2003; Panepucci et al. 2004). Taken together, it appears that MSCs from the subendothelial zone are more resemble, at least in part to the phenotypic profile of the perivascular layer. Furthermore, MSCs of this region expressed pluripotent markers, such as Nanog and Oct-4. In addition, these cells express markers of neural precursor cells such as nestin and PAX6, but they failed to differentiate into functional neurons. Also, the cells can be differentiated to adipoblast, osteoblast, chondroblast (Koh et al. 2008). These findings indicate that MSCs from the subendothelial layer are an important source of mesenchymal stem cells that could be used in cell therapy.

5.4 MSCs from the UC Lining (UCL)

After removing the other compartment of umbilical cord using razor blades, the subamnion region of UC lining membrane chopped into small pieces. These fragments placed in a cell culture dish containing growth media until MSCs migrate out of the tissue (explant method) (Gonzalez et al. 2010). In addition to common markers among different compartment of UC, cell populations derived from umbilical cord lining membrane were positive for the markers such as CD105 and negative for CD19 (Conconi et al. 2011). Some authors have reported that the cord lining or subamnion MSCs were positive for CD34, CD45 while others reported that they were negative (Gonzalez et al. 2010; Reza et al. 2011). Surprisingly the CD14 and CD106 expression which have role in immune response observed in UCLMSCs but not in other regions of UC. Furthermore, these cells expressed embryonic stem cell markers, such as Oct-4, SSEA4, and Nanog, but contradictory results were obtained about the presence of SOX-2. They were able to differentiate into osteogenic, adipogenic, chondrogenic, cardiogenic, and neurogenic lineages which are opening up promising perspectives in the cell therapy (Reza et al. 2011; Kita et al. 2010).

5.5 *MSCs from Whole Umbilical Cord*

In order to increase the yield of cell population, several researches have used whole UC instead of using compartments of UC as a source of MSCs (Tsagias et al. 2011). Entire UC with intact umbilical blood vessels was cut open followed by either an explant procedure or enzyme digestion. Indeed, these cells were positive to several markers detected in the other compartment of UC. The CD49a, CD80, CD133, and CD235a (glycophorin A) are the markers which were detected in the whole UC but not in other regions (Majore et al. 2011). Some discrepancies exist about CD106 and CD117 markers. Some authors have reported that MSCs isolated from whole UC were positive for CD106 and CD117 (Jo et al. 2008) while others reported that they were negative (Secco et al. 2009). Furthermore, cells were positive to several embryonic stem cell markers, such as Oct-4, SSEA-3, SSEA-4, Tra-1–60, and Tra-1–81, as both mRNA and protein (Jo et al. 2008). In addition to mesodermal lineage, these cells can differentiate into neuroectodermal and endodermal lineage under adequate stimulation. Their neural differentiation capacity supported by the expression of marker which have a role in neurite outgrowth such as nestin, nerve growth factor receptor (NGF), and CD56 (La et al. 2009; Yan et al. 2009). Furthermore, whole UC cells lack the expression of carcino-embryonic antigen (CEA), Eras which known as embryonic form of the RAS oncogene and CD86. Interestingly, these cells also expressed HLA-G which makes them an attractive candidate for clinical application (Fuks et al. 1975; Yasuda et al. 2007).

6 Clinical Application of UC-MSCs

Through the raise in the quality and quantity of experimental researches based on different stem cell (SC) lines (Rahbarghazi et al. 2012; Tehrani et al. 2014; Mohammadi et al. 2015; Ebrahimi-Barough et al. 2015; Geranmayeh et al. 2015), it seems that clinical application of SCs is optimistically becoming more functional. With the advances in clinical and experimental use of umbilical cord, this year faces 27th anniversary of the umbilical cord blood transplantation done first in France in a child with Fanconi anemia (Ballen et al. 2013). Since that time, multiple types of both related and unrelated UC transplant have been established in pediatric and adult patients, in which over 600,000 UC blood units stored globally and >30,000 UC blood transplantation have been performed (Appelbaum 2012). Additionally, prominent therapeutic effects of UC-MSCs have being increasingly proved in a number of diseases (Sun et al. 2010). Due to numerous disadvantages of invasive sampling and isolation, limited cell numbers and ethical constraints related with MSCs from bone marrow, adult organs, and fetus origin, stem cells from birth-associated tissues particularly UC-MSCs have been at the center of attention for a while (Bongso and Fong 2013). On the other hand, embryonic stem cells (ESCs) and newly found type of stem cell, induced pluripotent stem cells (iPSCs), encounter major impediments to clinical therapeutic trials (Bongso and Fong 2013). Enormous breadth of information, by in vivo or in vitro experiments, increased uncertainty and showed absence of useful safety indicator in clinical application by means of immunogenic and tumorigenic properties of either ESCs or iPSCs, although oftentimes human ESCs and majority of iPSCs line tested resulted efficiently for experimental tissue reconstitution (Zhao et al. 2011; Okita et al. 2011; Yamashita et al. 2013). In support of this uncertainty, study in chimeric mice by Yamashita et al. showed that certain human iPSCs displayed a pro-oncogenic status during cartilage differentiation (Yamashita et al. 2013). They also unveiled abnormality in five out of 21 iPSCs cell lines, originated by five different reprogramming methods, using three cellular sources. One key bottleneck to realize this discrepancy is related to nonabsolutely predictable target differentiation entity of both iPSCs and ESCs presumably due to their pro-oncogenic entity based on global gene expression patterns and epigenetic state (Guenther et al. 2010; Newman and Cooper 2010). In contrary, UC-MSCs do not trigger tumorigenesis (Weiss and Troyer 2006; Wang et al. 2013) or provoke immune responses (Weiss et al. 2006; Shawki et al. 2015), considered as a promising alternative source for stem cell. These cells could be collected painlessly in a very large quantity while having greater proliferation ability and long-lasting activity of stemness properties for numerous subsequent cell passages (Chen et al. 2014).

Of note, the initial step in appropriate UC associate cell therapy is to augment umbilical cord bio-bank to be collected. The term "bio-banks" stands for long-term repository storage of biological specimens for supporting and ascertaining of future scientific research (Artene et al. 2013). Two main components regarding on bio-banking system includes a: biologic material processing, consisting of sample collection, processing with long-time storing and b: database managing system. Database managing system per se relates with demographic and clinical data of each sample and bank stock (see Fig. 1) (Artene et al. 2013).

Over the past decades, both private and public cell banking systems have been established to provide a repository of either allergenic and autologous transplanting based on private and nonprivate use, geographical area, and investors (Gonzalez-Sanchez et al. 2013). It is accepted that different bio-banking production process of

Table 1 Clinical trial of UC-MSCs sorted by disease types by August-20-2015

Disease type	Number of clinical trials
Heart disease	6
Diabetes and related complications	8
Liver disease	12
Ulcerative colitis	2
Duchene muscular dystrophy	4
Spinal cord injury	4
Brain disease	10
Lung disease	10
Bone and cartilage disease	9
Skin disease	3
Graft versus host and cancer disease	6
Multiple sclerosis	3
Auto-immune disease	4
Other	19
Total	100

life-saving cells must be done by virtue of diversity of biological samples which ensued in-depth knowledge of costs, for example, fee-per-service, and banking facilities (Gonzalez-Sanchez et al. 2013). Although, plenty of bio-banks in industrialized countries, for example, USA and UK Stem Cell Banks, has also tight collaboration with academia and industry via active program of precompetitive research.

By August-20-2015, the public clinical trial database https://clinicaltrials.gov/ct2/results?term=umbilical+cord+mesenchymal+stem+cell&pg=1 displayed near to 100 clinical trials using UC-MSCs for a wide range of clinical applications (see Table 1). Most of the records show that the majority of clinical trials are in phase I, II, and a combined I/II studies while a small number of above-mentioned trials classified in phase 0, III, a mixture of II/III and IV studies (see Figs. 2 and 3). In addition, a survey on geographical distribution unveiled that East Asian countries, especially China, are pioneers in the field of UC-MSC clinical trials (see Table 2).

6.1 Immune-Modulatory and Reconstitution Effect of UC-MSCs

It is believed that GVHD, as an immunologic complication, occurred in large number of survivors (more than 60 %) from allogeneic hematopoietic cell transplantation (Ratanatharathorn et al. 2001). Corticosteroids are currently perceived as gold standard to relieve relative symptoms (Wang et al. 2012). The first experience of using MSCs for GVHD was the study by Le Blanc et al. who examined the immune-modulatory effects of haploidentical bone marrow MSCs transplant, an HLA-A, HLA-B, HLA-DRβ_1 identical, on 9-year-old boy with grade IV GVHD of liver and gastrointestinal apparatus (Le Blanc et al. 2004). In alignment with this study, administration of adult bone marrow MSCs to eight patients resulted in amelioration of six

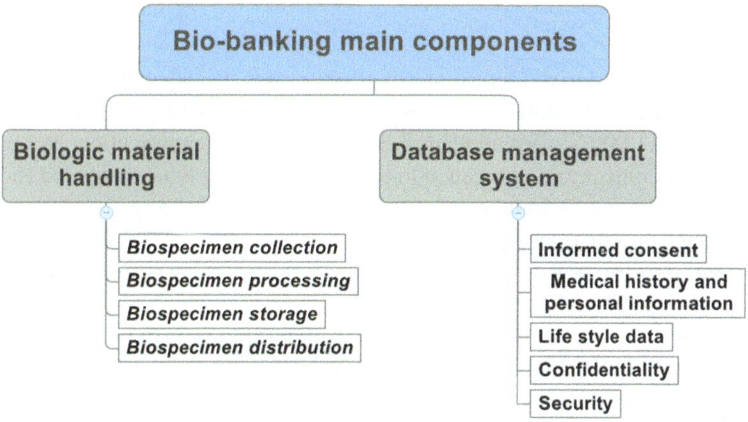

Fig. 2 Representative image of bio-banking main components

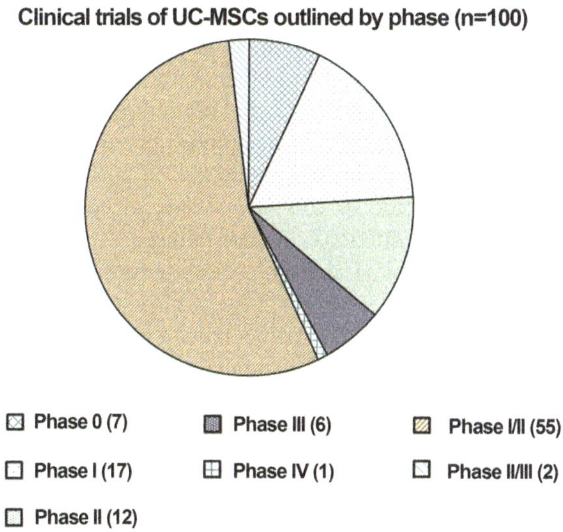

Fig. 3 An illustration of the UC-MSC clinical trials based on phase

Table 2 The number of conducted clinical trials using UC-MSCs in the world based on geographic region

Region name	Number of studies
World	100
Africa	1
Central America	8
East Asia	71
Europe	1
Middle East	4
North America	9
United States	9
South America	1
South Asia	1
Southeast Asia	2

people with GVHD-derived gastrointestinal symptoms (Ringdén et al. 2006). Based on public clinical trial database, we found that five out of one hundred clinical trials recorded for UC-MSCs were applied in immune-mediate complication. Noticeably, other relevant clinical trials pertained to immune-mediated issues included rheumatoid arthritis (two cases), immune reconstitution in HIV-infected patients (one case), autoimmune abnormalities (one case), and lupus nephritis (one case). All these experiments performed at different clinical phages with varying number of UC-MSCs dosage and in patients in different stages of the disease which resulted in various vital response rates. Regarding to comfort and plentiful access to birth associate tissues, it put them practical use and in the spotlight of biomedical research.

6.2 UC-MSCs and Cardiovascular Recovery

Despite overall outstanding progress of treatment strategies, cardiovascular problems, especially ischemic heart failure, remain extremely challenging and skeptical yet. According to the animal model of myocardial infarction generated by the ligation of the left anterior descending artery, many authorities confirmed improvement of autologous, allogeneic, xenogeneic transplantation on postinfarct myocardium function, although different degree of responses were, of course, reported with regarding to some recognized limitations of heart failure treatments (Rahbarghazi et al. 2014; Lim et al. 2006; Toma et al. 2002). Despite ostensible, shortcomings, multilineage differentiation potential of MSCs from different species to into functional cardiac tissue cells has been previously acclaimed as paramount milestone in this field. In initial step of preclinical experimental models of ischemic heart failure, many authorities discovered a potent high tendency of human UC-MSCs to preferentially home to damaged myocardium either systemic administration and local microinjection including intracoronary injection or direct implantation (Rahbarghazi et al. 2014; Erices et al. 2003). Although, UC-MSC-seeded implants and scaffolds have been newly investigated by different research groups (Murry et al. 2005). Up to now, two most preliminary prominent waves of clinical trials by different authorities were entered in clinical arena followed as application of skeletal myoblast and bone marrow trials in acute or chronic ischemic disease, respectively (Murry et al. 2005). They reported that phase II skeletal myoblast clinical trials are warranted. In addition, improvement in angina score, wall thickening, trends toward improved end diastolic volume, maximal oxygen consumption, and treadmill time (Murry et al. 2005). Through to August 2007, in a systematic review by Martin-Rendon et al. 13 trials with 14 comparisons, including a total of 811 participants, were included on evidence-based searching in MEDLINE, EMBASE, the Cochrane Library, and Current Controlled Trials Register databases (Martin-Rendon et al. 2008b). Regarding to random effects model analysis, they concluded that bone marrow autologous transplantation resulted in an improved left ventricular ejection fraction, marked reduction in left ventricular end-systolic volume and myocardial lesion area. Recently, Roura and colleagues showed that administration of UC-MSCs in afflicted people with idiopathic dilated cardiomyopathy could ameliorate endothelial dysfunction-related vascular complications of heart tissue (Roura et al. 2012). Of six cases registered for cardiovascular diseases regarding to public clinical trial database, UC-MSCs administration was subjected to following abnormalities including ischemic cardiomyopathy (five cases) and idiopathic dilated cardiopathy (one case). Despite fundamental issues remaining to be resolved for SC-based cardiac regeneration, for example, appropriate functional integration of introduced SCs into local myocardial milieu (Pijnappels et al. 2008); however, it seems that UC-MSCs are good candidate for different cell-based therapeutic purposes according to advantageous specifications described in earlier paragraphs.

6.3 Therapeutic Effects of UC-MSCs on Diabetes Mellitus 1 and 2

Many already up to date attempts have been focused to resuscitate endocrine pancreatic insufficiency due to β cell dysfunction peculiarly in diabetes mellitus type 1 and 2. A brief review on literature enunciated that the first milestone in cell-based therapies on diabetes has been accomplished with the use of islet transplantation from cadaveric donors (Shapiro et al. 2000). Current data from several SC-based in vitro and preclinical studies showed that various tissue-derived SCs, particularly UC-MSCs, could easily transdifferentiate into insulin-producing islet-like cells (Jiang et al. 2007; Mehrabi et al. 2015; Si et al. 2012). It was commonly perceived multiple mechanisms participated in in vivo animal models (Si et al. 2012). For example, MSC to β-cell differentiation, activation of β-cell function juxtaposed to MSCs, improvement of insulin sensitivity by an upregulated GLUT4 expression, promotion of phosphorylated IRS-1 and Akt content in insulin-dependent tissues are noteworthy (Si et al. 2012). Other underlying mechanisms MSCs exert their therapeutic effects during diabetic condition are MSC-derived exosome intercellular crosstalk with β cells, inhibition of immune responses, and acceleration of endogenous regeneration of pancreatic β cell (Tolar et al. 2010; Rahman et al. 2014; Aali et al. 2014; Berezin 2014). In eight projects of public clinical trial database, Phase I and II clinical trials by using UC-MSCs were conducted both on type 1, 2 diabetic patients and relevant complications, including diabetic foot root and peripheral arterial disease (Table 1). In other study by Li et al., a number of 15 diabetic patients with foot disease underwent UC-MSCs transplantation in which cells were injected quadriceps thigh muscles (Li et al. 2013). Four weeks posttransplantation, they found that diabetic immune deficiency was extenuated by a marked increase in number of $CD4^{+}CD25^{hi}FoxP3^{+}$ Treg/Th17 and $CD4^{+}CD25^{hi}FoxP3^{+}$ Treg/Th1 cell populations coincided with blood glucose reduction and required insulin injection. It seems logically that UC-MSCs could accelerate abrogating autoimmunity toward pancreatic β-cells.

6.4 UC-MSCs for Liver and Gastrointestinal Disorders

To date, different types of SCs have been eligibly exploited for gut and liver diseases. Intriguingly, we realized that the utmost number of UC-MSCs clinical trial was assigned to alimentary tract especially liver abnormalities ($n = 12$), although two cases were registered for gastrointestinal disease in public clinical trial database (Table 1). Decompensated primary biliary and liver cirrhosis, liver failure, autoimmune and HBV afflicted hepatitis, induction of liver transplant toleration, and active ulcerative colitis were included. One of the most challenging debates for liver cell-based therapy is cell-repopulating capacity of posttransplanted host liver (Wu and Tao 2012). It was found that hepatocyte-like cells differentiation of MSCs initiated under conditions favoring hepatocyte differentiation by virtue of albumin, alpha fetoprotein and urea production, cytochrome P_{450} enzyme, glycogen, and carbamoylphosphate synthetase activity (Wu and Tao 2012; Sarvandi et al. 2015). In addition, in some experiments the protective effect of MSCs secretome was evident on sinusoidal endothelial cells of radiation-induced hepatopathy (Chen et al. 2015). The locally direct injection of UC-MSC in carbon tetrachloride-induced rat hepatic cirrhosis model resulted in a marked reduction of liver fibrosis evidenced by total collagen deposition Sirius red stain (Tsai et al. 2009). While the upregulation of hepatic mesenchymal epithelial transition factor-phosphorylated type and hepatocyte growth factor was initiated postcell transplantation the content of levels of serum glutamic oxaloacetic transaminase, glutamic pyruvate transaminase, alpha-smooth muscle actin, and transforming growth factor-β_1 in the liver milieu (Tsai et al. 2009). Noticeably, the transplantation of exosomes derived from human UC-MSCs as adjuvant tool versus cell direct injection alone into mouse model of carbon tetrachloride-induced liver fibrosis in addition to an alleviated hepatic inflammation and collagen type I and III deposition, serum aspartate aminotransferase transforming growth factor-β_1, and phosphorylation Smad2 expression resulted in an increase of E-cadherin- and cytokeratin 18 positive cells concurrently with an reduced number of N-cadherin- and vimentin-positive cells (Li et al. 2012; Zhou et al. 2014). In human, long-term follow-up (near to a year) of 30 chronic hepatitis B patients with decompensated liver cirrhosis received UC-MSC transfusion contributed to an elimination of liver failure symptoms in comparison of 15 patients receiving saline alone (Zhang et al. 2012). In detail, Zhang et al. showed a greater level of improvement in the serum albumin and in total serum bilirubin levels. Totally, the therapeutic actions of MSCs, peculiarly UC-MSCs, on functional and structural aspects have been confirmed.

6.5 Neural Tissue Regeneration by UC-MSCs

Although not many encouraging experimental studies already been undertaken and are in primary stage, but both neuroprotective and neuroregenerative properties of different types of MSCs have been proved in multiple disorders affecting the brain, spinal cord, and even peripheral nervous system (Dalous et al. 2012; Phinney and Isakova 2005). Interestingly, it has become apparent that an innate neuroglial differentiation potential of UC-MSC samples confirmed by upregulation of Oct4, Nanog, Sox2, ABCG2, neuro-ectodermal marker nestin, glial fibrillary acidic protein (GFAP), and microtubule-associated protein-2 (MAP-2) represented by refractile cell body with extended neurite-like structures cell morphology in in vitro condition (Divya et al. 2012; Lee et al. 2004). Promising results have been already gained in numerous preclinical studies (Bauder and Ferguson 2012; Vogelaar et al. 2004). For example, xenogeneic administration of UC-MSC contributed into nerve regeneration evidenced by an upregulation of brain-derived neurotrophic factor and tyrosine kinase receptor B at the lesion site of a rat model of peripheral sciatic nerve crush injury (Sung et al. 2012). On the other hand, the severity of Wallerian degeneration was alleviated which simultaneously coincide with an increase in the number of regenerated fibers and thickness of myelin sheath (Gärtner et al. 2014). Because of autologous cell-based therapy would necessitate a proper nerve biopsy and a long period of culture times, therefore exploiting an array of different SC types, which supply a more accessible, suitable and off-the-shelf source, is plausible (Faroni et al. 2015; Mobarakeh et al. 2012). Of total 100 cases, 17 projects of public clinical trial database were allocated to nervous system associate diseases and abnormalities which mainly enrolled in brain diseases ($n=10$), spinal cord injury ($n=4$), and multiple sclerosis ($n=3$) in different clinical trial phases. Therefore, a multiple number of allogeneic, particularly in human and xenogeneic have been designed so far in in vivo condition (Cheng et al. 2014). Importantly, different methods of cell administration have been seen into target sites in nervous system with a possibly different degree of complications (Fransson et al. 2014; Guan et al. 2013). In a phase 1 clinical trial in nine patients with Alzheimer's disease, for instance, stereotactic accurate positioning of UC-MSCs in bilateral hippocampi and right precuneus regions showed highly feasible, safe, and well tolerated (Kim et al. 2015). Surprisingly, Hou et al. reported a fewer of unwanted adverse effects in allogeneic UC-MSC-treated individuals with multiple sclerosis as compared to autologous transplanted patient by bone marrow MSCs (Hou et al. 2013). Based on clinical trials undertaken, the rate of response to cell-based treatment seems to be dramatically different in accordance to the severity of the injury and the period of treatment. Intrathecal injection of 1×10^6/kg body weight in 22 patients with incomplete spinal cord injury attained 81.25 % while no hopeful response to treatment observed in six patients with complete spinal cord injury (Liu et al. 2013). In other study, among the eight cases with secondary progressive multiple scleroses, approximately six improved after intrathecal intravenous injections of UC-MSCs (Lu et al. 2013). In other type of central nervous system disorders, beneficial effects of UC-MSCs, with acceleration in healing process of the infarct cortex with subsequent functional recovery, have been elucidated in ischemic region of rat model of stroke (Lin et al. 2011). Of note, clinical trials based on different MSCs, peculiarly UC-MSC, transplantation for stroke is currently ongoing, although preliminary experiments on autologous intravenous administration of bone MSCs in patients with ischemic stroke revealed no stroke recurrence, adverse events near up to 1 year posttransplantation (Honmou et al. 2011). Even, 5-year-follow up showed no signs of venous thromboembolism, systemic malignancy, or systemic infection in stroke-afflicted people (Lee et al. 2010). Prior to define a routinely applied clinical treatment, it is logical that more preclinical and clinical studies are however essential.

6.6 UC-MSCs and Musculoskeletal Regeneration

In multiple preclinical experiments the regenerative capacity of MSCs has been proved (Ha et al. 2015). For instance, xenogeneic composite of human UC-MSCs juxtaposed with a hyaluronic acid hydrogel resulted in marked therapeutic effects in both rat and rabbit models (Ha et al. 2015; Gupta et al. 2012). A high degree of hyaline cartilage synthesis and regeneration was, 12 weeks posttransplantation, observed in pig that underwent interventional full-thickness chondral injury (Ha et al. 2015). The state of the art on UC-MSCs clinical trial and musculoskeletal regeneration, including cartilage and bone reconstitution was evident in public clinical trial database by nine patients with bone and cartilage and four with Duchene muscular dystrophy diseases received UC-MSC transplants. The most of experiments on animal models for bone regeneration focused on loading of different SC types and particularly UC-MSCs on scaffolds (Rosa and BacklyRania 2015). Rosa et al. recently acclaimed that the transplanted UC-MSCs triggered the bone healing procedure by stimulation of angiogenesis and bone formation at calvarial defects in bone defect mouse model (Rosa and BacklyRania 2015). Like the results of preclinical or clinical trials of other kind of disease or abnormalities, many further investigation must be undertaken to illuminate the safety and efficiency of different MSCs, especially UC-derived MSCs, on cartilage or bone healing. In accordance to cartilage and bone tissues consistency and inevitable effect of required matrix and scaffolds, it is essential to better know the possible key role and regenerative properties of environment enclosed by the MSCs.

6.7 UC-MSCs Versus Malignant Abnormalities

It is going without saying that both immunosuppressive and hematopoiesis capability of transplanted MSCs have been seen close together in many experiments (Zhi-Gang et al. 2008; Maitra et al. 2004). However, contradictory reports showed two-sided edge effects by both promotion and inhibition of cancer progression driven by MSCs (Zhang et al. 2013). The juxtaposition of UC-MSCs with normal lymphocytes or Jurkat leukemia cells induced the inhibition of cell proliferation and cell cycle progression in both cells and downregulated the HES-1 especially in cancer cells, nominated as classic transcriptional target of Notch signaling (Yuan et al. 2014; Xu et al. 2014). Some authorities reported the effective impact of human MSCs on immune cells by diminishing of tumor necrosis factor alpha, interferon gamma, and increasing interleukin-10 (Aggarwal and Pittenger 2005). In line with these statements, coinfusion of human MSCs with UC blood cells into sublethal irradiated nude mice suppresses alloantigen-specific activated T cells (Maitra et al. 2004). An in vitro study revealed that modulation of JNK and PI3K/AKT signaling, governed by UC-MSC, contributed to cell apoptosis in prostate cancer cell line (Han et al. 2014). Jing et al. demonstrated that the interleukin-15 production activity of UC-MSCs in with help of NK and $CD8^+$ T cells exerted antitumor activity on pancreatic tumor cells in syngeneic murine model (Jing et al. 2014). In contrary, breast cancer metastasis was mediated by UC-MSCs interleukin-8 and -6 which induced $CD44^+/CD24^-$ cell population (Ma et al. 2015). In our survey on public clinical trial database, two projects in patient with hematologic malignancies were found. Both cotransplantation of UC-MSCs with MPCs or UC-MSCs alone have been done in patients. Because of uncertainty on UC-MSCs behavior against different tumors, their application in cancerous-related niche is in its infancy.

7 Conclusion

UCMSCs are very valuable sources of stem cells that can be considered as an effective treatment in variety of clinical aspects. These cells are accessible source with fewer ethical concerns, lake of immunogenic rejection, high proliferation capacity without tumorigenicity properties, and placticity developmental flexibility. These characteristics make them a valuable source of stem cells for clinical application and cell-based therapies.

References

Aali E, Mirzamohammadi S, Ghaznavi H, Madjd Z, Larijani B, Rayegan S, Sharifi AM (2014) A comparative study of mesenchymal stem cell transplantation with its paracrine effect on control of hyperglycemia in type 1 diabetic rats. J Diabetes Metab Disord 13(1):76

Aggarwal S, Pittenger MF (2005) Human mesenchymal stem cells modulate allogeneic immune cell responses. Blood 105(4):1815–1822

Alt EU, Senst C, Murthy SN, Slakey DP, Dupin CL, Chaffin AE, Kadowitz PJ, Izadpanah R (2012) Aging alters tissue resident mesenchymal stem cell properties. Stem Cell Res 8(2):215–225

Appelbaum FR (2012) Pursuing the goal of a donor for everyone in need. N Engl J Med 367(16):1555

Artene S-A, Ciurea ME, Purcaru SO, Tache DE, Tataranu LG, Lupu M, Dricu A (2013) Biobanking in a constantly developing medical world. Scientific World Journal 2013:343275

Bagher Z, Ebrahimi-Barough S, Azami M, Mirzadeh H, Soleimani M, Ai J, Nourani MR, Joghataei MT (2015) Induction of human umbilical Wharton's jelly-derived mesenchymal stem cells toward motor neuron-like cells. In Vitro Cell Dev Biol Anim 51(9):987–994. doi:10.1007/s11626-015-9921-z

Bagher Z, Azami M, Ebrahimi-Barough S, Mirzadeh H, Solouk A, Soleimani M, Ai J, Nourani MR, Joghataei MT (2016a) Differentiation of Wharton's jelly-derived mesenchymal stem cells into motor neuron-like cells on three-dimensional collagen-grafted nanofibers. Mol Neurobiol 53(4):2397–2408. doi:10.1007/s12035-015-9199-x

Bagher Z, Ebrahimi-Barough S, Azami M, Safa M, Joghataei MT (2016b) Cellular activity of Wharton's Jelly-derived mesenchymal stem cells on electrospun fibrous and solvent-cast film scaffolds. J Biomed Mater Res A 104(1):218–226. doi:10.1002/jbm.a.35555

Bakhshi T, Zabriskie RC, Bodie S, Kidd S, Ramin S, Paganessi LA, Gregory SA, Fung HC, Christopherson KW 2nd (2008) Mesenchymal stem cells from the Wharton's jelly of umbilical cord segments provide stromal support for the maintenance of cord blood hematopoietic stem cells during long-term ex vivo culture. Transfusion 48(12):2638–2644

Baksh D, Yao R, Tuan RS (2007) Comparison of proliferative and multilineage differentiation potential of human mesenchymal stem cells derived from umbilical cord and bone marrow. Stem Cells 25:1384–1392

Ballen K (2010) Challenges in umbilical cord blood stem cell banking for stem cell reviews and reports. Stem Cell Rev Rep 6:8–14

Ballen KK, Gluckman E, Broxmeyer HE (2013) Umbilical cord blood transplantation: the first 25 years and beyond. Blood 122(4):491–498

Barry FP, Murphy JM, English K, Mahon BP (2005) Immunogenicity of adult mesenchymal stem cells: lessons from the fetal allograft. Stem Cells Dev 14:252–265

Batsali AK, Kastrinaki M, Papadaki HA, Pontikoglou C (2013) Mesenchymal stem cells derived from Wharton's jelly of the umbilical cord: biological properties and emerging clinical applications. Curr Stem Cell Res Ther 8:144–155

Bauder AR, Ferguson TA (2012) Reproducible mouse sciatic nerve crush and subsequent assessment of regeneration by whole mount muscle analysis. J Vis Exp (60). doi:10.3791/3606

Berezin AE (2014) Diabetes mellitus and cellular replacement therapy: expected clinical potential and perspectives. World J Diabetes 5(6):777

Bongso A, Fong CY (2013) The therapeutic potential, challenges and future clinical directions of stem cells from the Wharton's jelly of the human umbilical cord. Stem Cell Rev 9:226–240

Bongso A, Fong CY, Ng SC, Ratnam S (1994) Isolation and culture of inner cell mass cells from human blastocysts. Hum Reprod 9:2110–2117

Borhani-Haghighi M, Talaei-Khozani T, Ayatollahi M, Vojdani Z (2015) Wharton's Jelly-derived mesenchymal stem cells can differentiate into hepatocyte-like cells by HepG2 cell line extract. Iran J Med Sci 40(2):143–151

Broxmeyer HE, Douglas GW, Hangoc G, Cooper S, Bard J, English D, Arny M, Thomas L, Boyse EA (1989) Human umbilical cord blood as a potential source of transplantable hematopoietic stem/progenitor cells. Proc Natl Acad Sci U S A 86:3828–3832

Bustos ML, Huleihel L, Kapetanaki MG, Lino-Cardenas CL, Mroz L, Ellis BM, McVerry BJ, Richards TJ, Kaminski N, Cerdenes N, Mora AL, Rojas M (2014) Aging mesenchymal stem cells fail to protect because of impaired migration and antiinflammatory response. Am J Respir Crit Care Med 189(7):787–798

Can A, Karahuseyinoglu S (2007) Concise review: human umbilical cord stroma with regard to the source of fetus-derived stem cells. Stem Cells 25:2886–2895

Carlin R, Davis D, Weiss M, Schultz B, Troyer D (2006) Expression of early transcription factors Oct-4, Sox-2 and nanog by porcine umbilical cord (PUC) matrix cells. Reprod Biol Endocrinol 4:8

Castro-Malaspina H, Gay RE, Resnick G, Kapoor N, Meyers P, Chiarieri D, McKenzie S, Broxmeyer HE, Moore MA (1980) Characterization of human bone marrow fibroblast colony-forming cells (CFU-F) and their progeny. Blood 56:289–301

Che N, Li X, Zhou S, Liu R, Shi D, Lu L, Sun L (2012) Umbilical cord mesenchymal stem cells suppress B-cell proliferation and differentiation. Cell Immunol 274:46–53

Chen G, Yue A, Ruan Z, Yin Y, Wang R, Ren Y, Zhu L (2014) Human umbilical cord-derived mesenchymal stem cells do not undergo malignant transformation during long-term culturing in serum-free medium. PLoS One 9(6):e98565. doi:10.1371/journal.pone.0098565

Chen Y-X, Zeng Z-C, Sun J, Zeng H-Y, Zhang Z-Y (2015) Mesenchymal stem cell-conditioned medium prevents radiation-induced liver injury by inhibiting inflammation and protecting sinusoidal endothelial cells. J Radiat Res 56(4):700–708

Cheng H, Liu X, Hua R, Dai G, Wang X, Gao J, An Y (2014) Clinical observation of umbilical cord mesenchymal stem cell transplantation in treatment for sequelae of thoracolumbar spinal cord injury. J Transl Med 12(1):1–8

Cho PS, Messina DJ, Hirsh EL, Chi N, Goldman SN, Lo DP, Harris IR, Popma SH, Sachs DH, Huang CA (2008) Immunogenicity of umbilical cord tissue derived cells. Blood 111:430–438

Conconi MT, Di Liddo R, Tommasini M, Calore C, Parnigotto PP (2011) Phenotype and differentiation potential of stromal populations obtained from various zones of human umbilical cord: an overview. Open Tissue Eng Regen Med J 4:6–20

Covas DT, Siufi JL, Silva AR, Orellana MD (2003) Isolation and culture of umbilical vein mesenchymal stem cells. Braz J Med Biol Res 36(9):1179–1183

Dalous J, Larghero J, Baud O (2012) Transplantation of umbilical cord-derived mesenchymal stem cells as a novel strategy to protect the central nervous system: technical aspects, preclinical studies, and clinical perspectives. Pediatr Res 71(4-2):482–490

Ding DC, Chang YH, Shyu WC, Lin SZ (2015) Human umbilical cord mesenchymal stem cells: a new era for stem cell therapy. Cell Transplant 24(3):339–347. doi:10.3727/096368915X686841

Divya MS, Roshin GE, Divya TS, Rasheed VA, Santhoshkumar TR, Elizabeth KE, James J, Pillai RM (2012) Umbilical cord blood-derived mesenchymal stem cells consist of a unique population of progenitors co-expressing mesenchymal stem cell and neuronal markers capable of instantaneous neuronal differentiation. Stem Cell Res Ther 3(6):57

Dominici M, Le Blanc K, Mueller I, Slaper-Cortenbach I, Marini F, Krause D, Deans R, Keating A, Prockop D, Horwitz E (2006) Minimal criteria for defining multipotent mesenchymal stromal cells. The International Society for Cellular Therapy position statement. Cytotherapy 8(4):315–317

Ebrahimi-Barough S, Kouchesfahani HM, Ai J, Massumi M (2013) Differentiation of human endometrial stromal cells into oligodendrocyte progenitor cells (OPCs). J Mol Neurosci 51:265–273

Ebrahimi-Barough S, Javidan AN, Saberi H, Joghataei MT, Rahbarghazi R, Mirzaei E, Faghihi F, Shirian S, Ai A, Ai J (2015) Evaluation of motor neuron-like cell differentiation of hEnSCs on biodegradable PLGA nanofiber scaffolds. Mol Neurobiol 52(3):1704–1713

Ebrahimi-Barough S, Hoveizi E, Norouzi JA, Jafar Ai J (2015) Investigating the neuroglial differentiation effect of neuroblastoma conditioned medium in human endometrial stem cells cultured on 3D nanofibrous scaffold. J Biomed Mater Res A 103(8):2621–2627

Erices AA, Allers CI, Conget PA, Rojas CV, Minguell JJ (2003) Human cord blood-derived mesenchymal stem cells home and survive in the marrow of immunodeficient mice after systemic infusion. Cell Transplant 12(6):555–561

Farias VA, Linares-Fernández JL, Peñalver JL, Payá Colmenero JA, Ferrón GO, Duran EL, Fernández RM, Olivares EG, O'Valle F, Puertas A, Oliver FJ, Ruiz de Almodóvar JM (2011) Human umbilical cord stromal stem cell express CD10 and exert contractile properties. Placenta 32(1):86–95

Faroni A, Mobasseri SA, Kingham PJ, Reid AJ (2015) Peripheral nerve regeneration: experimental strategies and future perspectives. Adv Drug Deliv Rev 82:160–167

Fong CY, Richards M, Manasi N, Biswas A, Bongso A (2007) Comparative growth behaviour and characterization of stem cells from human Wharton's jelly. Rep Biomed Online 15(6):708–718

Fong CY, Chak LL, Biswas A, Tan JH, Gauthaman K, Chan WK, Bongso A (2011) Human Wharton's jelly stem cells have unique transcriptome profiles compared to human embryonic stem cells and other mesenchymal stem cells. Stem Cell Rev 7(1):1–16

Fransson M, Piras E, Wang H, Burman J, Duprez I, Harris RA, LeBlanc K, Magnusson PU, Brittebo E, Loskog AS (2014) Intranasal delivery of central nervous system-retargeted human mesenchymal stromal cells prolongs treatment efficacy of experimental autoimmune encephalomyelitis. Immunology 142(3):431–441

Friedenstein AJ, Piatetzky-Shapiro II, Petrakova KV (1966) Osteogenesis in transplants of bone marrow cells. J Embryol Exp Morphol 16(3):381–390

Friedman R, Betancur M, Boissel L, Tuncer H, Cetrulo C, Klingemann H (2007) Umbilical cord mesenchymal stem cells: adjuvants for human cell transplantation. Biol Blood Marrow Transplant 13(12):1477–1486

Fritsch MK, Singer DB (2008) Embryonic stem cell biology. Adv Pediatr 55:43–77

Fuks A, Banjo C, Shuster J, Freedman SO, Gold P (1975) Carcinoembryonic antigen (CEA): molecular biology and clinical significance. Biochim Biophys Acta 417(2):123–152

Gärtner A, Pereira T, Armada-da-Silva P, Amado S, Veloso A, Amorim I, Ribeiro J, Santos J, Bárcia R, Cruz P (2014) Effects of umbilical cord tissue mesenchymal stem cells (UCX®) on rat sciatic nerve regeneration after neurotmesis injuries. J Stem Cells Regen Med 10(1):14

Geranmayeh MH, Baghbanzadeh A, Barin A, Salar-Amoli J, Dehghan MM, Rahbarghazi R, Azari H (2015) Paracrine neuroprotective effects of neural stem cells on glutamate-induced cortical neuronal cell excitotoxicity. Adv Pharm Bull 5(4):515–521

Gluckman E, Broxmeyer HA, Auerbach AD, Friedman HS, Douglas GW, Devergie A (1989) Hematopoietic reconstitution in a patient with Fanconi's anemia by means of umbilical cord blood from a HLA matched sibling. N Engl J Med 321:1174–1178

Gonzalez R, Griparic L, Umana M, Burgee K, Vargas V, Nasrallah R, Silva F, Patel A (2010) An efficient approach to isolation and characterization of pre- and postnatal umbilical cord lining stem cells for clinical applications. Cell Transplant 19(11):1439–1449

Gonzalez-Sanchez MB, Lopez-Valeiras E, Morente MM, Fernández Lago O (2013) Cost model for biobanks. Biopreserv Biobanking 11(5):272–277

Gotherstrom C, Ringden O, Westgren M, Tammik C, Le Blanc K (2003) Immunomodulatory effects of human foetal liver-derived mesenchymal stem cells. Bone Marrow Transplant 32(3):265–272

Greider CW (1998) Telomerase activity, cell proliferation, and cancer. Proc Natl Acad Sci U S A 95(1):90–92

Guan J, Zhu Z, Zhao RC, Xiao Z, Wu C, Han Q, Chen L, Tong W, Zhang J, Han Q (2013) Transplantation of human mesenchymal stem cells loaded on collagen scaffolds for the treatment of traumatic brain injury in rats. Biomaterials 34(24):5937–5946

Guenther MG, Frampton GM, Soldner F, Hockemeyer D, Mitalipova M, Jaenisch R, Young RA (2010) Chromatin structure and gene expression programs of human embryonic and induced pluripotent stem cells. Cell Stem Cell 7(2):249–257

Gupta PK, Das AK, Chullikana A, Majumdar AS (2012) Mesenchymal stem cells for cartilage repair in osteoarthritis. Stem Cell Res Ther 3(4):25

Ha C-W, Park Y-B, Chung J-Y, Park Y-G (2015) Cartilage repair using composites of human umbilical cord blood-derived mesenchymal stem cells and hyaluronic acid hydrogel in a minipig model. Stem Cells Transl Med 4(9):1044–1051. doi:10.5966/sctm.2014-0264

Hamad A, Majda KY, Mohamed AF, Kazem B, Ashraf AM (2015) Multi-lineage differentiation of human umbilical cord Wharton's jelly mesenchymal stromal cells mediates changes in the expression profile of stemness markers. PLoS One 10(4):e0122465

Han I, Yun M, Kim E-O, Kim B, Jung M-H, Kim S-H (2014) Umbilical cord tissue-derived mesenchymal stem cells induce apoptosis in PC-3 prostate cancer cells through activation of JNK and downregulation of PI3K/AKT signaling. Stem Cell Res Ther 5(2):54

Henningson CT Jr, Stanislaus MA, Gewirtz AM (2003) Embryonic and adult stem cell therapy. J Allergy Clin Immunol 111(2):S745–S753

Honmou O, Houkin K, Matsunaga T, Niitsu Y, Ishiai S, Onodera R, Waxman SG, Kocsis JD (2011) Intravenous administration of auto serum-expanded autologous mesenchymal stem cells in stroke. Brain 134(Pt 6):1790–1807. doi:10.1093/brain/awr063

Hoogduijn MJ, Popp F, Verbeek R, Masoodi M, Nicolaou A, Baan C, Dahlke MH (2010) The immunomodulatory properties of mesenchymal stem cells and their use for immunotherapy. Int Immunopharmacol 10(12):1496–1500

Hoynowski SM, Fry MM, Gardner BM, Leming MT, Tucker JR, Black L, Sand T, Mitchell KE (2007) Charaterization and differentiation of equine umbilical cord derived matrix cells. Biochem Biophys Res Commun 362:347–353

Hsieh JY, Fu YS, Chang SJ, Tsuang YH, Wang HW (2010) Functional module analysis reveals differential osteogenic and stemness potentials in human mesenchymal stem cells from bone marrow and Wharton's jelly of umbilical cord. Stem Cells Dev 19:1895–1910

Ishige I, Nagamura-Inoue T, Honda MJ, Harnprasopwat R, Kido M, Sugimoto M, Nakauchi H, Tojo A (2009) Comparison of mesenchymal stem cells derived from arterial, venous, and Wharton's jelly explants of human umbilical cord. Int J Hematol 90(2):261–269

Jiang J, Au M, Lu K, Eshpeter A, Korbutt G, Fisk G, Majumdar AS (2007) Generation of insulin-producing islet-like clusters from human embryonic stem cells. Stem Cells 25(8):1940–1953

Jing W, Chen Y, Lu L, Hu X, Shao C, Zhang Y, Zhou X, Zhou Y, Wu L, Liu R, Fan K, Jin G (2014) Human umbilical cord blood-derived mesenchymal stem cells producing IL15 eradicate established pancreatic tumor in syngeneic mice. Mol Cancer Ther 13(8):2127–2137. doi:10.1158/1535-7163.mct-14-0175

Jo CH, Kim OS, Park EY, Kim BJ, Lee JH, Kang SB, Lee JH, Han HS, Rhee SH, Yoon KS (2008) Fetal mesenchymal stem cells derived from human umbilical cord sustain primitive characteristics during extensive expansion. Cell Tissue Res 334(3):423–433

Kadam SS, Tiwari S, Bhonde RR (2009) Simultaneous isolation of vascular endothelial cells and mesenchymal stem cells from the human umbilical cord. In Vitro Cell Dev Biol Anim 45(1-2):23–27

Kalaszczynska I, Ferdyn K (2015) Wharton's jelly derived mesenchymal stem cells: future of regenerative medicine? Recent findings and clinical significance. Biomed Res Int 2015:430847

Karahuseyinoglu S, Cinar O, Kilic E, Kara F, Akay GG, Demiralp DO, Tukun A, Uckan D, Can A (2007) Biology of stem cells in human umbilical cord stroma: in situ and in vitro surveys. Stem Cells 25:319–331

Kim DW, Staples M, Shinozuka K, Pantcheva P, Kang S, Borlongan CV (2013) Wharton's jelly-derived mesenchymal stem cells: phenotypic characterization and optimizing their therapeutic potential for clinical applications. Int J Mol Sci 14:11692–11712

Kim HJ, Seo SW, Chang JW, Lee JI, Kim CH, Chin J, Choi SJ, Kwon H, Yun HJ, Lee JM (2015) Stereotactic brain injection of human umbilical cord blood mesenchymal stem cells in patients with Alzheimer's disease dementia: a phase 1 clinical trial. Alzheimer's Dement Transl Res Clin Interv 1(2):95–102

Kita K, Gauglitz GG, Phan TT, Herndon DN, Jeschke MG (2010) Isolation and characterization of mesenchymal stem cells from the sub-amniotic human umbilical cord lining membrane. Stem Cells Dev 19(4):491–502

Koh SH, Kim KS, Choi MR, Jung KH, Park KS, Chai YG, Roh W, Hwang SJ, Ko HJ, Huh YM, Kim HT, Kim SH (2008) Implantation of human umbilical cord-derived mesenchymal stem cells as a neuroprotective therapy for ischemic stroke in rats. Brain Res 1229:233–248

La RG, Anzalone R, Corrao S, Magno F, Loria T, Lo Iacono M, Di Stefano A, Giannuzzi P, Marasà L, Cappello F, Zummo G, Farina F (2009) Isolation and characterization of Oct-4+/HLA-G+ mesenchymal stem cells from human umbilical cord matrix: differentiation potential and detection of new markers. Histochem Cell Biol 131(2):267–282

Le BK (2003) Immunomodulatory effects of fetal and adult mesenchymal stem cells. Cytotherapy 5(6):485–489

Le Blanc K, Rasmusson I, Sundberg B, Götherström C, Hassan M, Uzunel M, Ringdén O (2004) Treatment of severe acute graft-versus-host disease with third party haploidentical mesenchymal stem cells. The Lancet 363(9419):1439–1441

Lee OK, Kuo TK, Chen W-M, Lee K-D, Hsieh S-L, Chen T-H (2004) Isolation of multipotent mesenchymal stem cells from umbilical cord blood. Blood 103(5):1669–1675

Lee JS, Hong JM, Moon GJ, Lee PH, Ahn YH, Bang OY (2010) A long-term follow-up study of intravenous autologous mesenchymal stem cell transplantation in patients with ischemic stroke. Stem Cells 28(6):1099–1106

Li T, Yan Y, Wang B, Qian H, Zhang X, Shen L, Wang M, Zhou Y, Zhu W, Li W (2012) Exosomes derived from human umbilical cord mesenchymal stem cells alleviate liver fibrosis. Stem Cells Dev 22(6):845–854

Li X-Y, Zheng Z-H, Li X-Y, Guo J, Zhang Y, Li H, Wang Y-W, Ren J, Wu Z-B (2013) Treatment of foot disease in patients with type 2 diabetes mellitus using human umbilical cord blood mesenchymal stem cells: response and correction of immunological anomalies. Curr Pharm Des 19(27):4893–4899

Liang L, Yingfei G, Hongxia Z, Yaxin Y, Jinjin Z, Haiwei C, Lei W, Na L, Runmei L, Yunfeng X (2014) Aging increases the susceptivity of MSCs to reactive oxygen species and impairs their therapeutic potency for myocardial infarction. PLoS One 9(11):111850

Lim SY, Kim YS, Ahn Y, Jeong MH, Hong MH, Joo SY, Nam KI, Cho JG, Kang PM, Park JC (2006) The effects of mesenchymal stem cells transduced with Akt in a porcine myocardial infarction model. Cardiovasc Res 70(3):530–542. doi:10.1016/j.cardiores.2006.02.016

Lin Y-C, Ko T-L, Shih Y-H, Lin M-YA, Fu T-W, Hsiao H-S, Hsu J-YC, Fu Y-S (2011) Human umbilical mesenchymal stem cells promote recovery after ischemic stroke. Stroke 42(7):2045–2053

Liu J, Han D, Wang Z, Xue M, Zhu L, Yan H, Zheng X, Guo Z, Wang H (2013) Clinical analysis of the treatment of spinal cord injury with umbilical cord mesenchymal stem cells. Cytotherapy 15(2):185–191

Lu Z, Zhao H, Xu J, Zhang Z, Zhang X (2013) Human umbilical cord mesenchymal stem cells in the treatment of secondary progressive multiple sclerosis. J Stem Cell Res Ther 6:2

Ma F, Chen D, Chen F, Chi Y, Han Z, Feng X, Li X (2015) Human umbilical cord mesenchymal stem cells promote breast cancer metastasis by interleukin-8 and interleukin-6 dependent induction of CD44 (+)/CD24 (-) cells. Cell Transplant 24(12):2585–2599

Maitra B, Szekely E, Gjini K, Laughlin MJ, Dennis J, Haynesworth SE, Koc ON (2004) Human mesenchymal stem cells support unrelated donor hematopoietic stem cells and suppress T-cell activation. Bone Marrow Transplant 33(6):597–604

Majore I, Moretti P, Stahl F, Hass R, Kasper C (2011) Growth and differentiation properties of mesenchymal stromal cell populations derived from whole human umbilical cord. Stem Cell Rev 7(1):17–31

Marcus AJ, Woodbury D (2008) Fetal stem cells from extra-embryonic tissues: do not discard. J Cell Mol Med 12(3):730–742

Martin-Rendon E, Sweeney D, Lu F, Girdlestone J, Navarrete C, Watt SM (2008a) 5-Azacytidine-treated human mesenchymal stem/progenitor cells derived from umbilical cord, cord blood and bone marrow do not generate cardiomyocytes in vitro at high frequencies. Vox Sang 95(2):137–148

Martin-Rendon E, Brunskill SJ, Hyde CJ, Stanworth SJ, Mathur A, Watt SM (2008b) Autologous bone marrow stem cells to treat acute myocardial infarction: a systematic review. Eur Heart J 29(15):1807–1818

Mehrabi M, Mansouri K, Hosseinkhani S, Yarani R, Yari K, Bakhtiari M, Mostafaie A (2015) Differentiation of human skin-derived precursor cells into functional islet-like insulin-producing cell clusters. In Vitro Cell Dev Biol Anim 51:1–9

Mobarakeh ZT, Ai J, Yazdani F, Sorkhabadi SMR, Ghanbari Z, Javidan AN, Mortazavi-Tabatabaei SAR, Massumi M, Barough SE (2012) Human endometrial stem cells as a new source for programming to neural cells. Cell Biol Int Rep 19(1):7–14

Mohammadi E, Nassiri SM, Rahbarghazi R, Siavashi V, Araghi A (2015) Endothelial juxtaposition of distinct adult stem cells activates angiogenesis signaling molecules in endothelial cells. Cell Tissue Res 362:1–13

Murry CE, Field LJ, Menasché P (2005) Cell-based cardiac repair: reflections at the 10-year point. Circulation 112(20):3174–3183. doi:10.1161/circulationaha.105.546218

Nekanti U, Rao VB, Bahirvani AG, Jan M, Totey S, Ta M (2010a) Longterm expansion and pluripotent marker array analysis of Wharton's jelly-derived mesenchymal stem cells. Stem Cells Dev 19:117–130

Nekanti U, Mohanty L, Venugopal P, Balasubramanian S, Totey S, Ta M (2010b) Optimization and scale-up of Wharton's jelly derived mesenchymal stem cells for clinical applications. Stem Cell Res 5(3):244–254

Newman AM, Cooper JB (2010) Lab-specific gene expression signatures in pluripotent stem cells. Cell Stem Cell 7(2):258–262

Niknamasl A, Ostad SN, Soleimani M, Azami M, Salmani MK, Lotfibakhshaiesh N, Ebrahimi-Barough S, Karimi R, Roozafzoon R, Ai J (2014) A new approach for pancreatic tissue engineering: human endometrial stem cells encapsulated in fibrin gel can differentiate to pancreatic islet beta-cell. Cell Biol Int 38(10):1174–1182

Okita K, Nagata N, Yamanaka S (2011) Immunogenicity of induced pluripotent stem cells. Circ Res 109(7):720–721. doi:10.1161/RES.0b013e318232e187

Panepucci RA, Siufi JL, Silva WA Jr, Proto-Siquiera R, Neder L, Orellana M, Rocha V, Covas DT, Zago MA (2004) Comparison of gene expression of umbilical cord vein and bone marrow-derived mesenchymal stem cells. Stem Cells 22(7):1263–1278

Pappa KI, Anagnou NP (2009) Novel sources of fetal stem cells: where do they fit on the developmental continuum? Regen Med 4:423–433

Perdikogianni C, Dimitriou H, Stiakaki E, Martimianaki G, Kalmanti M (2008) Could cord blood be a source of mesenchymal stromal cells for clinical use? Cytotherapy 10(5):452–459

Phadnis SM, Ghaskadbi SM, Hardikar AA, Bhonde RR (2009) Mesenchymal stem cells derived from bone marrow of diabetic patients portrait unique markers influenced by the diabetic microenvironment. Rev Diabet Stud 6(4):260–270

Phinney DG, Isakova I (2005) Plasticity and therapeutic potential of mesenchymal stem cells in the nervous system. Curr Pharm Des 11(10):1255–1265

Pijnappels DA, Schalij MJ, Ramkisoensing AA, van Tuyn J, de Vries AAF, van der Laarse A, Ypey DL, Atsma DE (2008) Forced alignment of mesenchymal stem cells undergoing cardiomyogenic differentiation affects functional integration with cardiomyocyte cultures. Circ Res 103(2):167–176. doi:10.1161/circresaha.108.176131

Pittenger MF, Mackay AM, Beck SC, Jaiswal RK, Douglas R, Mosca JD, Moorman MA, Simonetti DW, Craig S, Marshak DR (1999) Multilineage potential of adult human mesenchymal stem cells. Science 284:143–147

Pountos I, Giannoudis PV, Jones E, English A, Churchman S, Field S, Ponchel F, Bird H, Emery P, McGonagle D (2011) NSAIDS inhibit in vitro MSC chondrogenesis but not osteogenesis: implications for mechanism of bone formation inhibition in man. J Cell Mol Med 15(3):525–534

Prasanna SJ, Jahnavi VS (2011) Wharton's jelly mesenchymal stem cells as off-the-shelf cellular therapeutics: a closer look into their regenerative and immunomodulatory properties. Open Tissue Eng Regen Med J 4:28–38

Rachakatla RS, Marini F, Weiss ML, Tamura M, Troyer D (2007) Development of human umbilical cord matrix stem cell-based gene therapy for experimental lung tumors. Cancer Gene Ther 14(10):828–835

Rahbarghazi R, Nassiri SM, Khazraiinia P, Kajbafzadeh A-M, Ahmadi SH, Mohammadi E, Molazem M, Zamani-Ahmadmahmudi M (2012) Juxtacrine and paracrine interactions of rat marrow-derived mesenchymal stem cells, muscle-derived satellite cells, and neonatal cardiomyocytes with endothelial cells in angiogenesis dynamics. Stem Cells Dev 22(6):855–865

Rahbarghazi R, Nassiri SM, Ahmadi SH, Mohammadi E, Rabbani S, Araghi A, Hosseinkhani H (2014) Dynamic induction of pro-angiogenic milieu after transplantation of marrow-derived mesenchymal stem cells in experimental myocardial infarction. Int J Cardiol 173(3):453–466

Rahman MJ, Regn D, Bashratyan R, Dai YD (2014) Exosomes released by islet-derived mesenchymal stem cells trigger autoimmune responses in NOD mice. Diabetes 63(3):1008–1020

Ratanatharathorn V, Ayash L, Lazarus H, Fu J, Uberti J (2001) Chronic graft-versus-host disease: clinical manifestation and therapy. Bone Marrow Transplant 28(2):121–129

Rennert RC, Sorkin M, Januszyk M, Duscher D, Kosaraju R, Chung MT, Lennon J, Radiya-Dixit A, Raghvendra S, Maan ZN, Hu MS, Rajadas J, Rodrigues M, Gurtner GC (2014) Diabetes impairs the angiogenic potential of adipose-derived stem cells by selectively depleting cellular subpopulations. Stem Cell Res Ther 5(3):79

Reza HM, Ng BY, Phan TT, Tan DT, Beuerman RW, Ang LP (2011) Characterization of a novel umbilical cord lining cell with CD227 positivity and unique pattern of P63 expression and function. Stem Cell Rev 7(3):624–638

Ringdén O, Uzunel M, Rasmusson I, Remberger M, Sundberg B, Lönnies H, Marschall H-U, Dlugosz A, Szakos A, Hassan Z (2006) Mesenchymal stem cells for treatment of therapy-resistant graft-versus-host disease. Transplantation 81(10):1390–1397

Riteau B, Moreau P, Menier C, Khalil-Daher I, Khosrotehrani K, Bras-Goncalves R, Paul P, Dausset J, Rouas-Freiss N, Carosella ED (2001) Characterization of HLA-G1, -G2, -G3, and -G4 isoforms transfected in a human melanoma cell line. Transplant Proc 33:2360–2364

Roobrouck VD, Ulloa-Montoya F, Verfaillie CM (2008) Self-renewal and differentiation capacity of young and aged stem cells. Exp Cell Res 314(9):1937–1944

Rosa T, BacklyRania E (2015) Transplanted umbilical cord mesenchymal stem cells modify the in vivo microenvironment enhancing angiogenesis and leading to bone regeneration. Stem Cells Dev 24(13):1570–1581

Roura S, Bagó JR, Soler-Botija C, Pujal JM, Gálvez-Montón C, Prat-Vidal C, Llucià-Valldeperas A, Blanco J, Bayes-Genis A (2012) Human umbilical cord blood-derived mesenchymal stem cells promote vascular growth in vivo. PLoS One 7(11):e49447

Salehinejad P, Alitheen NB, Ali AM, Omar AR, Mohit M, Janzamin E, Samani FS, Torshizi Z, Nematollahi-Mahani SN (2012) Comparison of different methods for the isolation of mesenchymal stem cells from human umbilical cord Wharton's jelly. In Vitro Cell Dev Biol Anim 48(2):75–83

Sarugaser R, Lickorish D, Baksh D, Hosseini MM, Davies JE (2005) Human umbilical cord perivascular (HUCPV) cells: a source of mesenchymal progenitors. Stem Cells 23(2):220–229

Sarugaser R, Hanoun L, Keating A, Stanford WL, Davies JE (2009) Human mesenchymal stem cells self-renew and differentiate according to a deterministic hierarchy. PLoS One 4(8):e6498

Sarvandi SS, Joghataei MT, Parivar K, Khosravi M, Sarveazad A, Sanadgol N (2015) In vitro differentiation of rat mesenchymal stem cells to hepatocyte lineage. Iran J Basic Med Sci 18(1):89

Secco M, Moreira YB, Zucconi E, Vieira NM, Jazedje T, Muotri AR, Okamoto OK, Verjovski-Almeida S, Zatz M (2009) Gene expression profile of mesenchymal stem cells from paired umbilical cord units: cord is different from blood. Stem Cell Rev 5(4):387–401

Seshareddy K, Troyer D, Weiss ML (2008) Method to isolate mesenchymal-like cells from Wharton's Jelly of umbilical cord. Methods Cell Biol 86:101–119

Shapiro AJ, Lakey JR, Ryan EA, Korbutt GS, Toth E, Warnock GL, Kneteman NM, Rajotte RV (2000) Islet transplantation in seven patients with type 1 diabetes mellitus using a glucocorticoid-free immunosuppressive regimen. N Engl J Med 343(4):230–238

Shawki S, Gaafar T, Erfan H, Khateeb EE, Sheikhah AE, Hawary RE (2015) Immunomodulatory effect of umbilical cord derived mesenchymal stem cells. Microbiol Immunol 59(6):348–356

Si Y, Zhao Y, Hao H, Liu J, Guo Y, Mu Y, Shen J, Cheng Y, Fu X, Han W (2012) Infusion of mesenchymal stem cells ameliorates hyperglycemia in type 2 diabetic rats: identification of a novel role in improving insulin sensitivity. Diabetes 61(6):1616–1625. doi:10.2337/db11-1141

Stefano F, Serena V, Lucia VF, Miriana JQ, Giampiero L, Alberto T, Erica V (2015) Wharton's jelly derived mesenchymal stromal cells: biological properties, induction of neuronal phenotype and current applications in neurodegeneration research. Acta Histochem 117(4-5):329–338

Stolzing A, Jones E, McGonagle D, Scutt A (2008) Age-related changes in human bone marrow-derived mesenchymal stem cells: consequences for cell therapies. Mech Ageing Dev 129(3):163–173

Subramanian A, Fong CY, Biswas A, Bongso A (2015) Comparative characterization of cells from the various compartments of the human umbilical cord shows that the Wharton's jelly compartment provides the best source of clinically utilizable mesenchymal stem cells. PLoS One 10(6):e0127992

Sun L, Wang D, Liang J, Zhang H, Feng X, Wang H, Hua B, Liu B, Ye S, Hu X (2010) Umbilical cord mesenchymal stem cell transplantation in severe and refractory systemic lupus erythematosus. Arthritis Rheum 62(8):2467–2475

Sung M-A, Jung HJ, Lee J-W, Lee J-Y, Pang K-M, Yoo SB, Alrashdan MS, Kim S-M, Jahng JW, Lee J-H (2012) Human umbilical cord blood-derived mesenchymal stem cells promote regeneration of crush-injured rat sciatic nerves. Neural Regen Res 7(26):2018

Taghizadeh RR, Cetrulo KJ, Cetrulo CL (2011) Wharton's jelly stem cells: future clinical applications. Placenta 32(4):311–315

Tehrani HJ, Parivar K, Ai J, Kajbafzadeh A, Rahbarghazi R, Hashemi M, Sadeghizadeh M (2014) Effect of dexamethasone, insulin and EGF on the myogenic potential on human endometrial stem cell. Iran J Pharm Res 13(2):659

Thomson JA, Itskovitz-Eldor J, Shapiro SS (1998) Embryonic stem cell lines derived from human blastocysts. Science 282:1145–1147

Tolar J, Le Blanc K, Keating A, Blazar BR (2010) Concise review: hitting the right spot with mesenchymal stromal cells. Stem Cells 28(8):1446–1455

Toma C, Pittenger MF, Cahill KS, Byrne BJ, Kessler PD (2002) Human mesenchymal stem cells differentiate to a cardiomyocyte phenotype in the adult murine heart. Circulation 105(1):93–98. doi:10.1161/hc0102.101442

Tong CK, Vellasamy S, Tan BC, Abdullah M, Vidyadaran S, Seow HF, Ramasamy R (2011) Generation of mesenchymal stem cell from human umbilical cord tissue using a combination enzymatic and mechanical disassociation method. Cell Biol Int 35(3):221–226

Troyer DL, Weiss ML (2008) Concise review: Wharton's jelly-derived cells are a primitive stromal cell population. Stem Cells 26:591–599

Tsagias N, Koliakos I, Karagiannis V, Eleftheriadou M, Koliakos GG (2011) Isolation of mesenchymal stem cells using the total length of umbilical cord for transplantation purposes. Transfus Med 21(4):253–261

Tsai PC, Fu TW, Chen YMA, Ko TL, Chen TH, Shih YH, Hung SC, Fu YS (2009) The therapeutic potential of human umbilical mesenchymal stem cells from Wharton's jelly in the treatment of rat liver fibrosis. Liver Transpl 15(5):484–495

Vogelaar CF, Vrinten DH, Hoekman MF, Brakkee JH, Burbach JPH, Hamers FP (2004) Sciatic nerve regeneration in mice and rats: recovery of sensory innervation is followed by a slowly retreating neuropathic pain-like syndrome. Brain Res 1027(1):67–72

Wang HS, Hung SC, Peng ST, Huang CC, Wei HM, Guo YJ, Fu YS, Lai MC, Chen CC (2004) Mesenchymal stem cells in the Wharton's jelly of the human umbilical cord. Stem Cells 22:1330–1337

Wang M, Yang Y, Yang D, Luo F, Liang W, Guo S, Xu J (2009) The immunomodulatory activity of human umbilical cord blood-derived mesenchymal stem cells in vitro. Immunology 126(2):220–232

Wang S, Qu X, Zhao RC (2012) Clinical applications of mesenchymal stem cells. J Hematol Oncol 5(1):19

Wang Y, Zhang Z, Chi Y, Zhang Q, Xu F, Yang Z, Meng L, Yang S, Yan S, Mao A, Zhang J, Yang Y, Wang S, Cui J, Liang L, Ji Y, Han ZB, Fang X, Han ZC (2013) Long-term cultured mesenchymal stem cells frequently develop genomic mutations but do not undergo malignant transformation. Cell Death Dis 4:e950. doi:10.1038/cddis.2013.480

Weiss ML, Troyer DL (2006) Stem cells in the umbilical cord. Stem Cell Rev 2(2):155–162

Weiss ML, Medicetty S, Bledsoe AR, Rachakatla RS, Choi M, Merchav S, Luo Y, Rao MS, Velagaleti G, Troyer D (2006) Human umbilical cord matrix stem cells: preliminary characterization and effect of transplantation in a rodent model of Parkinson's disease. Stem Cells 24(3):781–792

Weiss M, Anderson C, Medicetty S, SeshareddyK B, Weiss RJ, Vanderwerff I, Troyer D, Mcintosh KR (2008) Immune properties of human umbilical cord Wharton's jelly-derived cells. Stem Cells 26:2865–2874

Weissman IL, Anderson DJ, Gage F (2001) Stem and progenitor cells: origins, phenotypes, lineage commitments, and transdifferentiations. Annu Rev Cell Dev Biol 17:387–403

Welte K, Foeken L, Gluckman E, Navarrete C, Cord Blood Working Group of the World Marrow Donor Association (2010) International exchange of cord blood units: the registry aspects. Bone Marrow Transplant 45:825–831

Wetzig A, Alaiya A, Al-Alwan M, Pradez CB, Pulicat MS, Al-Mazrou A, Shinwari Z, Sleiman GM, Ghebeh H, Al-Humaidan H, Gaafar A, Kanaan I, Adra C (2013) Differential marker expression by cultures rich in mesenchymal stem cells. BMC Cell Biol 14:54

Wexler SA, Donaldson C, Denning-Kendall P, Rice C, Bradley B, Hows JM (2003) Adult bone marrow is a rich source of human mesenchymal 'stem' cells but umbilical cord and mobilized adult blood are not. Br J Haematol 121(2):368–374

Wu X-B, Tao R (2012) Hepatocyte differentiation of mesenchymal stem cells. Hepatobiliary Pancreat Dis Int 11(4):360–371

Xu Z, Sheng L, Ouyang G (2014) [Umbilical cord blood-derived mesenchymal stem cells inhibit proliferation of peripheral blood lymphocytes]. Xi Bao Yu Fen Zi Mian Yi Xue Za Zhi 30(9):968–971

Yamashita A, Liu S, Woltjen K, Thomas B, Meng G, Hotta A, Takahashi K, Ellis J, Yamanaka S, Rancourt DE (2013) Cartilage tissue engineering identifies abnormal human induced pluripotent stem cells. Sci Rep 3:1978

Yan Y, Xu W, Qian H, Si Y, Zhu W, Cao H, Zhou H, Mao F (2009) Mesenchymal stem cells from human umbilical cords ameliorate mouse hepatic injury in vivo. Liver Int 29(3):356–365

Yasuda K, Yashiro M, Sawada T, Ohira M, Hirakawa K (2007) ERas oncogene expression and epigenetic regulation by histone acetylation in human cancer cells. Anticancer Res 27(6B):4071–4075

Yoon JH, Roh EY, Shin S, Jung NH, Song EY, Chang JY, Kim BJ, Jeon HW (2013) Comparison of explant-derived and enzymatic digestion-derived MSCs and the growth factors from Wharton's jelly. Biomed Res Int 2013:428726

Yuan Y, Chen D, Chen X, Shao H, Huang S (2014) Human umbilical cord-derived mesenchymal stem cells inhibit proliferation but maintain survival of Jurkat leukemia cells in vitro by activating Notch signaling. Nan Fang Yi Ke Da Xue Xue Bao 34(4):441–447

Zhang Z, Lin H, Shi M, Xu R, Fu J, Lv J, Chen L, Lv S, Li Y, Yu S (2012) Human umbilical cord mesenchymal stem cells improve liver function and ascites in decompensated liver cirrhosis patients. J Gastroenterol Hepatol 27(s2):112–120

Zhang L, Xiang J, Li G (2013) The uncertain role of unmodified mesenchymal stem cells in tumor progression: what master switch? Stem Cell Res Ther 4(2):22

Zhao T, Zhang Z-N, Rong Z, Xu Y (2011) Immunogenicity of induced pluripotent stem cells. Nature 474(7350):212–215

Zhi-Gang Z, Wei-Ming L, Zhi-Chao C, Yong Y, Ping Z (2008) Immunosuppressive properties of mesenchymal stem cells derived from bone marrow of patient with hematological malignant diseases. Leuk Lymphoma 49(11):2187–2195

Zhou C, Yang B, Tian Y (2011) Immunomodulatory effect of human umbilical cord Wharton's jelly-derived mesenchymal stem cells on lymphocytes. Cell Immunol 272(1):33–38

Zhou R, Li Z, He C, Li R, Xia H, Li C, Xiao J, Chen Z-Y (2014) Human umbilical cord mesenchymal stem cells and derived hepatocyte-like cells exhibit similar therapeutic effects on an acute liver failure mouse model. PLoS One 9(8):e104392. doi:10.1371/journal.pone.0104392

Hou Z-l, Liu Y, Mao X-H, Wei C-y, Meng M-y, Liu Y-h, Zhuyun Yang Z, Zhu H, Short M, Bernard C, Xiao Z-c (2013) Transplantation of umbilical cord and bone marrow-derived mesenchymal stem cells in a patient with relapsing-remitting multiple sclerosis. Cell Adh Migr 7(5):404–407. doi:10.4161/cam.26941

Characteristics of Mesenchymal Stem Cells Derived from Amniotic Membrane: A Potential Candidate for Stem Cell-Based Therapy

Somaieh Kazemnejad, Manijeh Khanmohammadi, Amir-Hassan Zarnani, and Mohammad Reza Bolouri

1 Introduction

The stem cells are characterized by their capacity of self-renewal and transdifferentiation to various cell types if guided appropriately. These interesting properties have made stem cells an important component of tissue regeneration and regenerative medicine. Mesenchymal stem cells (MSCs) have been harvested successfully from a wide range of tissues such as bone marrow, adipose tissue, umbilical cord blood, menstrual blood, etc. (Phinney 2008; Lu et al. 2006; Gimble and Guilak 2003; Kazemnejad et al. 2012; Khanmohammadi et al. 2014). However, problems such as less availability, invasive methods for sample collection, and lower proliferation capacity in comparison with embryonic stem cells reduce the possibility to collect a large amount of cells, in an inexpensive and noninvasive way, and without being risky for the donor (Parolini et al. 2009). As such, amniotic membrane as postlabor medical waste has been recently recognized as appealing candidates for the derivation of MSCs.

The evidence delineates that presumptive adult stem cells derived from amniotic membrane retain highest proliferation capacity, longest telomere length, broadest differentiation, and extensive proliferative potential when compared with cells

S. Kazemnejad, Ph.D. (✉) • M. Khanmohammadi
Reproductive Biotechnology Research Center, Avicenna Research Institute, ACECR, Tehran, Iran
e-mail: kazemnejad_s@yahoo.com; s.kazemnejad@avicenna.ac.ir

A.-H. Zarnani
Department of Immunology, School of Public Health, Tehran University of Medical Sciences, Tehran, Iran

Immunology Research Center, Iran University of Medical Sciences, Tehran, Iran

M.R. Bolouri
Immunology Research Center, Iran University of Medical Sciences, Tehran, Iran

B. Arjmand (ed.), *Perinatal Tissue-Derived Stem Cells*, Stem Cell Biology and Regenerative Medicine, DOI 10.1007/978-3-319-46410-7_7

obtained from adult tissues (Kern et al. 2006; Kögler et al. 2004). MSCs obtained from human amnion have been shown to retain immunomodulatory properties as well as to strongly inhibit T lymphocyte proliferation (Magatti et al. 2008) and to survive when transplanted in immunocompetent animals without inducing any tumorigenic effect in vivo (Kubo et al. 2001).

The immunomodulatory capacity exerted by mesenchymal stem cells derived from human amniotic membrane is even greater than other cells population located in amniotic membrane named amniotic epithelial cells. Therefore, amniotic membrane-derived mesenchymal stem cells (AMSCs) might be useful in the clinic for autologous transplantation for fetuses and newborns, and after banking in later stages of life. In addition, due to immunosuppressive functions, these cells might be used for allogenic transplantation and in treatment of some autoimmune diseases (Kim et al. 2013a; Kang et al. 2012a; Rossi et al. 2012a). In this chapter, we will discuss characteristics and potential of AMSCs in terms of critical points for cell therapy like accessibility, proliferation and differentiation ability, ethical concern, tumorigenesis, and immunological consideration. Furthermore, the precise state of progressions in cell therapy of AMSCs for treatment of different disorders will be realized.

2 Main Issues Should Be Considered for Clinical Potential of Stem Cells

For clinical stem cell therapy, some challenges should be solved (Brooke et al. 2009): (1) Stem cell density in retrieved samples, (2) proliferation ability of stem cells in laboratory to dissolve demand of clinic, (3) differentiation potential of stem cell to the desired cell phenotype, (4) the bioreactive molecules such as cytokines or growth factors that can support the formation of the desired tissue, (5) the best biomarkers to characterize stem cells, (6) biosafety and tumorigenesis ability, (7) ethics, (8) immunological consideration, (9) the safety evaluation in preclinical and clinical studies, (10) the translation from laboratory to clinics by using good laboratory practice and good manufacturing practice.

For bone marrow and umbilical cord blood that are well-known stem cell sources and have longest histories compared to other sources, there are different regulation laws describing the procedure of authorization related to preparation, storage, and clinical use. Although these regulations help policy makers in cell therapy to define procedures and criteria for other newfound stem cells, before clinical application of these stem cells like AMSCs more researches are necessary to understand their properties and behavior upon transplantation. However, to investigate the application potential of AMSCs for cell therapy and regenerative medicine, here we have discussed the characteristics of these cells especially in comparison with other stem cells in terms of critical points for cell therapy.

3 Location of Amniotic Membrane Mesenchymal Stem Cells (AMSCs)

Human placenta plays an essential role in fetal development, nourishment, and immunological privilege. Two regions in the placental tissue can be distinguished: a maternal component, termed the decidua, which is derived from the endometrium, and a fetal component, which includes the amniotic membrane, the chorionic membrane, and the chorionic plate, from which villi extend and make intimate contact with the uterine decidua during pregnancy. These fetal membranes facilitate exchange of gases and wastes, provide a defense barrier, and support pregnancy and parturition. Amniotic membrane is the innermost layer of placenta and consists of a thin epithelial layer, a basement membrane, and an avascular stroma. This thin, avascular membrane, which lines the amniotic cavity and is bathed in amniotic fluid, is adjacent over the umbilical cord with the fetal skin (Benirschke and Kaufman 2000; Rennie et al. 2012; Toda et al. 2007a).

AMSCs are dispersed in an extracellular matrix in basement membrane largely composed of collagen and laminin, and a network of fibroblast-like mesenchymal cells (Ilancheran et al. 2009). These cells mostly originate from mesoderm layer and were developed from epiblast during the pregastrulation stages of embryogenesis (Ilancheran et al. 2009; Parolini et al. 2008) (Fig. 1).

4 Proliferation Potential of AMSCs Relative to Clinical Demand

The first report about AMSCs goes back to 2004, the year that In 't Anker et al. isolated and established the differentiation potential of these stem cells into osteogenic and adipogenic cells (In 't Anker et al. 2004). Nowadays, efficient protocols have been established for AMSCs isolation from term placenta and are generally based on the separation of the amniotic membrane from the chorionic membrane and subsequent enzymatic digestion (Parolini et al. 2008; In 't Anker et al. 2004; Marongiu et al. 2010).

Based on the reports, a typical human term amniotic membrane yields between 20 and 50×10^6 stem cells that are clonogenic and have great expansion potential in culture flask (Manuelpillai et al. 2011). It was found that human AMSCs could be expanded for at least 15 passages in culture (Soncini et al. 2007; Ilancheran et al. 2007). In a similar pattern with umbilical cord-MSCs, these cells exhibited a higher proliferation rate compared to MSCs derived from adult sources like bone marrow that their proliferative potential decrease as the donor's age increases (Alviano et al. 2007). Therefore, it has been suggested that human AMSCs be more primitive mesenchymal stem cells than those found in bone marrow (Taghizadeh et al. 2011).

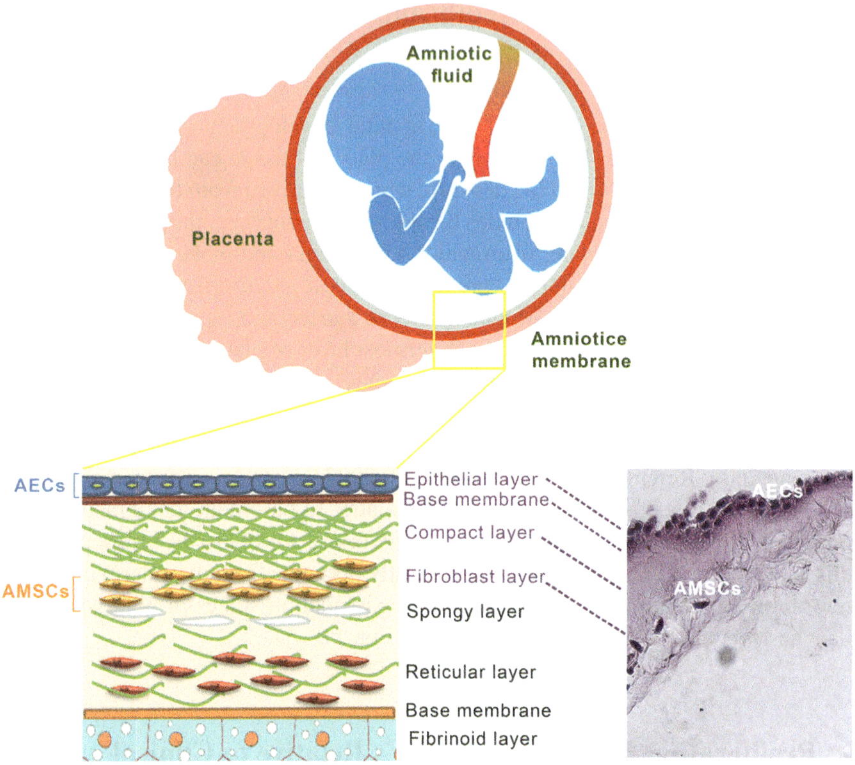

Fig. 1 Schematic illustration of the fetal membrane structure at term. The components of each layer and location of amniotic membrane stem cells have been shown

Interestingly, epithelial layer of amniotic membrane is composed of amniotic epithelial cells (AECs) that change from the cuboidal epithelial shape into elongated stromal-like cells after a few passages. The reason for these changes remains uncertain, but could be attributed to senescence, epigenetic modifications, and to the autocrine/paracrine effects of growth factors that induce an epithelial to mesenchymal transition. According to the studies, AECs express markers associated with mesenchymal and fibroblast cells and show low proliferation and differentiation potential (Stadler et al. 2008; Bilic et al. 2008a). It has been reported that proliferation ability and population doublings of AECs is significantly less than AMSCs. AECs can only proliferate for less than five population doublings during 2 months while human AMSCs expand more than 70 passages with higher population doublings compared to human AECs (Kim et al. 2013b; Kimura et al. 2012). This prolonged in vitro proliferation capacity of AMSCs is a notable point to obtain adequate cell numbers for manipulation in biology and medicine and their implant into the patient.

5 The Defined Biomarkers for Characterization of AMSCs

Human AMSCs express a repertoire of the surface and intracellular stem/progenitor markers (Fig. 2), suggesting that they could act as progenitors and differentiate into various cell types from each of the three germ layers (Manuelpillai et al. 2011). Considerable variation in percentages of different marker expression implies the dependence of levels and pattern of markers expression to the isolation protocol and passage number in culture (Díaz-Prado et al. 2010).

AMSCs share similar phenotypic characteristics with the stem cells derived from bone marrow (BMSCs) (Roubelakis et al. 2012a). These cells present characteristic

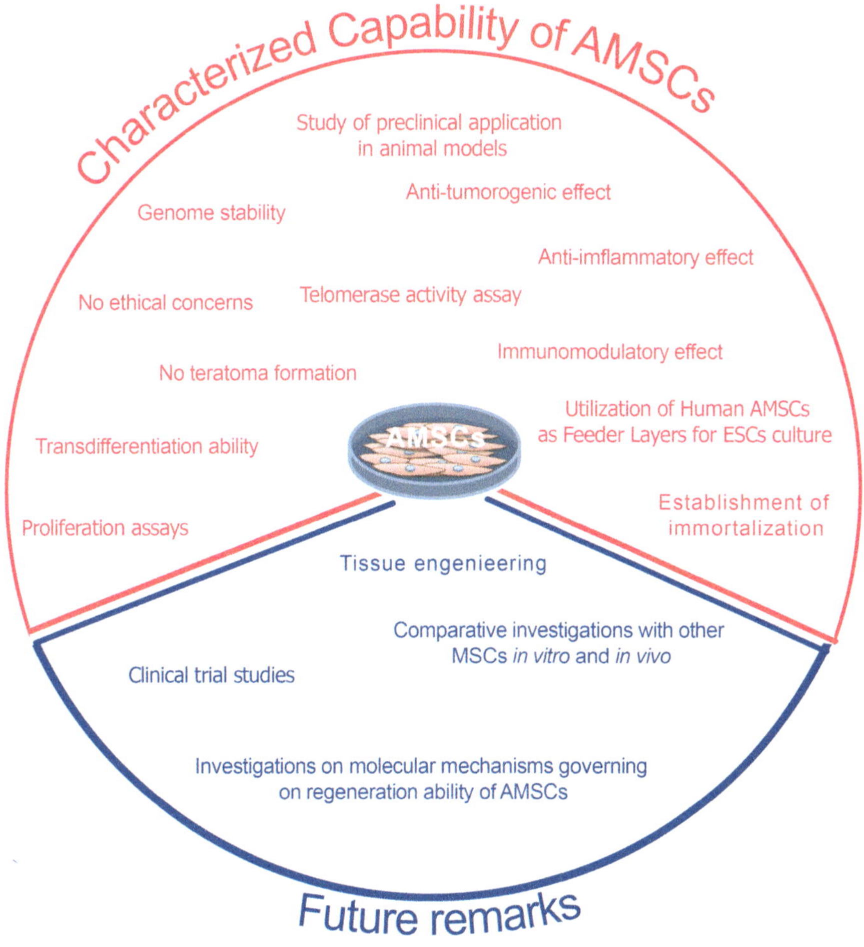

Fig. 2 The different characterized biomarkers expressing by amniotic membrane mesenchymal stem cells

fibroblast-like or spindle-shaped cells that express cell surface antigens markers associated to BMSCs such as adhesion and mesenchymal markers (STRO-1, CD13, CD73, CD90, CD29, CD44, CD49d, CD49e, CD56, CD166, CD105, and Vimentin) (Kmiecik et al. 2013), while they are negative for endothelial cell (CD31), hematopoietic (CD34, CD45), and monocyte (CD14) markers (Parolini et al. 2009; Kmiecik et al. 2013; Kamadjaja et al. 2014). In addition, these fetal cells express low levels of human leukocyte antigen class I (HLA-I), but do not express HLA-II, suggesting that these cells are proper candidate for cell therapy targets (Parolini et al. 2009; Kmiecik et al. 2013); a matter will be discussed in the following issues.

Some findings suggest that AMSCs have unique immunophenotypic characteristics that distinguish these cells from BMSCs. Since the amniotic membrane is derived from fetal origin, it is expected that amniotic cells contain pluripotent stem cells. In this line, several reports demonstrated that AMSCs express some pluripotency markers such as L-alkaline phosphatase or TRA-2-39, OCT-4, Nanog, SOX-2, and Rex-1 which is specifically expressed in embryonic stem cells and germ cells; GATA-4, which is a marker of definitive embryonic and visceral (extra-embryonic) endoderm; hepatocyte nuclear factor-3β (HNF-3β), which is a marker of definitive endoderm; nestin and musashi as a neural stem cell-specific marker. However, positivity for the SSEA-3 or SSEA-4 in these cells is still debated (Parolini et al. 2009; Kang et al. 2012b; Toda et al. 2007b).

According to the reports, the expression levels of embryonic markers including OCT3/4, SOX2, Klf4, c-Myc, Nanog, and Lin28 in AMSCs are much higher than those of BMSCs. On the other hand, the mean expression of Klf4 by AMSCs is high as much as induced pluripotent stem cells (iPSCs) suggesting that these stem cells are more similar to iPSCs compared to BMSC (Koike et al. 2014). These facts suggest that amniotic membrane cells possess pluripotency, and thus they would be a good candidate for an alternative cell source for cellular transplantation therapy.

6 Ethical and Legal Considerations

Amniotic membrane as a medical waste and part of the placenta routinely discarded postpartum with a rich source of MSCs, therefore, the ethical and legal consideration with this source is negligible. It does not require embryos destruction or intrusive procedures for MSCs recovery. Furthermore, harvesting of MSCs from the amniotic membrane is without the requirement for invasive procedures like bone marrow or adipose tissue aspiration and is completely safe (Koike et al. 2014; Díaz-Prado et al. 2011; Li et al. 2014).

These cells are immune privileged, thus have no prominent ethical concerns associated with human embryonic stem cells (ESCs) in cell therapy (Kim et al. 2012, 2013a; Paracchini et al. 2012). Indeed, the availability and high-yield in stem cell recovery from amniotic membrane makes this tissue as a truly exciting alternative source, and one that reveals new prospects of increasing the number of clinical application and its widespread use (Alviano et al. 2007). Moreover, this tissue offers an abundant source for bank development (Paracchini et al. 2012; Pianta et al. 2015). However, it is still in the ownership of the mother, therefore the use of human amniotic membrane in every research center must be approved by the ethics committee and written informed consents obtained from the mothers (Koike et al. 2014).

7 Biosafety of AMSCs (Tumor Formation, Genome Stability During In Vitro Culture)

Obtaining stem cell source should be done under sterile circumstances to avoid possibility of microbial contamination. Amniotic membrane has been retrieved from women who had cesarean section under sterilized condition in operation room. Therefore, the possibility of fungal, viral, or bacterial contamination in the primary culture of procured samples is lower than some stem cell sources like menstrual blood.

Another criterion to judge about biosafety of stem cell source is genome stability during in vitro culture. The evidence narrates that the AMSCs maintain genome stability and normal karyotype during multiple passages (Seo et al. 2013; Ghosh et al. 2015). Seo et al. indicated that isolated equine AMSCs at passage 5 have a normal karyotype with 64 chromosomes (Seo et al. 2013). Moreover, chromosomal spreads at passage 5 and passage 10 of buffalo AMSCs revealed normal number of chromosomes ($2n=50$) and no apparent changes in ploidy during culture (Ghosh et al. 2015). In another study, the karyotype of human AMSCs was analyzed at different culture passages (4 and 20) to monitor the occurrence of spontaneous chromosomal alterations during expansion. They indicated the chromosome number of human AMSCs was 46 and the G band karyotype analysis was normal type (Kang et al. 2012b; Tamagawa et al. 2004).

Tumorigenicity can be a main concern and a key obstacle in using stem cells as therapeutic purposes. Among various types of stem cells, the tumorigenicity is closely linked with the pluripotent stem cells like ESCs. Therefore, ESCs have a major hurdle to overcome before clinical translation, such their great potential for regenerative medicine along with the ethical concerns (Lee et al. 2009).

While some investigators have regarded AMSCs like pluripotent stem cells, there is no report about tumor induction of AMSCs. Amniotic membrane stem cells have never shown signs of aging and tumorigenicity even after propagation for more than 2 years in culture (Walther et al. 2009). Cat AMSCs at passage 4 were administered by intramuscular, subcutaneous, and intraperitoneal injection in immunodeficient (BALB/c-Nu) mice. No teratomas were observed by 45 days postinjection, and the transplanted animals survived without difficulty (Vidane et al. 2009). Moreover, in vivo teratoma formation and tumorigenicity of AMSCs have not been reported posttransplantation into human volunteers (Toda et al. 2007a). The low risk of tumor formation of these cells could be attributed to limited lifespan and no expression of telomerase. Some others have attributed these properties to DNA damage or shortened telomerase (Bilic et al. 2008b; Teng et al. 2013).

Some studies have demonstrated that AMSCs can inhibit the growth, proliferation, and invasion of tumor cells and promote their apoptosis with the release of the tumor necrosis factor-a (TNF-α), transforming growth factor-b (TGF-b), interleukins (ILs) or interferons (IFNs), granulocyte macrophage colony-stimulating factor (GM-CSF), neurotrophin 3, CCL18 (a chemokine), brain-derived neutrophic factor (BDNF), granulocyte chemotactic protein (GCP-2), and conserved dopamine neutrophic factors (Kang et al. 2012b; Li et al. 2011). In addition, these cells have anti-inflammatory functions and consequently potential antitumor effects. Therefore, their low risk of tumorigenicity and antitumor effects of AMSCs has great advantage for eventual therapeutic applications on account of their safety property (Kang et al. 2012b; Toda et al. 2007b).

8 Transdifferentiation Predisposition of AMSCs into Specialized Germinal Lineages

Already, a cocktail of growth factors, hormones, and/or other additives has been used to stimulate transdifferentiation of AMSCs in vitro. The differentiation potential of AMSCs into different lineages has been evaluated by assessment of morphological changes, expression of various lineage-specific genes, and tissue-specific functions. According to the reports (Table 1), AMSCs have a multilineage differentiation potential into cells derived from the three germinal cells lines (endodermal, mesodermal, and ectodermal lineages). However, there are inconsistent data about the levels and pattern of genes expression in terminally differentiated cells derived from AMSCs. This could be attributed to differences in procedures and stimulating factors in the induction media, utilization of extracellular matrices in culture condition, and/or donor age (Manuelpillai et al. 2011).

The first report about transdifferentiation ability of AMSCs was presented by In 't Anker et al. in 2004 that generated osteogenic and adipogenic lineages from these cells (In 't Anker et al. 2004). Later, other researchers confirmed the osteogenic and adipogenic differentiation capacity of AMSCs (Lindenmair et al. 2010; Chen et al. 2011; Kim et al. 2007; Tsai et al. 2007). In 2010, Lindenmair et al., by relying on osteogenic differentiation capability of AMSCs presented an interesting protocol substitute to current bone tissue engineering protocols. They differentiated AMSCs that were trapped in amniotic membrane matrix into the osteogenic lineage without prior stem cell isolation. Direct differentiation of trapped stem cells in amniotic membrane drawbacks the limitations of in vitro isolation and expansion of AMSCs for clinical use (Lindenmair et al. 2010). In other study, AMSCs were loaded in porcine gelatin microcarriers and cultured in osteogenic medium for bone tissue engineering purposes. The generated bone-like tissues were used as building blocks to fabricate a macroscopic bone construct in a cylindrical perfusion culture chamber. As a result a short-term perfusion culture (7 days) could effectively generate a macroscopic bone construct with high cell viability and uniform distribution of bone characteristic extracellular matrix (ECM) (Chen et al. 2011).

Table 1 Germinal lineages differentiated from amniotic membrane mesenchymal stem cells in vitro

Germinal lineage	Growth factors	Matrix	Results	Refs.
Osteogenic lineage	Dexamethasone, sodium-β-glycerophosphate, L-ascorbic acid-2-phosphate, mercaptoethanol	CultiSphermicrocarriers (macroporous porcine gelatin microcarriers)	Development of bone-characteristic ECM, significant ALP activity and great mineralization	Chen et al. 2011
	Osteogenic stimulatory kit (Stem Cell Technologies, Cologne, Germany) or osteogenic medium containing dexamethasone, 1,25-dihydroxy-vitamin D3, β-glycerophosphate	Amniotic membrane	Increase of calcium contents, ALP activity and expression levels of selected genes involved in osteogenesis such as RUNX2, alkaline phosphatase, BMP-2, BMP-4	Lindenmair et al. 2010
	Dexamethasone, β-glycerol phosphate, ascorbic acid-phosphate	Cell culture plate	Mineralization of accumulated calcium shown by von Kossa staining	Kim et al. 2007
	Bullekit Osteogenic differentiation medium	One-well chamber slide	High calcium deposition	Díaz-Prado et al. 2010
	Bullekit Osteogenic differentiation medium	One-well chamber slide	High percentage of alizarin red-positive-stained cells	Díaz-Prado et al. 2011
	Bullekit Osteogenic differentiation medium (Cambrex) or in NH OsteoDiff Medium (MiltenyiBiotec)	Glass chamber slides	Calcium deposition shown by positive staining for alizarin red, upregulation of osteopontin	Soncini et al. 2007
Chondrogenic lineage	Sodium pyruvate, ITS + 1, dexamethasone, L-ascorbic acid-2-phosphate, TGF-β3	Polypropylene conical tubes	The appearance of abundant extracellular matrix and type II collagen after 3 weeks'	Alviano et al. 2007
	Dexamethasone, L-ascorbic acid, sodium pyruvate, ITS + 1, TGF-β3	Six-well plate	Formation of collagen fibers and GAGs revealed by Alcian blue staining	Xu et al. 2014
	L-ascorbic acid, MTG, knockout serum, dexamethasone, transferrin, RA, TGF-β	Micropellet	Detection of collagen II and proteoglycans throughout the newly formed matrix, weak immunodetection for aggrecan and collagen I	Díaz-Prado et al. 2010
	L-ascorbic acid, dexamethasone, transferrin, RA, TGF-β3	Micropellet	Detection of collagen II and proteoglycans throughout the newly formed matrix, weak immunodetection for aggrecan	Díaz-Prado et al. 2011
	Dexamethasone, L-ascorbic acid 2-phosphate, sodium pyruvate, proline, ITS, TGF-β1	Glass chamber slides	Positive staining of toluidine blue staining	Soncini et al. 2007
	Insulin, TGF-β1, L-ascorbic acid	Cell culture plate	Positive alcian blue staining of the proteoglycan and detection of type II collagen	Portmann-Lanz et al. 2006

(continued)

Table 1 (continued)

Germinal lineage	Growth factors	Matrix	Results	Refs.
Adipogenic lineage	Dexamethasone, isobutyl-1-methylxanthine, human insulin, indomethacin	Cell culture plate	Presence of intracellular lipid droplets shown by Oil Red O staining	Kim et al. 2007
	Bullekit Adipogenic differentiation medium	One-well chamber slide	AMSCs showed positive staining, with single adipocyticmultivacuolar cells secreting lipid droplets	Díaz-Prado et al. 2010
	Bullekit Adipogenic differentiation medium	One-well chamber slide	Positive staining characterized by single adipocyticmultivacuolar cells secreting lipid droplets	Díaz-Prado et al. 2011
	Bullekit Adipogenic differentiation medium (Cambrex)	Glass chamber slides	Positive staining for Oil Red O staining	Soncini et al. 2007
Endothelial lineage	VEGF and 1 % hypoxic	Type I collagen-coated dish	The gene expression of endothelial markers such as KDR, VCAM, Flt-1and vWF and positive immunocytochemical staining for KDR and VE-cadherin proteins was observed after 2 week of culture with VEGF under hypoxic conditions as compared with that in the control condition	Maruyama et al. 2013
	VEGF	Gel matrix	AMSCs spontaneously form capillary-like structures when cultured in semisolid medium with VEGF. Also the endothelial commitment of the cells was demonstrated by the expression of vWF	Alviano et al. 2007
Myogenic lineage	MCDB-201, ITS + 1, dexamethasone, L-ascorbic acid-2-phosphate, bFGF, VEGF, IGF-1	Chamber slides	Gene expression of MyoD, myogenin and desmin as a skeletal protein by RT-PCR and immunohistochemical methods	Alviano et al. 2007
	Horse serum, hydrocortisone, dexamethasone	Cell culture plate	The development of long multinucleated cells forming precursors of myotubes, positive expression of the transcription factor MyoD1 and skeletal muscle myosin heavy chain	Portmann-Lanz et al. 2006

Neurogenic lineage	Dexamethasone, linoleic acid, PDGF, bFGF	Cell culture plate	Positive immunohistochemical staining for neurofilament M	Sun et al. 2014
	DMSO, butylated hydroxyanisole, KCl, valproic acid and hydrocortisone in N2 medium plus 1 × N2 supplement	Cell culture plate	Immunoreactivity against the neuronal marker, Neu N, and the oligodendrocyte marker, Gal C	Kim et al. 2007
	DMSO, BHA	Six-well plate	Up regulation of NSE and GFAP	Li et al. 2012
	Nerve induction media	Cover glasses coated with membranous matrix in 24-well plates	Morphological changes, increase of NSE and SYN expression	Hu et al. 2013
	IBMX, insulin	Cell culture plate	The upregulation of Tuj1, MAP-2 (both neuron markers), NeuN and GFAP genes, positive for the dopamine neuron marker tyrosine hydroxylase	Chang et al. 2010
	CIM containing bFGF, NGF and RA	Cell culture plate	Neuron-like morphological characteristics, demonstrated by sharp, elongated, tri-, or multipolar cells, upregulation of NSE gene, no expression of nestin in end stage differentiated cells	Zeng et al. 2011
	All-trans RA	Cell culture plate	Generation of neuron-like structures, positive staining for CD133, Nestin and NF200	Portmann-Lanz et al. 2006
	EGF, bFGF, N2, butylatedhydroxyanisole, IBMX, all-trans RA	Cell culture plate	The formation of retracted bodies, long processes and neuron network-like structures, upregulation of neuron-specific genes such as NSE, NF-M, β-tubulin isotype III, (TUJ1), and GFAP	Tamagawa et al. 2008
	BHA, 10 M forskolin, 2% DMSO, 5 U/mL heparin, 5 nM K252a, 25 mMKCl, 2 mM valproic acid, 1- N2 supplement, 10 ng/mL bFGF, and 10 ng/mL (PDGF-BB)	Cell culture plate	Morphologically change to bipolar or multipolar cells with branched process, upregulation of nestin, Musashi1, Tuj1 and NF-M and also GFAP	Li et al. 2008b

(continued)

Table 1 (continued)

Germinal lineage	Growth factors	Matrix	Results	Refs.
Hepatogenic lineage	Beta-mercaptoethanol, EGF, dexamethaone	Type 1 collagen-coated culture dishes	Positive expression of CK7, albumin, CK19 and weakly alpha1-antitrypsin, no alpha-fetoprotein expression	Paracchini et al. 2012
	HGF, bFGF, oncostatin M, dexamethasone	Type I collagen-coated plastic dishes	Upregulation of albumin, α-FP, CK18, α1 antitrypsin and HNF-4α	Tamagawa et al. 2007
Cardiomyogenic lineage	Cardiac lysate from porcine hearts	3D culture with Fibrin scaffolds	The enhanced fluorescence intensity of cTnT and presence of cTnI mRNA shown by in situ hybridization, the upregulation of cTnT, CX43, β-MHC, Mef2c, GATA-4 and HCN2 judged by immunocytochemistry, RT-PCR and western blot analysis, higher levels of these markers in cells differentiated by cardiac lysate than those of cells maintained in standard adherent cultures or those in 3D cultures without cardiac lysate	Lin et al. 2013
	Activin, bFGF or Heart explants for coculture	Plastic dishes	The upregulation of Nkx2.5, ANP and α-MHC, positive immunostaining for Nkx2.5 and β-MHC	Zhao et al. 2005

AMSCs amniotic membrane mesenchymal stem cells, *ECM* extracellular matrix, *ALP* alkaline phosphatase, *BMP-2* Bone morphogenetic proteins-2, *MTG* Monotioglycerol, *TGF-b* transforming growth factor-b, *RA* Retinoic Acid, *GAG* glycosaminoglycans, *VEGF* vascular endothelial growth factor, *VCAM* vascular cell adhesion molecule, *vWF* von Willebrand Factor, *bFGF* basic fibroblast growth factor, *IGF-1* Insulin-like growth factor-1, *PDGF* platelet-derived growth factor, *NSE* neuron-specific enolase, *GFAP* glial fibrillary acidic protein, *BHA* butylatedhydroxyanisole, *DMSO* Dimethyl sulfoxide, *SYN* synaptophysin, *IBMX* dibutyryl cyclic AMP, 3-isobutyl-1-methyl-xanthine, *MAP-2* microtubule-associated protein-2, *CIM* Cytokine induction medium, *NGF* Nerve growth factor, *EGF* Epidermal growth factor, *NF-M* neurofilament-medium, *α-FP* α-fetoprotein, *HNF* hepatocyte nuclear factor

Chondrogenic differentiation of AMSCs was carried out in different culture conditions such as micro pellet or two dimensional (2D) culture in presence of different growth factors and cytokines and demonstrated by the presence of glycoseaminoglycans and collagen in the extracellular matrix using histological, molecular, and biochemical assays (Soncini et al. 2007; Díaz-Prado et al. 2010). While the best strategies to utilization of stem cells for cartilage regeneration are incorporation of the cells into suitable three-dimensional (3D) scaffolds, cartilage tissue engineering using AMSCs and 3D matrix such as fibrous structures, porous sponges, woven or nonwoven meshes, and hydrogels has been remained to work.

AMSCs are also able to differentiate toward the skeletal myogenic lineage under physiological culture conditions without the addition of demethylating drugs (Alviano et al. 2007). This ability has been shown by expression of myogenic transcription factors such as Myo D and Myogenin and the protein expression of desmin (Portmann-Lanz et al. 2006). Alviano et al. (2007) confirmed these results and also was the first to demonstrate the angiogenic differentiation potential of these cells (Alviano et al. 2007). This latter study revealed that human AMSCs after culture in induction media with vascular endothelial growth factor (VEGF) expressed endothelial-specific markers such as the receptors of the VEGF 1 and 2 (FLT-1, KDR), intercellular adhesion molecule 1 (ICAM-1), as well as the appearance of CD34 and von Willebrand Factor (vWF) positive cells. It is notable that AMSCs spontaneously form capillary-like structures when cultured in semisolid medium (Matrigel system), however, this characteristic will be improved by exposure to angiogenic factor VEGF (Alviano et al. 2007). In 2013, Maruyama et al. demonstrated culture under hypoxic condition reinforces the endothelial differentiation of AMSCs through VEGF. They indicated upregulation of endothelial genes such as KDR, vascular cell adhesion molecule (VCAM), FLT-1, and vWF in differentiated cells by VEGF under hypoxic conditions as compared to the cells induced with VEGF under normoxic (Pirjali et al. 2013).

One main target of researches on AMSCs is future utilization of these cells for curing of cardiac diseases. Therefore, there are some reports about differentiation potential of these cells into cardiomyocytes. Some growth factors like basic fibroblast growth factor (bFGF) or activin A are required to stimulate differentiation of AMSCs into cardiomyocytes. The human AMSCs without stimulation did not express cardiac markers such as Nkx2.5, ANP, and or α-MHC, while AMSCs stimulation with bFGF or activin A led to expression of these markers (Zhao et al. 2005). The efficiency of 3D culture system in development of AMSCs into cardiomyocytes has been shown by higher expression levels of cardiac markers in differentiated cells in 3D culture compared to those in 2D culture. In this line, Lin et al. developed fibrin scaffold for differentiation of AMSCs into cardiomyocyte-like cells in the presence of porcine cardiac lysate. The results of AMSCs culture in the presence of porcine cardiac lysate indicated a significantly higher expression of cardiac markers in these cells compared to AMSCs culture without cardiac lysate. Specifically, AMSCs cultured in 3D supplemented medium with cardiac lysates displayed more upregulation levels of Mef2c, GATA-4, and HCN-2 mRNA as compared with cells grown under adherent conditions or cells grown in 3D culture with bovine serum (Lin et al. 2013).

A tendency of AMSCs to neuronal differentiation has been established by the observation that these cells express neuronal and glial markers such as Nestin (NES), Musashi1, Tuj1, NF-M, glial fibrillary acidic protein (GFAP), Neural cell adhesion molecule (NCAM), neuron-specific enolase, neurofilament medium, microtubule-associated protein (MAP)-2, Neu-N. The expression of some of these markers enhanced after culture of AMSCs in specific neural-induction media (Parolini et al. 2009). One group investigated differentiation of human AMSCs into motor neuron precursor cells by combination of ECM and multicell factors in a multistep induction process. After stimulation, the cells connected with adjacent cells to form oriented net. Besides the morphological changes, expression levels of neuron-specific enolase (NSE) and synaptophysin (SYN) increased and GFAP expression decreased. In the group without ECM, NSE expression increased, while the expression of Nestin and SYN did not change implying that ECM enhanced the effectiveness of used growth factors and cytokines (Hu et al. 2013). Other researchers confirmed differentiation ability of AMSCs into neuronal-like cells, which was identified by neuronal-specific markers and their ability to secrete dopamine. Based on their results, the developed neuronal-like cells expressed Tuj1, MAP-2 (both neuron markers), NeuN (neuron nuclear marker), and GFAP (astroglial marker). These cells were also positive for the dopamine neuron marker tyrosine hydroxylase (TH). Moreover compared to undifferentiated cells, nestin expression was downregulated, and the expression of Tuj1, MAP-2, and GFAP was upregulated. So far, there is no knowledge about comparative differentiation potential of these cells with other stem cells; however, it seems that AMSCs could be a great candidate for future cell therapy of neuronal degenerative disorders (Chang et al. 2010).

AMSCs could also differentiate into endoderm-derived cells, such as hepatocytes. In this regard, Tamagawa et al. showed that AMSCs culture on type I collagen-coated dishes in differentiation medium causes significant expression of hepatic markers including albumin, CK18, α1 antitrypsin, and HNF 4α. Furthermore, the storage of glycogen was clearly detected in the cells following the induction of hepatocyte differentiation (Tamagawa et al. 2007).

These studies suggested that AMSCs possess some characteristics that introduce them as a proper cell population for cell therapy, although molecular mechanisms governing on transdifferentiation ability of these cells are remained to be clarified. Nonetheless, the multilineage differentiation potential is not exclusive to AMSCs, such that MSCs derived from other sources like bone marrow (Chivu et al. 2009), umbilical cord blood (Sanberg et al. 2011), adipose tissue (Choi et al. 2010), and menstrual blood (Khanjani et al. 2015) exhibit this potency in appropriate inductive condition. While the differences in transdifferentiation capability of other newfound stem cell sources like menstrual blood in reference to bone marrow have been well established (Khanjani et al. 2014; Darzi et al. 2012; Rahimi et al. 2014), comparative studies about transdifferentiation ability of AMSCs in comparison with well-known MSCs such as bone marrow have been kept to settle.

9 Immunomodulatory Features of AMSCs

One hallmark of the cells derived from amniotic membranes is their ability to engraft different tissues following allogeneic (Marcus et al. 2008) and xenogeneic (Bailo et al. 2004) transplantation indicating the presence of potent in situ immunomodulatory properties and low immunogenicity of these cell populations. It has been shown that transplantation of human amnion-derived cells in neonatal swine and rats resulted in cell microchimerism in various organs and tissues without inducing lymphocyte proliferation responses (Bailo et al. 2004). In line with preclinical settings, there are plenty of reports taking advantages of immunomodulatory properties of amniotic-derived cells in clinical practice. It is for several years that human amniotic membrane is successfully utilized as an immune tolorable biomaterial for surgical dressing, burn treatment (Singh and Chacharkar 2011; Ghieh et al. 2015), and tissue grafting in surface ocular diseases (Palamar et al. 2014).

Mechanisms involved in maternal tolerance toward semiallogeneic fetus have been the center of many researches during the past half century (Erlebacher 2013; Jeddi-Tehrani et al. 2009; Zarnani et al. 2008; Shojaeian et al. 2007) and mesenchymal cells resident within the fetal membranes may be considered as functionally effective immunomodulatory cells for establishment of immune tolerance at the feto-maternal interface. Such assumption is supported by high expression levels of HLA-G, a molecule actively involved in immune suppression of local T and natural killer (NK) cell populations (Lynge Nilsson et al. 2014), experiments showing active suppression of T cell proliferation by cells isolated from amniotic and chorionic membranes (Wolbank et al. 2007) and also long-term survival of engrafted allogeneic and xenogeneic amniotic membrane cells (Khanjani et al. 2014).

Such observations have been the basis for elucidation of different immunomodulatory aspects of AMSCs with this hope that these findings could dramatically expand their therapeutic potential clinical applications and in this context those immunomodulatory properties directly relevant to the mechanisms of graft immunologic tolerance have been the main focus of recent researches (Insausti et al. 2014). Although there is a considerable body of evidence on immunomodulatory effects of either placental or amniotic fluid-derived stem cells, data on immunomodulation by MSCs directly derived from mesenchymal layer of the amniotic membrane is relatively scarce. In the following part, we will try to have a general overview on immunomodulatory effects of AMSCs on components of adaptive and innate immune system.

One important feature of AMSCs directly relevant to their immunomodulatory properties is their immunophenotype explained earlier. Besides expression of surface markers of mesenchymal origin (Parolini et al. 2008; In 't Anker et al. 2004; Marongiu et al. 2010; Manuelpillai et al. 2011; Soncini et al. 2007; Roubelakis et al. 2012b), a feature that these cells share with human amniotic epithelial stem cells (Parolini et al. 2008; Tabatabaei et al. 2014), AMSCs lack MHC-II and costimulatory molecules CD80, CD86, CD40, and CD40 ligand pointing to the inability of these cells to act as antigen-presenting cells (Parolini et al. 2008; Wu et al. 2014). Moreover, AMSCs express very low levels of MHC-I and high levels of HLA-G indicating their tolerogenic property and low propensity to induce allogeneic immune responses (Parolini et al. 2008; Ge et al. 2012). HLA-G is mainly recognized by T lymphocytes, NK cells, and abolish their responsiveness to activating signals (Shojaeian et al. 2007; Favier et al. 2010). Indeed, interaction of this molecule with its cognate receptors on dendritic cells leads to induction of tolerogenic phenotype in this cells type suggesting that this might be a key mechanism through which tolerance to allogeneic AMSCs is established. More importantly, the level of HLA-G expression in AMSCs is increased following exposure to IFN-γ treatment (Chang et al. 2006). If such finding is found to be the case in vivo, it may be considered as an added benefit when expression of HLA-G in transplanted AMSCs is increased following homing of these cells to the damaged tissue of interest, where high concentration of proinflammatory cytokines exists. More importantly, AMSCs express ligands for programmed cell death receptor, PD-L1 and PD-L2 after IFN-γ stimulation (Kronsteiner et al. 2011; Kang et al. 2012c). This phenomenon could potentially result in inhibition of activated T cells which are positive PD1 (Carter et al. 2002; Parry et al. 2005). The immunophenotypic features of AMSCs presented earlier clearly imply that this cell type might exert negative regulatory action on different subsets of the immune cells.

In a comprehensive study by Pianta et al., it was demonstrated that conditioned medium (CM) of human AMSCs markedly suppressed proliferation of effector/memory CD4+ and CD8+ T cells stimulated either with allogeneic mixed lymphocyte reaction (MLR) or by T cell receptor (TCR) stimulation using anti-CD3/CD28 (Pianta et al. 2015). These findings reinforce on the concept that soluble factors released by these cells could potentially suppress T cell proliferation (Kronsteiner et al. 2011). Importantly, no effect on proliferation of naïve T cells was observed (Pianta et al. 2015), a finding which is in sharp contrast with antiproliferative effects of bone marrow (BM)-derived MSCs (Krampera et al. 2003). Inhibitory effect of unfractionated AMSCs and their CM on proliferation of TCR- or allogeneic MLR-stimulated peripheral blood mononuclear cells has also been reported earlier (Magatti et al. 2008). Such inhibitory effect is exerted in both cell contact-dependent and independent manner (Magatti et al. 2008; Kronsteiner et al. 2011). Such antiproliferative capacity is not restricted to AMSCs as MSCs derived from menstrual blood (Nikoo et al. 2012), bone marrow (Krampera et al. 2003), placenta (Chang et al. 2006; Li et al. 2007), adipose (Wolbank et al. 2007), and fetal tissues (Weiss et al. 2008) have been shown to possess the same inhibitory function. Notably and in contrast to BMSCs which attain antiproliferative ability after stimulation with activating substances such as IL-1β, TNF-α, or IFN-γ, antiproliferative effect of AMSCs has an intrinsic nature without need for prior stimulation with aforesaid activating mediators (Kamadjaja et al. 2014). Such difference may point to the heterogeneity of MSCs with respect to their functionality. Nonetheless, it has been observed that pretreatment of AMSCs with IFN-γ could enhance their antiproliferative effects on stimulated Peripheral Blood Mononuclear Cells (PBMCs) and T cells in conjunction with upregulation of inhibitory costimulatory molecules PD-L1 and PD-L2 (Kronsteiner et al. 2011).

In an attempt to delineate the factors and mechanisms responsible for the immunoregulatory activities of human AMSCs, Rossi et al. recently showed that nonproteinaceous small compounds with chemical–physical nature of prostaglandins are mainly responsible for antiproliferative property of AMSCs, since protease treatment of CM of AMSCs did not significantly reduce the antiproliferative capacity of these cells on CD3-stimulated T cells (Rossi et al. 2012b). Indeed, these small molecules were involved neither directly nor indirectly in the production of other inhibitory molecules such as indoleamine 2,3-dioxygenase (IDO) and nitric oxide (NO) synthase (Rossi et al. 2012b). Nonetheless, upregulation of IDO and NO in AMSCs following coculture with stimulated PBMCs is a matter of debate as it was also reported that production of hepatocyte growth factor (HGF), TGF-b, prostaglandin E2 (PGE2), and IDO increased significantly in human AMSCs cocultured with PBMCs (Kang et al. 2012c). Such discrepancy may in part be attributed to the different stimulation strategy employed in these two reports (Kronsteiner et al. 2011).

Besides inhibitory activity on T cell proliferation, AMSCs were also found to skew development of T cell subsets and related cytokine profile. It has been recently reported that CM of human AMSCs significantly reduced the expression of TH1 (T-bet+CD119+) and TH17 (RORγt+CD161+) markers, while had no effect on TH2 cells (GATA3+CD193+/GATA3+CD294+cells). Such inhibitory effect was concomitant with a significant decrease of TH17 (IL-17A, IL-22) and TH1-related (TNFα,γ, IL-1β) (Pianta et al. 2015) Surprisingly, they also found that such inhibitory action on cytokine production by T cells is not restricted to TH1 or TH17 cells and cytokines related to the TH2 (IL-5, IL-6) and TH9 (IL-9) are also negatively affected by culture supernatant of human AMSCs. Nonetheless, it increased production of IL-10 and IL-13 (Pianta et al. 2015). An increase in production of IL-10 in secondary MLR by fetal MSCs (derived from either amniotic membrane or amniotic fluid) has been also reported by Rolelen et al., where they observed that inhibition of MLR response is restored by anti-IL-10 indicating that inhibitory effect of AMSCs is in part mediated through IL-10 (Roelen et al. 2009). By contrast, it was reported that coculturing PBMCs and IFN-γ-treated human AMSCs caused downregulation of IL-13 and IL-10 (Kronsteiner et al. 2011). Whether or not such discrepancy could be attributed to the differential action of AMSCs in cell–cell contact-dependent or -independent conditions and also whether pretreatment with IFN-γ is a prerequisite for induction of IL-10 and IL-13 by AMSCs need to be elucidated.

Regulatory T cells (Tregs) have a potent immunomodulatory activity and maintain immunological tolerance to self-antigens and transplanted tissues (Josefowicz et al. 2012). One hallmark of AMSCs is their ability to induce Tregs. In line with this notion, it was observed that CM from human AMSCs markedly induced Tregs as shown by proliferation of CD25+FOXP3+ cells in the CD4+ population. Indeed, induction of Tregs was documented by increased secretion of TGF-b and IL-13 in the coculture of allogeneic-activated T cells with CM-human AMSCs (Pianta et al. 2015). Similar effect has also been attributed to placental-derived MSCs showing that these cells are able to threefold increase in the proportion of Tregs in PHA-stimulated T cells (Chang et al. 2006).

Besides fundamental immunoregulatory effects that AMSCs exert on adaptive arm of immune system, these cells also have profound regulatory activity on innate immune system and in this context dendritic cells (DCs) as the most important immune cells linking the innate and adaptive immune systems have drawn much attention due to their central regulatory role in induction of immunity or tolerance (Albert et al. 2001). The latter is basically influenced by maturation and activation state of DCs and also the nature of microenvironment (Lutz and Schuler 2002; Steinman et al. 2003). Therefore, factors controlling DCs maturation state and/or cytokine microenvironment affect the outcome of a graft after transplantation. In a study by Magatti et al., it was reported that mesenchymal cells derived from amniotic membrane cause G0 arrest and inhibit differentiation and maturation of peripheral blood monocytes as judged by downregulation of CD1a, MHC-II, CD83, and CD80. Such effect resulted in diminished allostimulatory potential of these cells on allogeneic T cells and impaired production of such proinflammatory cytokines and chemokines and TNF-α, CXCL10, CXCL9, and CCL5 (Magatti et al. 2009). Interestingly, AMSCs predominantly exerted their inhibitory effect in the course of monocyte to immature DCs differentiation, while had only minimal effect of transition of immature DCs to mature DCs (Magatti et al. 2009). Such differential effect has also been reported for MSCs derived from menstrual blood (Bozorgmehr et al. 2014). Obviously, AMSCs share this impairing activity on DCs generation and maturation with MSCs originated from other sources including menstrual blood (Bozorgmehr et al. 2014), bone marrow (Beyth et al. 2005; Chen et al. 2007; Ramasamy et al. 2007; Li et al. 2008a), fetal lung and bone morrow (Nauta et al. 2006), and amniotic epithelial cells (Banas et al. 2014). AMSCs were also found to produce high levels of TH2-related cytokines, CCL2, CXCL8, and IL-6, the latter is known to inhibit differentiation of monocytes to DCs (Menetrier-Caux et al. 1998). One other mechanism through which AMSCs induce generation of tolerogenic DCs stems from their high expression levels of HLA-G which is a well-established pathway known to arrest maturation and activation of DCs (Ristich et al. 2005). The tolerized DCs could in turn induce Tregs leading to control of ongoing immune responses.

NK cells, as effector cells of innate immune system responsible for cytolysis of virally infected or cancer-modified cells have also been studied for their being affected by modulatory action of AMSCs. In terms of transplantation, NK cells are involved in both graft tolerance and rejection depending on their activation status (Murphy et al. 2011). Such opposing effects are mainly dictated by fine tuning of cell surface receptors signaling in an either activatory or inhibitory manner (Vivier et al. 2008). It was reported that AMSCs could significantly inhibit NK-mediated cytolysis of K562 cells through downregulation of NKp30, NKp44, NKp46, NKG2D, and CD69, which are involved in the NK cytotoxicity. In parallel coculture of AMSCs with NK cells resulted in an increased expression levels of IL-10 and PGE2 (Li et al. 2015), two known mediators responsible for immunoregulatory function of MSCs derived from other sources (Kyurkchiev et al. 2014). Interestingly, the same authors showed that reduced NK cytotoxicity following interaction with AMSCs is reversible implying that decreased NK cytotoxicity is not due to damage of NK cells treated with AMSCs (Li et al. 2015). These results are in accordance with those published earlier showing that BMSCs inhibit cytokine-induced proliferation and effectors function of freshly isolated NK through inhibitory activity of IDO and PGE2 and downregulation of activating signals (Spaggiari et al. 2008).

Although there are a limited number of publications on immunomodulatory effects of AMSCs, the published data confirm the concept that this cell type as with MSCs from other sources exhibit strong immunomodulation on different arms of immune system. Taken together, the present data suggest that AMSCs exert immunomodulatory effects through inhibition of T cell proliferation, skewing cytokine profile toward TH2 immunity, inhibition of generation, and maturation of DCs and NK cells. Such features along with their low immunogenic profile make AMSCs an attractive source for cell transplantation approaches, controlling graft rejection and modulation of inflammatory processes in autoimmune diseases.

10 Regenerative Properties of AMSCs: Preclinical Studies

Regenerative medicine is a novel field based on the cell therapy approach to generate biological substitutes and improve tissue functions (Toda et al. 2007a). As we described here, AMSCs are an attractive and accessible stem cell source with the immunomodulatory characteristics, anti-inflammatory and nontumorigenicity effect, thus introduce them as suitable candidate for cell therapy and regenerative medicine application (Maruyama et al. 2013; Resca et al. 2015). For cell administration into the patient, it is required that clinically high cell numbers with appropriate quality be achieved within a very limited time (Pirjali et al. 2013). This circumstance will be fulfilled via the prolonged capacity for proliferation of AMSCs in vitro. Preclinical studies in small animal models have provoked great promise to restore functions of injured tissues in spinal cord injury, lung and liver fibrosis, severe colitis, cerebral ischemia, diabetes, and myocardial infarction posttransplantation of AMSCs. The most studies on therapeutic potential of AMSCs were carried out in treatment of neurological disorders (Table 2). In particular, stroke has been a major target disease for testing the efficiency of transplantation of AMSCs. In a study, EGFP (enhanced green fluorescence protein)-labeled and BDNF overexpressing human AMSCs were transplanted into the ischemic rat brain after middle cerebral artery occlusion. It was observed that the graft integrated and migrated in the rat brain and could express MAP-2 protein after transplantation. In addition, the graft significantly decreases behavioral dysfunction and infarct size (Tao et al. 2012). Significant recovery in neurological behavior was also detected in the focal cerebral ischemia model treated with BrdU-labeled AMSCs transplantation compared with the controls (Li et al. 2012).

In another study, effects of intravenous administration of human AMSCs into a SOD1 mouse G93A models for treatment of Amyotrophic lateral sclerosis (ALS) between 12 and 16 weeks were investigated. Human AMSCs transplantation through systemic delivery can ameliorate the phenotype and prolong the lifespan of

Table 2 Studies on regeneration ability of AMSCs in treatment of different disorders in animal model

Disorder type	Disease model	Species	Condition of transplanted cells	Detection time	Results	Refs.
Neurological	MCAO	Rat	EGFP-labeled and BDNF overexpressing AMSCs	3 weeks	The transplanted cells expressed the neuronal marker MAP-2 *and the neuronal progenitor marker* Nestin. In addition the graft significantly decreases behavioral dysfunction and infarct size	Tao et al. 2012
	MCAO	Rat	BrdU-labeled AMSCs induced with BHA	8 weeks	Significant recovery in neurological behavior symptoms was detected in the MCAO model treated with cells transplantation compared with the control group that received PBS	Li et al. 2012
	ALS	hSOD1^{G93A} mouse	–	16 weeks	These cells delayed disease progression, extended survival and improved motor behavior. Also transplanted cells prevented motor neuron loss and decreased neuroinflammation	Sun et al. 2014
	T11 spinal cord clip compressive injury	Rat	GFP-labeled AMSCs	8 weeks	The combination of methylprednisolone and AMSCs could inhibit effectively the apoptosis of both the endogenous cells and the grafted cells	Gao et al. 2014
	Sciatic nerve injury	Mice	DiI dye-labeled AMSCs	4 weeks	The higher expression of angiogenic factors was detected in AMSCs injected into the mice sciatic nerve injury. Also AMSCs increased blood perfusion and vascularization of nerves	Li et al. 2014

Cardiological	Hindlimb ischemia	Mice	DiI dye-labeled AMSCs	8 weeks	The mice that received the AMSCs had normal hindlimb appearance with no autoamputation compared to mice that received PBS after severe ischemic damage. Additionally the analysis showed that the cutaneous blood flow and vascular density were increased in the intramuscular injection of cultured AMSCs than PBS	Kim and Choi 2011
	Myocardial infarction	NOD/ SCID mice	DiI dye-labeled AMSCs	4 weeks	AMSCs transplantation through enhanced paracrine and engraftment factors ameliorated left ventricular function, capillary density, angiogenic cytokine levels, Ang-1 and VEGF-A levels in affected tissue	Kim et al. 2013b
	Ischaemic hindlimb	Mice	DiI dye-labeled AMSCs	4 weeks	AMSCs had considerable therapeutic effects on the capillary density and blood perfusion in the ischaemic hindlimb adduct or muscles after transplantation	Kim et al. 2012
	Chronic myocardial ischemia	Porcine	GFP-labeled AMSCs	4 weeks	The results of echocardiography and histological assessment indicated that left ventricular ejection fraction was significantly improved and left ventricular dilatation was well attenuated in animlas after AMSCs transplantation compared to normal saline injection. Moreover percentage of fibrosis tissue in impaired myocardial reduced	Kimura et al. 2012
	Myocardial infarction model	Rat	PKH67-labeled AMSCs were co cultivated with neonatal rat heart explants	2 months	Labeled AMSCs survived in the scar tissue for at least 2 months and differentiated into cardiomyocyte-like cells. The human-specific gene-β-2-microglobulin also can be found in cases	Zhao et al. 2005
	Myocardial infarction model	Rat	EGFP-labeled AMSCs	4 weeks	Transplanted AMSCs survived and transdifferentiated into cardiomyocytes in myocardial infarction, and improved impaired left ventricular fractional shortening measured by echocardiogram and also decreased myocardial fibrosis area significantly	Tsuji et al. 2010

(continued)

Table 2 (continued)

Disorder type	Disease model	Species	Condition of transplanted cells	Detection time	Results	Refs.
Colitis	DSS-induced colitis model	Rat	–	5 days	AMSCs transplantation significantly ameliorated the disease activity index score, weight loss, colon shortening and the histological colitis score. The AMSCs transplantation significantly decreased expression levels of TNF-α, IL-1β and MIF in the rectums of AMSCs-treated rats. In addition, the infiltration of monocytes/macrophages and serum levels of MCP-1 was markedly decreased in treated animals	Onishi et al. 2015
Liver cirrhosis	CCL4-induced liver cirrhosis	Mouse	DiI dye-labeled AMSCs	4 weeks	The transplantation of AMSCs significantly decreased hepatic fibrosis and progression of CCL4-induced cirrhosis, thereby providing a new approach for the treatment of fibrotic liver disease	Zhang et al. 2011
Diabetes	Diabetes	Mice	AMSCs were differentiated into functional pancreatic lineage and encapsulated in PU-PVmacrocapsules	2 weeks	Transplantation of fabricated structures could regulate insulin secretion and gave back glucose level	Kadam et al. 2010
Lung fibrosis	Bleomycin-induced lung fibrosis	Mouse	–	2 weeks	Findings indicated a clear decrease in neutrophil infiltration and a significant reduction in the severity of bleomycin-induced lung fibrosis in mice treated with AMSCs	Cargnoni et al. 2009

MCAO Middle cerebral artery occlusion, *EGFP* enhanced green fluorescence protein, *BDNF* brain derived neurotrophic factor, *AMSCs* amniotic membrane mesenchymal stem cells, *MAP-2* microtubule-associated protein-2, *BHA* Butylatedhydroxyanisole, *SOD1* copper/zinc superoxide dismutase 1, *ALS* Amyotrophic lateral sclerosis, *PBS* phosphate buffered saline, *DiI* 1, 1–dioctadecyl-3, 3, 3′, 3′-tetra-methylindocarbocyanine, *Ang-1* Angiopoetin-1, *VEGF-A* vascular endothelial growth factor, *TNF-α* transforming growth factor-α, *IL-1* interleukin-1, *MIF* macrophage migration inhibitory factor, *MCP-1* monocyte chemoattractant protein-1, *DSS* dextran sulfate sodium, *PU-PV* polyurethane-polyvinyl pyrrolidone

ALS mice. These cells delay disease progression, extend survival, and improve motor behavior. In addition, transplanted cells prevent motor neuron loss and decrease neuroinflammation (Sun et al. 2014).

Spinal cord injury is one of the common destructive injuries that could take profits from stem cell therapy using AMSCs. One study in 2014 was undertaken to assess the combinatory therapy of administering clinical dose of methylprednisolone (MP) and grafting AMSCs after rat spinal cord injury. According to data, AMSCs or MP treatment alone was not effective to inhibit the endogenous cells and the grafted cells apoptosis, while, the numbers of apoptotic cells and the levels of the bax and caspase-3 were typically reduced in the combination therapy group. Therefore, the combination therapy sounds to be a potential strategy for reducing secondary damage and promoting functional recovery following spinal cord injury (Gao et al. 2014).

Furthermore, it has demonstrated that AMSCs transplantation is a promising alternative therapeutic option to treat peripheral neuropathy. In this line, the higher expression of angiogenic factors was detected after AMSCs transplantation into the mice with sciatic nerve injury. In addition, AMSCs increase blood perfusion and vascularization of nerves (Li et al. 2014).

Besides neurological disorders, efficiency of AMSCs in treatment of cardiovascular diseases has been documented. Some studies reported that AMSCs possess high angio-vasulogenic properties. The angiogenic potential of the human AMSCs was first assessed in a mouse hind limb ischemia model. The mice that received the AMSCs had normal hind limb appearance with no autoamputation compared to mice that received phosphate-buffered saline (PBS) after severe ischemic damage. Additionally the analysis showed that the cutaneous blood flow and vascular density were increased in the intramuscular injection of cultured AMSCs compared to PBS group (Kim and Choi 2011).

Furthermore, the effectiveness of AMSCs transplantation in improvement of cardiac function and angiogenesis after myocardial infarction has been indicated. In the study by Kim et al., in 2013, AMSCs were directly transplanted into the border sections of ischemic heart tissue postmyocardial infarction in NOD/SCID mice that resulted to ameliorate left ventricular function, capillary density, angiogenic cytokine levels, angiopoietin (Ang)-1, and VEGF-A levels in affected tissue. They suggested that AMSCs contain chemotactic capabilities that can improve ischemic heart through enhanced paracrine and engraftment factors (Kim et al. 2013b).

In addition, angio-vasculogenic effect of AMSCs has been shown in ischemic hind limbs of mice. It has been suggested that AMSCs posttransplantation spontaneously differentiate into vascular-like structures and express endothelial-specific genes and proteins including Ang-1 and VEGF-A involve in neovascularization (Kim et al. 2012). Likewise, the safety and efficacy of porcine AMSCs transplantation in a swine model of chronic myocardial ischemia was reported by Kimura et al. Based on the results of echocardiography and histology, left ventricular ejection fraction was significantly improved and left ventricular dilatation was well attenuated and also percentage of fibrosis tissue in impaired myocardial reduced in animals after AMSCs administration compared to normal saline injection (Kimura et al. 2012).

Moreover, it has been indicated that AMSCs regenerate heart tissue, improve impaired left ventricular fractional shortening, and decrease myocardial fibrosis area after ischemic injury in rat model. It was suggested that the in vivo niche induces the cardiac differentiation of transplanted PKH26-labeled human AMSCs in scar tissue of infracted heart (Zhao et al. 2005). These results open the possibility that AMSCs could be a highly promising cell population for cell therapy of human ischemic cardiovascular diseases.

There are some reports implying efficiency of AMSCs administration in treatment of different diseases related to digestive tract. AMSCs transplantation significantly ameliorated the disease activity index score, weight loss, colon shortening, and the histological colitis score in rats with severe colitis. Moreover, the AMSCs transplantation significantly decreased expression levels of TNF-α, IL-1β, and macrophage migration inhibitory factor (MIF) in the rectums as well as the infiltration of monocytes/macrophages and serum levels of monocyte chemoattractant protein-1 (MCP-1) (Onishi et al. 2015).

Human AMSCs have also been evaluated as a treatment for liver fibrosis. Zhang et al. have reported that the transplantation of AMSCs significantly decreased hepatic fibrosis and progression of CCL4-induced cirrhosis, thereby providing a new approach for the treatment of fibrotic liver disease (Zhang et al. 2011). Furthermore, AMSCs are able to restore normoglycemic experimental diabetic mice. In one study, AMSCs were differentiated into functional pancreatic lineage and encapsulated in polyurethane–polyvinylpyrrolidone macrocapsules. Transplantation of these structures into diabetic mice could regulate insulin secretion and give back glucose level (Kadam et al. 2010).

In addition, these cells are applicable for the restoration of tissue damage associated with inflammatory and fibrotic degeneration. In this line, a clear decrease in neutrophil infiltration and a significant reduction in the severity of bleomycin-induced lung fibrosis in mice treated with AMSCs were reported (Cargnoni et al. 2009).

Therefore, the current findings support the feasibility of AMSCs in clinical cell therapy of various diseases especially neurological and cardiovascular disorders, and hold promise in application of these cells as an "off-the-shelf" product. While low immunogenicity, ethical restrictions, and pluripotency of AMSCs make these cells suitable for clinical application, already, there are no reports about clinical cell therapy using AMSCs. Maybe to exert beneficial effects of AMSCs, more evaluation in larger animal models especially in comparison with other conventional stem cells such as BMSCs is required. Nevertheless, molecular mechanisms of AMSCs effectiveness in improvement and regeneration of injured tissues have been remained to investigate. However, these cells may produce progenitor cells to generate differentiated cells in vivo. On the other hand, the in vivo regenerative ability of these cells might be attributed to paracrine signaling impact of growth factors, cytokines, and other trophic and anti-inflammatory factors secreted by AMSCs on injured or damaged tissues.

11 Conclusion and Remarks

In this chapter, we have delineated that mesenchymal stem cells isolated from human amniotic membrane have characteristics that would introduce them as a suitable candidate for cell therapy goals. Unlimited availability of fetal membranes and their easy procurement, which are routinely discarded postpartum, allows isolating large number of stem cells from this tissue with short population doubling time. In addition, nonteratogenicity, anti-inflammatory and antifibrotic effects and immunomodulatory properties, less ethical controversy, and great multipotency make AMSCs extremely attractive and useful for stem cells therapy even in allogeneic cellular therapies and open a wide perspective of potential clinical applications (Fig. 3). Although, there is no report about clinical trial of AMSCs administration for treatment of different diseases, the findings of animal studies indicate that human AMSCs can survive in grafts and may produce progenitor cells to generate adult cells. The molecular mechanisms responsible for their regenerative and immunomodulatory effects have been remained to investigate. However, the most probable mechanism may be through the release of cytokines and other growth-promoting factors. Furthermore, to translate the results of animal studies into clinical phase, further long-term researches in large animal models of diverse diseases are required

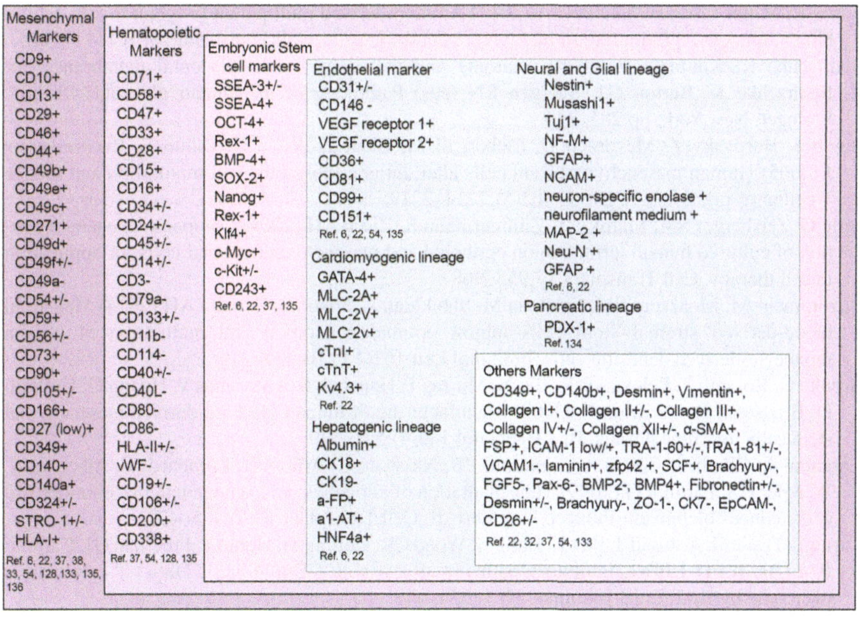

Fig. 3 A comprehensive diagram about past, present, and future works on AMSCs. By having a glance look at this descriptive image, perspective of studies on AMSCs application in the field of regenerative medicine will be assigned

to assure safety of AMSCs administration in clinical step. Along with stem cell therapy, one application of these cells could be tissue engineering that have been scantly proceed. In a comprehensive view, gathered information about characteristics of AMSCs in terms of critical points for cell therapy would allow researchers and clinicians to get better insights for choosing suitable cell source for cell therapy purposes.

Study Finding/Competing Interest The authors indicate no potential conflict of interest.

References

Albert ML, Jegathesan M, Darnell RB (2001) Dendritic cell maturation is required for the cross-tolerization of CD8+ T cells. Nat Immunol 2(11):1010–1017

Alviano F, Fossati V, Marchionni C, Arpinati M, Bonsi L, Franchina M, Lanzoni G, Cantoni S, Cavallini C, Bianchi F, Tazzari PL, Pasquinelli G, Foroni L, Ventura C, Grossi A, Bagnara GP (2007) Term amniotic membrane is a high throughput source for multipotent mesenchymal stem cells with the ability to differentiate into endothelial cells in vitro. BMC Dev Biol 7:11

Bailo M, Soncini M, Vertua E, Signoroni PB, Sanzone S, Lombardi G, Arienti D, Calamani F, Zatti D, Paul P, Albertini A, Zorzi F, Cavagnini A, Candotti F, Wengler GS, Parolini O (2004) Engraftment potential of human amnion and chorion cells derived from term placenta. Transplantation 78(10):1439–1448

Banas R, Miller C, Guzik L, Zeevi A (2014) Amnion-derived multipotent progenitor cells inhibit blood monocyte differentiation into mature dendritic cells. Cell Transplant 23(9):1111–1125

Benirschke K, Kaufman P (2000) Anatomy and pathology of the placental membranes. In: Benirschke K, Burton GJ, Baergen RN (eds) Pathology of the human placenta, 4th edn. Springer, New York, pp 281–334

Beyth S, Borovsky Z, Mevorach D, Liebergall M, Gazit Z, Aslan H, Galun E, Rachmilewitz J (2005) Human mesenchymal stem cells alter antigen-presenting cell maturation and induce T-cell unresponsiveness. Blood 105(5):2214–2219

Bilic G, Zeisberger SM, Mallik AS, Zimmermann R, Zisch AH (2008) Comparative characterization of cultured human term amnion epithelial and mesenchymal stromal cells for application in cell therapy. Cell Transplant 17:955–968

Bozorgmehr M, Moazzeni SM, Salehnia M, Sheikhian A, Nikoo S, Zarnani AH (2014) Menstrual blood-derived stromal stem cells inhibit optimal generation and maturation of human monocyte-derived dendritic cells. Immunol Lett 162(2 Pt B):239–246

Brooke G, Rossetti T, Pelekanos R, Ilic N, Murray P, Hancock S, Antonenas V, Huang G, Gottlieb D, Bradstock K, Atkinson K (2009) Manufacturing of human placenta-derived mesenchymal stem cells for clinical trials. Br J Haematol 144(4):571–579

Cargnoni A, Gibelli L, Tosini A, Signoroni PB, Nassuato C, Arienti D, Lombardi G, Albertini A, Wengler GS, Parolini O (2009) Transplantation of allogeneic and xenogeneic placenta-derived cells reduces bleomycin-induced lung fibrosis. Cell Transplant 18(4):405–422

Carter L, Fouser LA, Jussif J, Fitz L, Deng B, Wood CR, Collins M, Honjo T, Freeman GJ, Carreno BM (2002) PD-1:PD-L inhibitory pathway affects both CD4(+) and CD8(+) T cells and is overcome by IL-2. Eur J Immunol 32(3):634–643

Chang CJ, Yen ML, Chen YC, Chien CC, Huang HI, Bai CH, Yen BL (2006) Placenta-derived multipotent cells exhibit immunosuppressive properties that are enhanced in the presence of interferon-gamma. Stem Cells 24(11):2466–2477

Chang YJ, Hwang SM, Tseng CP, Cheng FC, Huang SH, Hsu LF, Hsu LW, Tsai MS (2010) Isolation of mesenchymal stem cells with neurogenic potential from the mesoderm of the amniotic membrane. Cells Tissues Organs 192(2):93–105

Chen L, Zhang W, Yue H, Han Q, Chen B, Shi M, Li J, Li B, You S, Shi Y, Zhao RC (2007) Effects of human mesenchymal stem cells on the differentiation of dendritic cells from CD34+ cells. Stem Cells Dev 16(5):719–731

Chen M, Wang X, Ye Z, Zhang Y, Zhou Y, Tan WS (2011) A modular approach to the engineering of a centimeter-sized bone tissue construct with human amniotic mesenchymal stem cells-laden microcarriers. Biomaterials 32(30):7532–7542

Chivu M, Dima SO, Stancu CI, Dobrea C, Uscatescu V, Necula LG, Bleotu C, Tanase C, Albulescu R, Ardeleanu C, Popescu I (2009) In vitro hepatic differentiation of human bone marrow mesenchymal stem cells under differential exposure to liver-specific factors. Transl Res 154:122–132

Choi YS, Matsuda K, Dusting GJ, Morrison WA, Dilley RJ (2010) Engineering cardiac tissue in vivo from human adipose-derived stem cells. Biomaterials 31:2236–2242

Darzi S, Zarnani AH, Jeddi-Tehrani M, Entezami K, Mirzadegan E, Akhondi MM, Talebi S, Khanmohammadi M, Kazemnejad S (2012) Osteogenic differentiation of stem cells derived from menstrual blood versus bone marrow in the presence of human platelet releasate. Tissue Eng Part A 18(15-16):1720–1728

Díaz-Prado S, Muiños-López E, Hermida-Gómez T, Rendal-Vázquez ME, Fuentes-Boquete I, de Toro FJ, Blanco FJ (2010) Multilineage differentiation potential of cells isolated from the human amniotic membrane. J Cell Biochem 111:846–857

Díaz-Prado S, Muiños-López E, Hermida-Gómez T, Rendal-Vázquez ME, Fuentes-Boquete I, de Toro FJ, Blanco FJ (2011) Isolation and characterization of mesenchymal stem cells from human amniotic membrane. Tissue Eng Part C Methods 17(1):49–59

Dobreva MP, Pereira PN, Deprest J, Zwijsen A (2010) On the origin of amniotic stem cells: of mice and men. Int J Dev Biol 54(5):761–777

Erlebacher A (2013) Mechanisms of T cell tolerance towards the allogeneic fetus. Nat Rev Immunol 13(1):23–33

Favier B, Lemaoult J, Lesport E, Carosella ED (2010) ILT2/HLA-G interaction impairs NK-cell functions through the inhibition of the late but not the early events of the NK-cell activating synapse. FASEB J 24(3):689–699

Gao S, Ding J, Xiao HJ, Li ZQ, Chen Y, Zhou XS, Wang JE, Wu J, Shi WZ (2014) Anti-inflammatory and anti-apoptotic effect of combined treatment with methylprednisolone and amniotic membrane mesenchymal stem cells after spinal cord injury in rats. Neurochem Res 39(8):1544–1552

Ge X, Wang IN, Toma I, Sebastiano V, Liu J, Butte MJ, Reijo Pera RA, Yang PC (2012) Human amniotic mesenchymal stem cell-derived induced pluripotent stem cells may generate a universal source of cardiac cells. Stem Cells Dev 21(15):2798–2808

Ghieh F, Jurjus R, Ibrahim A, Geagea AG, Daouk H, El Baba B, Chams S, Matar M, Zein W, Jurjus A (2015) The use of stem cells in burn wound healing: a review. Biomed Res Int 2015:684084

Ghosh K, Kumar R, Singh J, Gahlawat SK, Kumar D, Selokar NL, Yadav SP, Gulati BR, Yadav PS (2015) Buffalo (Bubalus bubalis) term amniotic-membrane-derived cells exhibited mesenchymal stem cells characteristics in vitro. In Vitro Cell Dev Biol Anim 51(9):915–921

Gimble JM, Guilak F (2003) Adipose-derived adult stem cells: isolation, characterization and differentiation potential. Cytotherapy 5:362–369

Hu W, Guan FX, Li Y, Tang YJ, Yang F, Yang B (2013) New methods for inducing the differentiation of amniotic-derived mesenchymal stem cells into motor neuron precursor cells. Tissue Cell 45(5):295–305

Ilancheran S, Michalska A, Peh G, Wallace EM, Pera M, Manuelpillai U (2007) Stem cells derived from human fetal membranes display multilineage differentiation potential. Biol Reprod 77:577–588

Ilancheran S, Moodley Y, Manuelpillai U (2009) Human fetal membranes: a source of stem cells for tissue regeneration and repair? Placenta 30:2–10

In 't Anker PS, Scherjon SA, Kleijburg-van der Keur C, de Groot-Swings GM, Claas FH, Fibbe WE, Kanhai HH (2004) Isolation of mesenchymal stem cells of fetal or maternal origin from human placenta. Stem Cells 22(7):1338–1345

Insausti CL, Blanquer M, García-Hernández AM, Castellanos G, Moraleda JM (2014) Amniotic membrane-derived stem cells: immunomodulatory properties and potential clinical application. Stem Cells Cloning 7:53–63

Jeddi-Tehrani M, Abbasi N, Dokouhaki P, Ghasemi J, Rezania S, Ostadkarampour M, Rabbani H, Akhondi MA, Fard ZT, Zarnani AH (2009) Indoleamine 2,3-dioxygenase is expressed in the endometrium of cycling mice throughout the oestrous cycle. J Reprod Immunol 80(1-2):41–48

Josefowicz SZ, Lu LF, Rudensky AY (2012) Regulatory T cells: mechanisms of differentiation and function. Annu Rev Immunol 30:531–564

Kadam SS, Sudhakar M, Nair PD, Bhonde RR (2010) Reversal of experimental diabetes in mice by transplantation of neo-islets generated from human amnion-derived mesenchymal stromal cells using immuno-isolatory macrocapsules. Cytotherapy 12(8):982–991

Kamadjaja DB, Purwati, Rantam FA, Ferdiansyah, Pramono C (2014) The osteogenic capacity of human amniotic membrane mesenchymal stem cell (hAMSC) and potential for application in maxillofacial bone reconstruction in vitro study. J Biomed Sci Eng 7:497–503

Kang JW, Koo HC, Hwang SY, Kang SK, Ra JC, Lee MH, Park YH (2012a) Immunomodulatory effects of human amniotic membrane-derived mesenchymal stem cells. J Vet Sci 13(1):23–31

Kang NH, Hwang KA, Kim SU, Kim YB, Hyun SH, Jeung EB, Choi KC (2012b) Potential antitumor therapeutic strategies of human amniotic membrane and amniotic fluid-derived stem cells. Cancer Gene Ther 19(8):517–522

Kazemnejad S, Akhondi MM, Soleimani M, Zarnani AH, Khanmohammadi M, Darzi S (2012) Characterization and chondrogenic differentiation of menstrual blood-derived stem cells on a nanofibrous scaffold. Int J Artif Organs 35(1):55–66

Kern S, Eichler H, Stoeve J, Kluter H, Bieback K (2006) Comparative analysis of mesenchymal stem cells from bone marrow, umbilical cord blood, or adipose tissue. Stem Cells 24:1294–1301

Khanjani S, Khanmohammadi M, Zarnani AH, Akhondi MM, Ahani A, Ghaempanah Z, Naderi MM, Eghtesad S, Kazemnejad S (2014) Comparative evaluation of differentiation potential of menstrual blood- versus bone marrow-derived stem cells into hepatocyte-like cells. PLoS One 9(2):e86075

Khanjani S, Khanmohammadi M, Zarnani AH, Talebi S, Edalatkhah H, Eghtesad S, Nikokar I, Kazemnejad S (2015) Efficient generation of functional hepatocyte-like cells from menstrual blood-derived stem cells. J Tissue Eng Regen Med 9(11):E124–E134

Khanmohammadi M, Khanjani S, Edalatkhah H, Zarnani AH, Heidari-Vala H, Soleimani M, Alimoghaddam K, Kazemnejad S (2014) Modified protocol for improvement of differentiation potential of menstrual blood-derived stem cells into adipogenic lineage. Cell Prolif 47(6):615–623

Kim HG, Choi OH (2011) Neovascularization in a mouse model via stem cells derived from human fetal amniotic membranes. Heart Vessels 26(2):196–205

Kim J, Kang HM, Kim H, Kim MR, Kwon HC, Gye MC, Kang SG, Yang HS, You J (2007) Ex vivo characteristics of human amniotic membrane-derived stem cells. Cloning Stem Cells 9(4):581–594

Kim SW, Zhang HZ, Kim CE, An HS, Kim JM, Kim MH (2012) Amniotic mesenchymal stem cells have robust angiogenic properties and are effective in treating hindlimb ischaemia. Cardiovasc Res 93(3):525–534

Kim KS, Kim HS, Park JM, Kim HW, Park MK, Lee HS, Lim DS, Lee TH, Chopp M, Moon J (2013a) Long-term immunomodulatory effect of amniotic stem cells in an Alzheimer's disease model. Neurobiol Aging 34(10):2408–2420

Kim SW, Zhang HZ, Kim CE, Kim JM, Kim MH (2013b) Amniotic mesenchymal stem cells with robust chemotactic properties are effective in the treatment of a myocardial infarction model. Int J Cardiol 168(2):1062–1069

Kimura M, Toyoda M, Gojo S, Itakura Y, Kami D, Miyoshi S, Kyo S, Ono M, Umezawa A (2012) Allogeneic amniotic membrane-derived mesenchymal stromal cell transplantation in a porcine model of chronic myocardial ischemia. J Stem Cells Regen Med 8(3):171–180

Kmiecik G, Niklińska W, Kuć P, Pancewicz-Wojtkiewicz J, Fil D, Karwowska A, Karczewski J, Mackiewicz Z (2013) Fetal membranes as a source of stem cells. Adv Med Sci 58(2):185–195

Kögler G, Sensken S, Airey JA, Trapp T, Müschen M, Feldhahn N, Liedtke S, Sorg RV, Fischer J, Rosenbaum C, Greschat S, Knipper A, Bender J, Degistirici O, Gao J, Caplan AI, Colletti EJ, Almeida-Porada G, Müller HW, Zanjani E, Wernet P (2004) A new human somatic stem cell from placental cord blood with intrinsic pluripotent differentiation potential. J Exp Med 200:123–135

Koike C, Zhou K, Takeda Y, Fathy M, Okabe M, Yoshida T, Nakamura Y, Kato Y, Nikaido T (2014) Characterization of amniotic stem cells. Cell Reprogram 16(4):298–305

Krampera M, Glennie S, Dyson J, Scott D, Laylor R, Simpson E, Dazzi F (2003) Bone marrow mesenchymal stem cells inhibit the response of naive and memory antigen-specific T cells to their cognate peptide. Blood 101(9):3722–3729

Kronsteiner B, Wolbank S, Peterbauer A, Hackl C, Redl H, van Griensven M, Gabriel C (2011) Human mesenchymal stem cells from adipose tissue and amnion influence T-cells depending on stimulation method and presence of other immune cells. Stem Cells Dev 20(12):2115–2126

Kubo M, Sonoda Y, Muramatsu R, Usui M (2001) Immunogenicity of human amniotic membrane in experimental xenotransplantation. Invest Ophthalmol Vis Sci 42(7):1539–1546

Kyurkchiev D, Bochev I, Ivanova-Todorova E, Mourdjeva M, Oreshkova T, Belemezova K, Kyurkchiev S (2014) Secretion of immunoregulatory cytokines by mesenchymal stem cells. World J Stem Cells 6(5):552–570

Lee AS, Tang C, Cao F, Xie X, van der Bogt K, Hwang A, Connolly AJ, Robbins RC, Wu JC (2009) Effects of cell number on teratoma formation by human embryonic stem cells. Cell Cycle 8:2608–2612

Li C, Zhang W, Jiang X, Mao N (2007) Human-placenta-derived mesenchymal stem cells inhibit proliferation and function of allogeneic immune cells. Cell Tissue Res 330(3):437–446

Li YP, Paczesny S, Lauret E, Poirault S, Bordigoni P, Mekhloufi F, Hequet O, Bertrand Y, Ou-Yang JP, Stoltz JF, Miossec P, Eljaafari A (2008a) Human mesenchymal stem cells license adult CD34+ hemopoietic progenitor cells to differentiate into regulatory dendritic cells through activation of the Notch pathway. J Immunol 180(3):1598–1608

Li W, He H, Chen YT, Hayashida Y, Tseng SC (2008b) Reversal of myofibroblasts by amniotic membrane stromal extract. J Cell Physiol 215(3):657–664

Li L, Tian H, Yue W, Zhu F, Li S, Li W (2011) Human mesenchymal stem cells play a dual role on tumor cell growth in vitro and in vivo. J Cell Physiol 226:1860–1867

Li F, Miao ZN, Xu YY, Zheng SY, Qin MD, Gu YZ, Zhang XG (2012) Transplantation of human amniotic mesenchymal stem cells in the treatment of focal cerebral ischemia. Mol Med Rep 6(3):625–630

Li Y, Guo L, Ahn HS, Kim MH, Kim SW (2014) Amniotic mesenchymal stem cells display neurovascular tropism and aid in the recovery of injured peripheral nerves. J Cell Mol Med 18(6):1028–1034

Li J, Koike-Soko C, Sugimoto J, Yoshida T, Okabe M, Nikaido T (2015) Human amnion-derived stem cells have immunosuppressive properties on NK Cells and monocytes. Cell Transplant 24(10):2065–2076

Lin X, Li HY, Chen LF, Liu BJ, Yao Y, Zhu WL (2013) Enhanced differentiation potential of human amniotic mesenchymal stromal cells by using three-dimensional culturing. Cell Tissue Res 352(3):523–535

Lindenmair A, Wolbank S, Stadler G, Meinl A, Peterbauer-Scherb A, Eibl J, Polin H, Gabriel C, van Griensven M, Redl H (2010) Osteogenic differentiation of intact human amniotic membrane. Biomaterials 31(33):8659–8665

Lu LL, Liu YJ, Yang SG et al (2006) Isolation and characterization of human umbilical cord mesenchymal stem cells with hematopoiesis-supportive function and other potentials. Haematologica 91:1017–1026

Lutz MB, Schuler G (2002) Immature, semi-mature and fully mature dendritic cells: which signals induce tolerance or immunity? Trends Immunol 23(9):445–449

Lynge Nilsson L, Djurisic S, Hviid TV (2014) Controlling the immunological crosstalk during conception and pregnancy: HLA-G in reproduction. Front Immunol 5:198

Magatti M, De Munari S, Vertua E, Gibelli L, Wengler GS, Parolini O (2008) Human amnion mesenchyme harbors cells with allogeneic T-cell suppression and stimulation capabilities. Stem Cells 26:182–192

Magatti M, De Munari S, Vertua E, Nassauto C, Albertini A, Wengler GS, Parolini O (2009) Amniotic mesenchymal tissue cells inhibit dendritic cell differentiation of peripheral blood and amnion resident monocytes. Cell Transplant 18(8):899–914

Manuelpillai U, Moodley Y, Borlongan CV, Parolini O (2011) Amniotic membrane and amniotic cells: potential therapeutic tools to combat tissue inflammation and fibrosis? Placenta 32(Suppl 4):S320–S325

Marcus AJ, Coyne TM, Black IB, Woodbury D (2008) Fate of amnion-derived stem cells transplanted to the fetal rat brain: migration, survival and differentiation. J Cell Mol Med 12(4):1256–1264

Marongiu F, Gramignoli R, Sun Q, Tahan V, Miki T, Dorko K, Ellis E, Strom SC (2010) Isolation of amniotic mesenchymal stem cells. Curr Protoc Stem Cell Biol Chapter 1:Unit 1E.5

Maruyama N, Kokubo K, Shinbo T, Hirose M, Kobayashi M, Sakuragawa N, Kobayashi H (2013) Hypoxia enhances the induction of human amniotic mesenchymal side population cells into vascular endothelial lineage. Int J Mol Med 32(2):315–322

Menetrier-Caux C, Montmain G, Dieu MC, Bain C, Favrot MC, Caux C, Blay JY (1998) Inhibition of the differentiation of dendritic cells from CD34(+) progenitors by tumor cells: role of interleukin-6 and macrophage colony-stimulating factor. Blood 92(12):4778–4791

Murphy SP, Porrett PM, Turka LA (2011) Innate immunity in transplant tolerance and rejection. Immunol Rev 241(1):39–48

Nauta AJ, Kruisselbrink AB, Lurvink E, Willemze R, Fibbe WE (2006) Mesenchymal stem cells inhibit generation and function of both CD34+-derived and monocyte-derived dendritic cells. J Immunol 177(4):2080–2087

Nikoo S, Ebtekar M, Jeddi-Tehrani M, Shervin A, Bozorgmehr M, Kazemnejad S, Zarnani AH (2012) Effect of menstrual blood-derived stromal stem cells on proliferative capacity of peripheral blood mononuclear cells in allogeneic mixed lymphocyte reaction. J Obstet Gynaecol Res 38(5):804–809

Onishi R, Ohnishi S, Higashi R, Watari M, Yamahara K, Okubo N, Nakagawa K, Katsurada T, Suda G, Natsuizaka M, Takeda H, Sakamoto N (2015) Human amnion-derived mesenchymal stem cell transplantation ameliorates dextran sulfate sodium-induced severe colitis in rats. Cell Transplant 24(12):2601–2614

Palamar M, Kaya E, Egrilmez S, Akalin T, Yagci A (2014) Amniotic membrane transplantation in surgical management of ocular surface squamous neoplasias: long-term results. Eye (Lond) 28(9):1131–1135

Paracchini V, Carbone A, Colombo F, Castellani S, Mazzucchelli S, Gioia SD, Degiorgio D, Seia M, Porretti L, Colombo C, Conese M (2012) Amniotic mesenchymal stemcells: a new source for hepatocyte-like cells and induction of CFTR expression by coculture with cystic fibrosis airway epithelial cells. J Biomed Biotechnol 2012:575471

Parolini O, Alviano F, Bagnara GP, Bilic G, Bühring HJ, Evangelista M, Hennerbichler S, Liu B, Magatti M, Mao N, Miki T, Marongiu F, Nakajima H, Nikaido T, Portmann-Lanz CB, Sankar V, Soncini M, Stadler G, Surbek D, Takahashi TA, Redl H, Sakuragawa N, Wolbank S, Zeisberger S, Zisch A, Strom SC (2008) Concise review: isolation and characterization of cells from human term placenta: outcome of the first international workshop on placenta derived stem cells. Stem Cells 26:300–311

Parolini O, Soncini M, Evangelista M, Schmidt D (2009) Amniotic membrane and amniotic fluid-derived cells: potential tools for regenerative medicine? Regen Med 4(2):275–291

Parry RV, Chemnitz JM, Frauwirth KA, Lanfranco AR, Braunstein I, Kobayashi SV, Linsley PS, Thompson CB, Riley JL (2005) CTLA-4 and PD-1 receptors inhibit T-cell activation by distinct mechanisms. Mol Cell Biol 25(21):9543–9553

Phinney DG (2008) Isolation of mesenchymal stem cells from murine bone marrow by immunodepletion. Methods Mol Biol 449:171–186

Pianta S, BonassiSignoroni P, Muradore I, Rodrigues MF, Rossi D, Silini A, Parolini O (2015) Amniotic membrane mesenchymal cells-derived factors skew T cell polarization toward Treg and downregulate Th1 and Th17 cells subsets. Stem Cell Rev 11(3):394

Pirjali T, Azarpira N, Ayatollahi M, Aghdaie MH, Geramizadeh B, Talai T (2013) Isolation and characterization of human mesenchymal stem cells derived from human umbilical cord Wharton's jelly and amniotic membrane. Int J Organ Transplant Med 4(3):111–116

Portmann-Lanz CB, Schoeberlein A, Huber A, Sager R, Malek A, Holzgreve W, Surbek DV (2006) Placental mesenchymal stem cells as potential autologous graft for pre- and perinatal neuroregeneration. Am J Obstet Gynecol 194(3):664–673

Rahimi M, Zarnani AH, Mohseni-Kouchesfehani H, Soltanghoraei H, Akhondi MM, Kazemnejad S (2014) Comparative evaluation of cardiac markers in differentiated cells from menstrual blood and bone marrow-derived stem cells in vitro. Mol Biotechnol 56(12):1151–1162

Ramasamy R, Fazekasova H, Lam EW, Soeiro I, Lombardi G, Dazzi F (2007) Mesenchymal stem cells inhibit dendritic cell differentiation and function by preventing entry into the cell cycle. Transplantation 83(1):71–76

Rennie K, Gruslin A, Hengstschläger M, Pei D, Cai J, Nikaido T, Bani-Yaghoub M (2012) Applications of amniotic membrane and fluid in stem cell biology and regenerative medicine. Stem Cells Int 2012:721538

Resca E, Zavatti M, Maraldi T, Bertoni L, Beretti F, Guida M, La Sala GB, Guillot PV, David AL, Sebire NJ, De Pol A, De Coppi P (2015) Enrichment in c-Kit improved differentiation potential of amniotic membrane progenitor/stem cells. Placenta 36(1):18–26

Ristich V, Liang S, Zhang W, Wu J, Horuzsko A (2005) Tolerization of dendritic cells by HLA-G. Eur J Immunol 35(4):1133–1142

Roelen DL, van der Mast BJ, In 't Anker PS, Kleijburg C, Eikmans M, van Beelen E, de Groot-Swings GM, Fibbe WE, Kanhai HH, Scherjon SA, Claas FH (2009) Differential immunomodulatory effects of fetal versus maternal multipotent stromal cells. Hum Immunol 70(1):16–23

Rossi D, Pianta S, Magatti M, Sedlmayr P, Parolini O (2012) Characterization of the conditioned medium from amniotic membrane cells: prostaglandins as key effectors of its immunomodulatory activity. PLoS One 7(10):e46956

Roubelakis MG, Trohatou O, Anagnou NP (2012) Amniotic fluid and amniotic membrane stem cells: marker discovery. Stem Cells Int 2012:107836

Sanberg PR, Eve DJ, Willing AE, Garbuzova-Davis S, Tan J, Sanberg CD, Allickson JG, Cruz LE, Borlongan CV (2011) The treatment of neurodegenerative disorders using umbilical cord blood and menstrual blood-derived stem cells. Cell Transpl 20:85–94

Seo MS, Park SB, Kim HS, Kang JG, Chae JS, Kang KS (2013) Isolation and characterization of equine amniotic membrane-derived mesenchymal stem cells. J Vet Sci 14(2):151–159

Shojaeian J, Moazzeni SM, Nikoo S, Bozorgmehr M, Nikougoftar M, Zarnani AH (2007) Immunosuppressive effect of pregnant mouse serum on allostimulatory activity of dendritic cells. J Reprod Immunol 75(1):23–31

Singh R, Chacharkar MP (2011) Dried gamma-irradiated amniotic membrane as dressing in burn wound care. J Tissue Viability 20(2):49–54

Soncini M, Vertua E, Gibelli L, Zorzi F, Denegri M, Albertini A, Wengler GS, Parolini O (2007) Isolation and characterization of mesenchymal cells from human fetal membranes. J Tissue Eng Regen Med 1:296–305

Spaggiari GM, Capobianco A, Abdelrazik H, Becchetti F, Mingari MC, Moretta L (2008) Mesenchymal stem cells inhibit natural killer-cell proliferation, cytotoxicity, and cytokine production: role of indoleamine 2,3-dioxygenase and prostaglandin E2. Blood 111(3):1327–1333

Stadler G, Hennerbichler S, Lindenmair A, Peterbauer A, Hofer K, van Griensven M, Gabriel C, Redl H, Wolbank S (2008) Phenotypic shift of human amniotic epithelial cells in culture is associated with reduced osteogenic differentiation in vitro. Cytotherapy 10:743–752

Steinman RM, Hawiger D, Nussenzweig MC (2003) Tolerogenic dendritic cells. Annu Rev Immunol 21:685–711

Sun H, Hou Z, Yang H, Meng M, Li P, Zou Q, Yang L, Chen Y, Chai H, Zhong H, Yang ZZ, Zhao J, Lai L, Jiang X, Xiao Z (2014) Multiple systemic transplantations of human amniotic mesenchymal stem cells exert therapeutic effects in an ALS mouse model. Cell Tissue Res 357(3):571–582

Tabatabaei M, Mosaffa N, Nikoo S, Bozorgmehr M, Ghods R, Kazemnejad S, Rezania S, Keshavarzi B, Arefi S, Ramezani-Tehrani F, Mirzadegan E, Zarnani AH (2014) Isolation and partial characterization of human amniotic epithelial cells: the effect of trypsin. Avicenna J Med Biotechnol 6(1):10–20

Taghizadeh RR, Cetrulo KJ, Cetrulo CL (2011) Wharton's jelly stem cells: future clinical applications. Placenta 32(Suppl 4):S311–S315

Tamagawa T, Ishiwata I, Saito S (2004) Establishment and characterization of a pluripotent stem cell line derived from human amniotic membranes and initiation of germ layers in vitro. Hum Cell 17(3):125–130

Tamagawa T, Oi S, Ishiwata I, Ishikawa H, Nakamura Y (2007) Differentiation of mesenchymal cells derived from human amniotic membranes into hepatocyte-like cells in vitro. Hum Cell 20(3):77–84

Tamagawa T, Ishiwata I, Ishikawa H, Nakamura Y (2008) Induced in vitro differentiation of neural-like cells from human amnion-derived fibroblast-like cells. Hum Cell 21(2):38–45

Tao J, Ji F, Liu B, Wang F, Dong F, Zhu Y (2012) Improvement of deficits by transplantation of lentiviral vector-modified human amniotic mesenchymal cells after cerebral ischemia in rats. Brain Res 1448:1–10

Teng Z, Yoshida T, Okabe M, Toda A, Higuchi O, Nogami M, Yoneda N, Zhou K, Kyo S, Kiyono T, Nikaido T (2013) Establishment of immortalized human amniotic mesenchymal stem cells. Cell Transplant 22(2):267–278

Toda A, Okabe M, Yoshida T, Nikaido T (2007) The potential of amniotic membrane/amnion-derived cells for regeneration of various tissues. J Pharmacol Sci 105:215–228

Tsai MS, Hwang SM, Chen KD, Lee YS, Hsu LW, Chang YJ, Wang CN, Peng HH, Chang YL, Chao AS, Chang SD, Lee KD, Wang TH, Wang HS, Soong YK (2007) Functional network analysis of the transcriptomes of mesenchymal stem cells derived from amniotic fluid, amniotic membrane, cord blood, and bone marrow. Stem Cells 25(10):2511–2523

Tsuji H, Miyoshi S, Ikegami Y, Hida N, Asada H, Togashi I, Suzuki J, Satake M, Nakamizo H, Tanaka M, Mori T, Segawa K, Nishiyama N, Inoue J, Makino H, Miyado K, Ogawa S, Yoshimura Y, Umezawa A (2010) Xenografted human amniotic membrane-derived mesenchymal stem cells are immunologically tolerated and transdifferentiated into cardiomyocytes. Circ Res 106(10):1613–1623

Vidane AS, Souza AF, Sampaio RV, Bressan FF, Pieri NC, Martins DS, Meirelles FV, Miglino MA, Ambrósio CE (2009) Cat amniotic membrane multipotent cells are nontumorigenic and are safe for use in cell transplantation. Stem Cells Cloning 7:71–78

Vivier E, Tomasello E, Baratin M, Walzer T, Ugolini S (2008) Functions of natural killer cells. Nat Immunol 9(5):503–510

Walther G, Gekas J, Bertrand OF (2009) Amniotic stem cells for cellular cardiomyoplasty: promises and premises. Catheter Cardiovasc Interv 73:917–924

Weiss ML, Anderson C, Medicetty S, Seshareddy KB, Weiss RJ, VanderWerff I, Troyer D, McIntosh KR (2008) Immune properties of human umbilical cord Wharton's jelly-derived cells. Stem Cells 26(11):2865–2874

Wolbank S, Peterbauer A, Fahrner M, Hennerbichler S, van Griensven M, Stadler G, Redl H, Gabriel C (2007) Dose-dependent immunomodulatory effect of human stem cells from amniotic membrane: a comparison with human mesenchymal stem cells from adipose tissue. Tissue Eng 13(6):1173–1183

Wu W, Lan Q, Lu H, Xu J, Zhu A, Fang W, Ge F, Hui G (2014) Human amnion mesenchymal cells negative co-stimulatory molecules PD-L1 expression and its capacity of modulating microglial activation of CNS. Cell Biochem Biophys 69(1):35–45

Xu M, Zhang B, Liu Y, Zhang J, Sheng H, Shi R, Liao L, Liu N, Hu J, Wang J, Ning H, Liu T, Zhang Y, Chen H (2014) The immunologic and hematopoietic profiles of mesenchymal stem cells derived from different sections of human umbilical cord. Acta Biochim Biophys Sin (Shanghai) 46(12):1056–1065

Zarnani AH, Moazzeni SM, Shokri F, Salehnia M, Dokouhaki P, Ghods R, Mahmoodi AR, Jeddi-Tehrani M (2008) Microenvironment of the feto-maternal interface protects the semiallogenic fetus through its immunomodulatory activity on dendritic cells. Fertil Steril 90(3):781–788

Zeng G, Wang G, Guan F, Chang K, Jiao H, Gao W, Xi S, Yang B (2011) Human amniotic membrane-derived mesenchymal stem cells labeled with superparamagnetic iron oxide nanoparticles: the effect on neuron-like differentiation in vitro. Mol Cell Biochem 357(1-2):331–341

Zhang D, Jiang M, Miao D (2011) Transplanted human amniotic membrane-derived mesenchymal stem cells ameliorate carbon tetrachloride-induced liver cirrhosis in mouse. PLoS One 6(2):e16789

Zhao P, Ise H, Hongo M, Ota M, Konishi I, Nikaido T (2005) Human amniotic mesenchymal cells have some characteristics of cardiomyocytes. Transplantation 79(5):528–535

Amniotic Fluid: A Source of Stem Cells for Therapeutic Use and Modeling of Human Genetic Diseases

Somaieh Kazemnejad, Manijeh Khanmohammadi, Abolfazl Shirazi, Shaghayegh Arasteh, Sayeh Khanjani, and Mehdi Aleahmad

Abbreviations

AF	Amniotic fluid
AFSCs	Amniotic fluid stem cells
AFSCs-ECs	Amniotic fluid stem cell-derived endothelial cells
ASCs	Adult stem cells
BMSCs	Bone marrow mesenchymal stem cells
BMP-4	Bone morphogenetic protein-4
ESCs	Embryonic stem cells
GDNF	Glia cell line-derived neurotrophic factor
GVHD	Graft versus host disease
HGFR	Hepatocyte growth factor receptor
iPSC	Induced pluripotent stem cells
ICAM-1	Intercellular adhesion molecule-1
MDSC	Muscle-derived stem cells
MSCs	Mesenchymal stem cells
NRVM	Neonatal rat ventricular myocytes
NCAM	Neural cell adhesion molecule
PGA	Polyglycolic acid
SSEA-1	Stage-specific embryonic antigens
SPCL	Starch-poly(ε-caprolactone)
TB4	Thymosin-beta-4

S. Kazemnejad, Ph.D. (✉) • M. Khanmohammadi • A. Shirazi • S. Arasteh • S. Khanjani
Reproductive Biotechnology Research Center, Avicenna Research Institute, ACECR, Tehran, Iran
e-mail: kazemnejad_s@yahoo.com; s.kazemnejad@avicenna.ac.ir

M. Aleahmad
Department of Immunology, School of Public Health, Tehran University of Medical Sciences, Tehran, Iran

B. Arjmand (ed.), *Perinatal Tissue-Derived Stem Cells*, Stem Cell Biology and Regenerative Medicine, DOI 10.1007/978-3-319-46410-7_8

Tra-1-60	Tumor-rejection antigens-1-60
Tra-1-81	Tumor-rejection antigens-1-81
VEGF	Vascular endothelial growth factor

1 Amniotic Fluid: Novel Source of Stem Cell with Therapeutic Application

Amniotic fluid (AF) that is delimited by a membrane called amnion is constituted by around 98 % of water. This fluid is produced between day 7 and 14 postfertilization and its volume elevates by different mechanisms including fetal urine production, and oral, nasal, tracheal and pulmonary fluid secretions, and fetal swallowing. Amniotic fluid mediates a dynamic environment, which allows the fetus to grow and move inside the uterus and prevents the fetus from attachment to the developing fetal membranes. In addition, it acts as a vehicle for the exchange of proteins, carbohydrates, lipids, electrolytes, enzymes, hormones, and growth factors between the mother and embryo, and is vital for the development of some organs such as lungs (Pozzobon et al. 2014).

Due to contact of amniotic fluid to the fetus during development and its important roles, this fluid has been applied in order to detect fetal abnormalities since 6 decades ago (Gholizadeh-Ghalehaziz et al. 2015). Although in humans adequate DNA can be extracted from amniotic fluid from 8 weeks' gestation onward and these samples are suitable for prenatal diagnosis, amniocentesis is usually performed after 15 weeks of gestation since earlier sampling is associated with abnormalities of limb development (Odibo et al. 2008).

Human amniotic fluid samples are readily available, noncontroversial, and routinely obtained during the second trimester of pregnancy by scheduled amniocenteses under ultrasonographic guidance with a less than 1 % rate of fetal loss together with facile storage at minimum cost and use in the standard evaluation of fetal pulmonary maturity, metabolic diseases, fetal infections, and intrauterine infections. These tests have recently been complemented by applying chromosomal microarray as a more efficient prenatal genetic screening tool to detect fetal abnormalities (Rennie et al. 2012).

It is noticeable that AF cells procured after prenatal diagnostic testing provide clinically important information about the fetus (Kim et al. 2014). These cells could be stored in cell banks and used in disease research, drug screening, and genetic disorders (Gholizadeh-Ghalehaziz et al. 2015).

The AF cells are mainly composed of subpopulations of adherent cells that vary in proportion according to gestational age and calcified based on their growth, morphology, behavior, and biochemical characteristics that are derived from the three germ layers. About 33.7 % of AF cells are composed of epithelioid (E-type) cells that are cuboidal to columnar and are derived from the fetal skin and urine. Amniotic fluid (AF-type) cells are originating from the membranes and trophoblast (60.8 % of AF cells), and fibroblastic (F-type) cells are generated mainly from fibrous connective tissue and dermal fibroblasts (5.5 %) (Pozzobon et al. 2014; Gholizadeh-Ghalehaziz et al. 2015; Loukogeorgakis and De Coppi 2016). During the early days of amniotic fluid cell culture, both E-and AF-type cells appear. AF-type cells are the dominant cell type and coexpress keratins and vimentins and have a fibroblastoid morphology that remain throughout the culture period, while E-type cells soon disappear. F-type cells usually appear during late primary culture and possess fibroblastoid morphology, phenotypic, and differentiation characteristics similar to bone marrow mesenchymal stem cells (BMSCs) (Eslaminejad and Jahangir 2012; Roubelakis et al. 2012).

Amniotic fluid stem cells (AFSCs) can be easily obtained from a small amount of second trimester AF without destroying a human embryo, thus eliminating many ethical controversies associated with the application of human embryonic stem cells (ESCs) and necessity for application of invasive methods for bone marrow aspiration (Roubelakis et al. 2012; Koike et al. 2014; Iordache et al. 2016). Isolation and identification of AF cells goes back to the early 1990s (Streubel et al. 1995). However, first progenitor cells derived from amniotic fluid were reported in 1993 (Torricelli et al. 1992). Later, the scientists could indicate the existence of a small population of expanding cells in human amniotic fluid expressing pluripotent stem cells and/or mesenchymal stem cells (MSCs) markers with multilineage differentiation capacity (De Coppi et al. 2007). The isolation and culture of amniotic fluid MSCs from weeks 16 to 20 of pregnancy was a great achievement in 2004 that made human amniotic fluid stem cells fascinating for future application to replace dysfunctional cells in myriad of diseases, including Parkinson's and Alzheimer's diseases, heart diseases, diabetes, stroke, spinal cord injuries, and burns (Kim et al. 2014; Eslaminejad and Jahangir 2012).

2 Characteristics of Stem Cell Population in Amniotic Fluid

The great self-renewal potential of AFSCs mediates a great opportunity for expansion of these cells appropriate for clinical demand. According to the reports, in each milliliter of AF, there is an average of 100,000 cells with 60–90 % viability (Dziadosz et al. 2016) that generate a large number of cells ($2–4 \times 10^8$) from a single cell clone after only nine passages (Rennie et al. 2013). AFSCs divide and double in an average of 36 h with high proliferative capacity without feeder layers (Loukogeorgakis and De Coppi 2016; Dziadosz et al. 2016; Edwards and Hollands 2007). These cells maintain their karyotypic stability and long telomeres and remain diploid after 250–300 generations and have normal G1 and G2 cell cycle checkpoints (Rennie et al. 2012; Kim et al. 2014). In addition, there is no report about tumor formation or malignant change or other abnormalities after implantation of these cells to animal models (Pozzobon et al. 2014; Martinelli et al. 2016; Sessarego et al. 2008). These data imply the safety of these cells for clinical application.

AFSCs are a stem cell population with intermediate characteristics between embryonic and adult stem cells (ASCs). Human AFSCs were identified for expressing of pluripotency markers such as OCT-4, alkaline phosphatase, c-kit, SOX-2, Nanog, KLF-4, WDR-5 (a key factor that interact with OCT-4), c-MYC, Rex-1, cyclin A, and SSEA-4 (stage-specific embryonic antigens-4). In addition, there are some controversial reports about expression of SSEA-3, SSEA-1 (stage-specific embryonic antigens), Tra-1-60, and Tra-1-81 (tumor-rejection antigens) (Eslaminejad and Jahangir 2012; Iordache et al. 2016; Dziadosz et al. 2016). Moreover, they present typical cell surface mesenchymal markers such as CD15, CD44, CD29, CD9, CD73, CD90, CD105, CD166, CD49e, CD58, CD133, vimentin, intercellular adhesion molecule-1 (ICAM-1), and the histocompatibility protein HLA-class I. However, they do not express markers of the hematopoietic lineage such as CD45, CD34, CD14, CD11b, CD79, T-cell coreceptors markers of CD8, and CD4, the endothelial marker such as CD31, CD144, von Willebrand, KDR, the epithelial marker of CD326 (Eslaminejad and Jahangir 2012; Iordache et al. 2016; Dziadosz et al. 2016; Edwards and Hollands 2007; Martinelli et al. 2016; Mareschi et al. 2009), and cell surface markers associated with rejection including the histocompatibility protein HLA-class II, CD80, CD86, and CD40 (Fig. 1). This result suggests that they lie somewhere between ESCs and ASCs on the developmental continuum and could reduce the risk of rejection and graft versus host disease (GVDH) (Sessarego et al. 2008; Chun et al. 2015).

In addition, costaining of more than 75 % of AFSCs with at least two of the surface proteins such as SOX-17 endodermal marker, SM_{22a} mesodermal marker (smooth muscle), and Tubb-3 ectodermal marker (neuronal) reflect their broad differentiation potency (Pozzobon et al. 2014).

Another important feature of these cells is that they express markers of neural lineage, such as Nestin (an intermediate filament protein expressed by neural stem or progenitor cells), TUBB3 (a neuronal cytoskeletal dimer), NEFH (a marker of neurofilament located primarily in the cytoplasm of mature neurons), GFAP (a structure element of fibrillary astrocytes), NEUNA60 (a marker for early neurons), GALC (a marker for oligodendrocytes), Brn-2 (pou3f2), and Neurofilament, suggesting that distinct populations of AFSCs may have capacity to differentiate into neural lineage (Rennie et al. 2012; Tsai et al. 2006).

However, they did not express C-MET or hepatocyte growth factor receptor (HGFR), ABCG2 or CDw338, neural cell adhesion molecule (NCAM) or CD56, cytokeratin, and bone morphogenetic protein-4 (BMP-4) (Joo et al. 2012).

Therefore, AFSCs are an attractive cell source due to a number of fascinating features, including no legal considerations associated with their collection, reduced donor damage, lack of ethical concerns and restrictions, capacity of in vitro expansion and self-renewal in culture, lack of immune reactions when administered in vivo due to the absence of MHC-II surface antigens and immunosuppressive and immunomodulatory activities, absence of teratocarcinomas formation when implanted into recipients, lack of karyotypic abnormalities and maintenance of long telomeres despite the high proliferative potential, broad differentiation potency, and high tolerating cryopreservation (Chun et al. 2015; Joo et al. 2012). These characterizations represent a great alternative source to embryonic and adult stem cells for use in regenerative medicine and clinical applications.

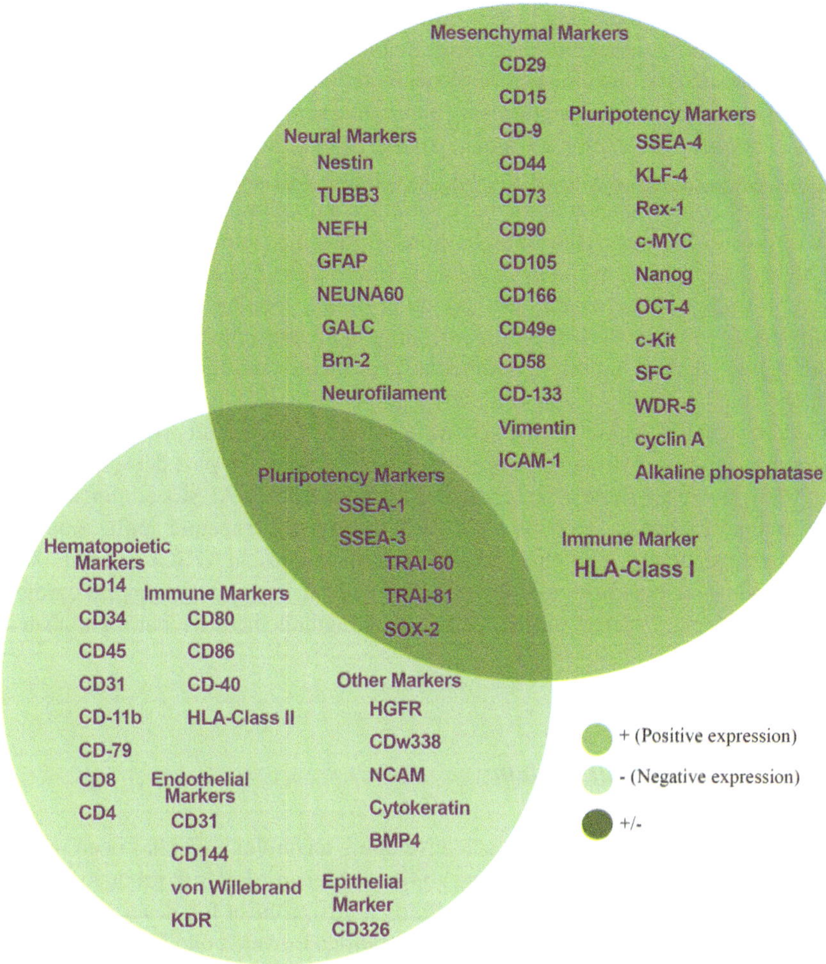

Fig. 1 The expression pattern of pluripotent, mesenchymal, hematopoietic, and immunologic markers in AFSCs

3 Amniotic Fluid Stem Cells As a Candidate Source for Regenerative Medicine

Regenerative medicine involves the use of living cells to repair and restore normal function to damaged tissues. Due to stem cells capacities for self-renewal and differentiation into diverse mature progeny, they are viewed as promising candidates for use in cell-based therapies over the past few years. However, the source of stem cells, in order to maximize the safety and efficacy of tissue engineering and cell-based therapies, is clearly of great importance (Sessarego et al. 2008).

Characterizations, including high proliferation rate and multilineage differentiation capacity into adipogenic, osteogenic, chondrogenic, myogenic, neurogenic, epithelial, hepatocyte, insulin secreting pancreatic β-cells, and endothelial cell lineages, have raised AFSCs as a promising alternative source of cells for use in regenerative therapies. The major advantage of AFSCs is that potential cells can be obtained as early as the second trimester of gestation, providing a chance/time window for autologous transplantation back into the same fetus as a cell or cell-based gene therapy. In addition, AFSCs can be stored for further postnatal use with autologous stem cell origin, pre- and postnatal autologous transplantation (Shaw 2014).

Several studies indicated that AFSCs exhibit significant plasticity and can be differentiated in vitro toward different lineages, including adipogenic, osteogenic, chondrogenic, skeletal muscle, endothelial, neurogenic, and hepatogenic lineages (Iordache et al. 2016; Martinelli et al. 2016; Joo et al. 2012).

In addition, AFSCs have been extensively examined in a variety of experimental models of injuries and diseases in preclinical studies over the past decade (Rennie et al. 2013). However, there is no report about clinical trial of AFSCs administration for treatment of diseases. Hereby, to get insights about therapeutic applications of AFSCs for regenerative medicine, we reviewed the studies on AFSCs in preclinical disease models such as musculoskeletal, neurological, respiratory system, cardiovascular, and urinary system disorders, and diaphragmatic hernia repair in neonates or adults.

3.1 Musculoskeletal System

Bone regeneration using stem cells is a developing technology that has been investigated by different groups. Already, the ability of human AFSCs to produce mineralized matrix in conditioned porous scaffolds holds promise for generation of an applicable artificial construct for bone regeneration (Edwards and Hollands 2007; Martinelli et al. 2016; Mareschi et al. 2009). In this regard, it has been shown that three-dimensional scaffolds containing AFSCs could produce highly mineralized bone tissue 8 weeks following transplantation into mice (Murphy and Atala 2013). In addition, Maraldi et al. demonstrated the capacity of these cells in repairing of critical size femoral rat defect after seeding preconditioned AFSCs in starch-poly-(e caprolactone) (SPCL) scaffolds. Moreover, AFSCs could restore critical size cranial bone defects in immunocompromised rats after implantation of silk fibroin scaffolds seeded with predifferentiated AFSCs (Maraldi et al. 2011; Riccio et al. 2012). On the other hand, several studies presented evidence about this matter that mesenchymal cells derived from amniotic fluid could be used for repair of cartilage defects and tendon injuries (Rodrigues et al. 2012; Kunisaki et al. 2006). Kunisaki et al. studied the potential of ovine AFSCs/GFP^{+}-seeded polyglycolic acid polymer mesh (PGA) to repair either partial or full circumferential tracheal defects in allogeneic fetal lambs. They concluded that engineered cartilaginous grafts containing AFSCs may become a viable alternative for tracheal reconstruction (Kunisaki et al. 2006).

These reports demonstrate that AFSCs are a valuable stem cell population for future therapies of osteochondral defects. Moreover, the ability of AFSCs to differentiate into skeletal myogenic cells persuaded several preclinical studies to evaluate the therapeutic potential of xenogeneic and allogeneic AFSCs in vivo as a tool for the treatment of degenerative skeletal muscle disorders such as Duchenne muscular dystrophy and trauma. The incorporation of differentiated AF type of human AFSCs into skeletal myogenic cells in regeneration of irradiated tibialis anterior muscle of mice has been reported by a group of Chinese researchers. Moreover, tail vein transplantation of allogeneic AFSCs in mouse muscular dystrophies model enhanced the muscle strength and animals' survival rate (Piccoli et al. 2012). In addition, it has been demonstrated that GFP-labeled allogeneic AFSCs could represent a valuable tool for preclinical studies, shown by 1-month survival within the host tissue and enhancement of early phase tendon healing in ovine Achilles tendon defect (Colosimo et al. 2013).

3.2 *Neural System*

A major goal of regenerative medicine is to ameliorate irreversible destruction of brain tissue by utilizing stem cells to control the process of neurogenesis. Aside from the common mesenchymal lineages, cultured AFSCs have also been successfully differentiated into neuron-like cells (Tsai et al. 2006; Cheng et al. 2010; Pan et al. 2007; Prasongchean et al. 2011).

Animal models with human AFSCs have been used to study neurodegenerative diseases. In fact, the transplanted stem cells migrated toward areas of damaged neural tissue and resulted in symptomatic improvement (Dziadosz et al. 2016). In this line, when AFSCs injected into lateral cerebral ventricles of newborn mice, cells participated in the growth of the central nervous system by 1-month postinjection (Edwards and Hollands 2007). In addition, when AFSCs induced to neural lineage in presence of specific growth factor and implanted in rat, cells contributed to healing of neuronal degenerative (Eslaminejad and Jahangir 2012). Pan et al. showed that injection of embedding rat AFSCs in fibrin glue to the crushed sciatic nerve of rat model could regenerate nerve injury (Pan et al. 2007). Likewise, transplantation of matrigel containing induced human AFSCs with glia cell line-derived neurotrophic factor (GDNF) into the injured sciatic nerve of rats promoted nerve regeneration (Cheng et al. 2010). It has been indicated that 70 % of grafted AFSCs into the lateral cerebral of mice model with neurodegeneration could survive; therefore AFSCs are recommended for novel therapies in diseases of the central nervous system (Joo et al. 2012). Some authors suggested that regenerative effect of AMSCs is associated with the secretion of paracrine factors rather than the ability of AFSCs to fully differentiate into neuronal cells. In this line, it has been indicated that grafting of differentiated rat $GFP^{+}c\text{-}kit^{+}$ AFSCs into neuronal cells at the site of an extensive thoracic crush injury in E2.5 Chick embryos caused significant hemorrhage reduction and survival increase (Prasongchean et al. 2011). Furthermore, AFSCs have the potency to heal ischemic stroke. It has been shown that intracerebroventricular administration of AFSCs had a significant neuroprotective effect and reduced behavioral brain disorders in mouse model of ischemic stroke (Murphy and Atala 2013). Therefore, it is hoped that the beneficial effects of AFSCs gradually transit into clinical application and pave the way for potential treatments of diseases such as stroke, Parkinson, Alzheimer disease, and spinal injuries.

3.3 Urinary System

Stem cell-based therapy is a promising avenue for curing kidney injury; however, the investigations for finding out the most suitable stem cell source have been continued. AFSCs may represent a promising candidate to treat kidney injury as they demonstrate renoprotective effects postinjury via the secretion of promitotic, anti-apoptotic, anti-inflammatory, and immunomodulatory factors. These cells could stimulate the proliferation of tubular cells via the local release of factors, including interleukin-6, VEGF, and stromal cell-derived factor-1. It is demonstrated that infusion of human AFSCs in cisplatin-treated mice improves renal function and limited tubular damage. In addition, human AFSCs provide a protective effect by decreasing creatinine and BUN blood levels (Perin et al. 2010). The mechanisms governing the curing and protective effects of AFSCs are remained to clarify, however, it is indicated that AFSCs pretreatment with glial cell line-derived neurotrophic factor (GDNF) could enhance stem cell homing to the tubule interstitial compartment and ameliorate renal function and tubular injury (Rota et al. 2012). Moreover, it has been reported that AFSCs provide a protective effect through inhibition of the renin–angiotensin system, ameliorating acute tubular necrosis in mouse model (Sedrakyan et al. 2012). Furthermore, the antiapoptotic activity of AFSCs against renal tubular cells has been reported more potent in comparison with BMSCs, but stimulatory activity for the proliferation of renal tubular cells is more dominant in BMSCs. In fact, AFSCs and BMSCs express distinct sets of paracrine factors, impressing their activities in vivo (Hauser et al. 2010).

Recently, a three-dimensional chimeric organoid composed of AFSCs and mouse embryonic kidney cells was developed that could generate vascularized glomeruli and tubular structures after engraftment (Xinaris et al. 2016). Generating kidney organoids using these cells could offer promising prospects for therapeutic purposes of kidney injury.

Urinary incontinence, which is characterized by involuntary leakage of urine and has profound effects on quality of life (Corcos et al. 2002), might occur as a result of the damage of the external urethral sphincter and associated nerves after vaginal delivery, weight gain, diagnosis with diabetes, or other conditions which stretch the pelvic floor muscles (Rortveit et al. 2001; Deng 2011). Muscle-derived stem cells (MDSC) exhibit therapeutic potential for regeneration of striated muscle (Cannon et al. 2003; Yokoyama et al. 2000); however, its usage is restricted by complications associated with biopsy (Chun et al. 2012). Application of stem cells from amniotic fluid eliminates some limitations associated with the application of MDSCs. Preclinical studies demonstrated the safety and efficacy of AFSCs in urinary incontinence mouse model by cells injection to external urethral sphincter (Chun et al. 2012, 2014; Choi et al. 2015; Kim et al. 2012). Considering the significant role of innervation and angiogenesis in regeneration of urethral sphincter, a research group proposed that combined cell therapy might be more effective (Chun et al. 2014). Triple cell combination of early differentiated AFSCs into muscle, neuron, and endothelial progenitor cells showed synergistic effects in mice urethral sphincter regeneration and improved urodynamic function and formation of new striated muscle fibers and neuromuscular junctions at the cell injection site compared to single-cell or double-cell combinations.

3.4 Cardiovascular System

Nowadays, stem cell therapies have been fascinated as a possible treatment approach for fatal cardiovascular disease that does not respond to current medical therapies. There are some preclinical studies on cell therapy of ischemic and nonischemic cardiomyopathy using AFSCs. Chiavegato et al. (2007) was the first group investigating the ability of AFSCs to differentiate into cardiomyocytes (Chiavegato et al. 2007). It is demonstrated that systemic injection of human AFSCs diminishes the skeletal muscle atrophy in damaged cardiac rat by ameliorating apoptosis and expression of pro-inflammatory cytokines (Castellani et al. 2013). In addition, systemic injection of human AFSCs has therapeutic potential in acute myocardial infarction, which may be mediated through paracrine effectors such as the actin monomer-binding protein thymosin b4 (Tb4) that previously shown to be both cardioprotective and proangiogenic (Bollini et al. 2011). In this regard, direct cocultures of AFSCs and neonatal rat ventricular myocytes (NRVM) have been a good strategy for in vitro pretreatment of AFSCs for further in vivo myocardial improvement (Yeh et al. 2010; Guan et al. 2011).

Amniotic fluid represents an attractive fetal cell source for pediatric cardiovascular tissue engineering. In this line, a heart valve leaflet construct composed of $CD133^+$ AFSCs and a biodegradable polymer was developed, which showed near-native behavior under low-pressure conditions (Schmidt et al. 2007).

Furthermore, AFSCs have previously shown angiogenic potential and could be used in the prevascularization of engineered constructs and the treatment of ischemic disease. Besides the studies demonstrating endothelial differentiation ability of AFSCs (Piccoli et al. 2012; Colosimo et al. 2013), AFSC-derived ECs (AFSCs-ECs) have been regarded as a cell source for therapeutic angiogenesis in a mouse hind limb ischemia model. It is suggested that matrix metalloproteinase of MMP-3 and MMP-9 might activate angiogenesis by regulating vascular endothelial growth factor (VEGF) expressions, which are significantly higher in AFSCs-ECs in comparison with control groups (Liu et al. 2013).

3.5 Skin

Evidence has emerged that AFSCs hold great potential in skin regeneration and the stem cell-based therapy of chronic wounds. Beside some reports about epithelial commitment of AFSCs, there is some evidence about effectiveness of AFSCs in wound healing. Yoon et al. (2010) provided the first evidence on the potential of cultured AFSCs under the conditioned medium in wound healing. It is supposed that AFSCs significantly enhanced wound healing by improving dermal fibroblasts proliferation and migration via the TGF-β/SMAD2 pathway (Yoon et al. 2010). Recently, Sun et al. showed that AFSCs provide considerable advantages in epidermal regeneration via secretion of B7H4 that creates a moderate inflammation microenvironment to promote intentional excisional wounds (Sun et al. 2015).

3.6 Respiratory System

Respiratory diseases occur due to varied reasons with similar resultant of chronic inflammation, fibrosis, scaring, and consequently loss of functional lung tissue. In an attempt to investigate the potential applications of AFSCs in patients with congenital diaphragmatic hernia or prematurity, it is demonstrated that early prenatal administration of AFSCs in a rat nitrofen model of pulmonary hypoplasia improves lung growth, bronchial motility, and innervations (Pederiva et al. 2013). Moreover, upon intravascular injection of AFSCs into the nude mice with hyperoxia-induced pulmonary injury, these cells migrated to the lung and expressed the human pulmonary epithelial differentiation markers (Carraro et al. 2008). Recently, proof of AFSCs efficacy in improvement of pulmonary injury impelled studies in engineering of respiratory tissues by integrating AFSCs with synthetic scaffolds or decellularized tissues including lung (Vadasz et al. 2014).

4 Modeling of Human Genetic Diseases Using AFSCs

Despite continuous increase in our knowledge about the genetic basis of a number of congenital and late-onset human diseases, so far a large majority of these conditions still remain untreatable. This is largely due to the lack of information about the precise sequence of early molecular events occurring during tissue development and underlying the pathogenesis of the disease. The use of animal models for the study of the consequences of gene mutations during development, although able to provide useful information, does not produce results which can be entirely translated to humans. Indeed, due to the anatomical and physiological differences between the two species, animal models often are not able to completely represent the pathological mechanisms underlying human diseases. Additionally, the differences in specific molecular pathways connecting genotype to phenotype between human and animal models raise another challenge. In this regard, the use of alternative models to study human genetic diseases is more appreciated. In recent years, great interest has been devoted to the use of human cells for disease modeling and the best ones are so far the human pluripotent stem cells; ESCs and induced Pluripotent Stem cells (iPS), harboring naturally occurring disease-causing mutations and genomic aberration (Kobold et al. 2015). Since ESCs are obtained by the destruction of embryos, there are serious ethical objections that have yet to be resolved. On the other hand, despite major advances in iPS technology, reprogrammed cells often have an imperfectly cleared epigenetic memory of the source cells and are vulnerable to genomic instability (Ohi et al. 2011). Moreover, the protocol of iPS generation still suffers from technical limitations, low efficiency, high costs, and lengthy procedure. Due to the limitations associated with ESCs and iPS cells, much effort has been directed at finding an alternative source of cells that are able to bypass the aforementioned criticisms for use in regenerative medicine. AFSCs represent a potential alternative novel source of stem cells for modeling of human genetic diseases that rule out the above-mentioned drawbacks. In fact, by means of prenatal diagnosis using amniocentesis, the amniotic fluid collected for genetic testing can be used for the isolation, culture, and differentiation of AFSCs. Additionally, banking of clonal AFSCs lines derived from pregnancies with specific genetic aberrations might be considered as a promising tool to model in vitro pathogenic phenotypes (Rosner et al. 2014). However, gene modeling by AFS cells would involve only genetic diseases currently investigated in prenatal diagnosis, such as chromosomal abnormalities or monogenic disease in at-risk families, but not multifactorial disease or late onset monogenic diseases. To drawback this limitation, the novel emerging technology for genome editing, also known as CRISPR (Clustered Regularly Interspaced Short Palindromic Repeats)/Cas9 system, is developing that could generate disease models for both monogenic and complex genetic disorders, enabling creation of knockout cells in vitro. This technology mediates the procurement of AFS clones with induced mutations for testing the modes of action and the efficacy of various drug candidates (Antonucci et al. 2016). The potential ways for utilization of AFSCs in modeling of genetic diseases have been demonstrated in Fig. 2.

5 In Utero Therapy of Congenital Disorders Using AFSCs

Over a third of all pediatric hospital admissions are related to congenital diseases. Progressions in prenatal screening and molecular diagnosis have mediated the determination of life-threatening genetic diseases early in gestation. In utero transplantation with stem cells could cure affected human fetuses using allogeneic hematopoietic stem cells or MSCs transplantation. However, it has been limited to fetuses with severe immunologic defects and more recently osteogenesis imperfecta. The in utero treatment of congenital diseases using stem cell or gene therapy may overcome the necessity of postnatal treatment and reduce future costs and therefore impress the attitude of congenital diseases. AFSCs have a great potency for prenatal or postnatal treatment of disorders. Recently, preclinical autologous transplantation of transduced AFSCs has been achieved in fetal sheep using minimally invasive

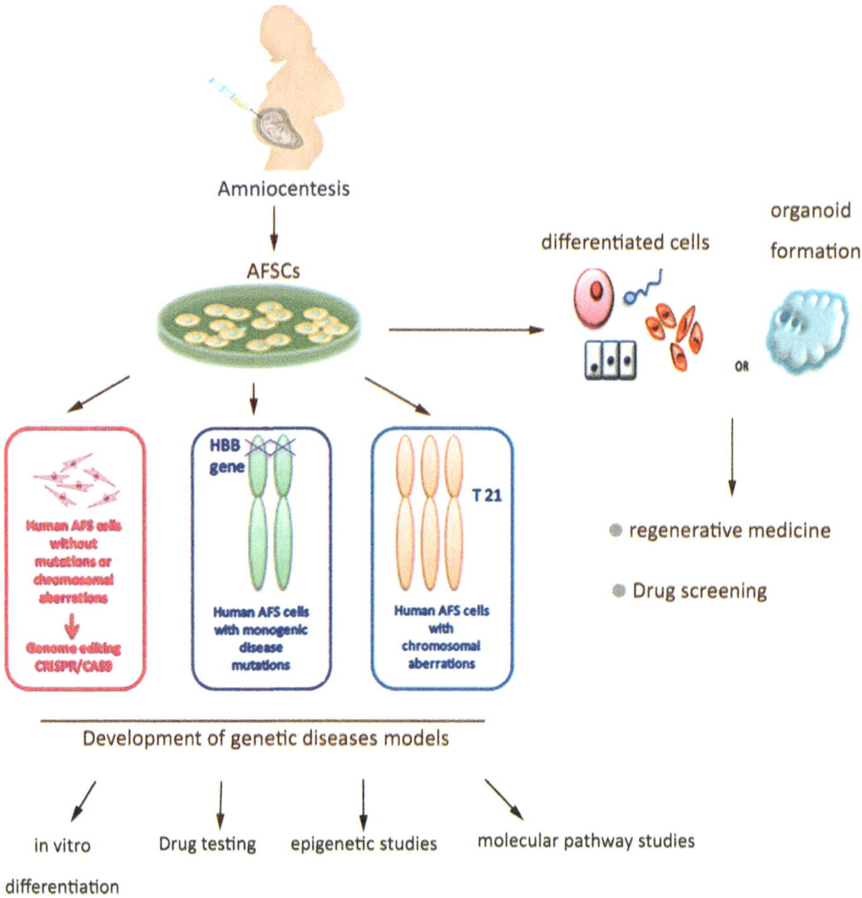

Fig. 2 Schematic diagram of potential benefits of AFSCs in regenerative medicine and the modeling of human genetic diseases (Adopted from Antonucci et al. (2016) with modification)

ultrasound-guided injection techniques (Ramachandra et al. 2014). Clinically relevant levels of transgenic protein were expressed in the blood of transplanted lambs for at least 6 months. These results bring hope and hold promise for prenatal treatment of genetic disorders in near future.

6 Potential of AFSCs for Future Basic Science

At present, AFSCs are widely accepted as a new powerful tool for basic research as well as for the establishment of new stem cell-based therapy concepts. It is possible to generate monoclonal genomically stable AFSCs lines harboring high proliferative potential without raising ethical issues. Stem cells are very useful tools to study the molecular and cellular regulation of differentiation processes (Gundacker et al.

2012). One approach to learn more about the role of a specific gene for a certain differentiation process is to knockdown the endogenous expression of the gene of interest. Such an approach allows clarifying the role of modulated gene expression for the cell potential to differentiate into a specific lineage. Application of siRNA-mediated gene silencing in AFSCs has already been tested for a variety of different genes (Rosner et al. 2010). Another very interesting aspect for future basic research is the banking of AFSCs lines carrying naturally occurring mutations, which are of relevance for certain human pathological phenotypes. In medical genetics, the future development of new prophylactic and therapeutic strategies directly depend on a better understanding of the mechanisms by which naturally occurring genetic variation contributes to a disease (Zhu et al. 2011). Moreover, AFSCs, with their potential proliferation, differentiation, and 3D organoid formation, could represent an interesting source to produce physiologically relevant systems to adequately test pharmaceutical agents before their administration to patients (Fig. 2). Taken together, it seems that generation and banking of normal human AFSCs lines, AFSCs lines with chromosomal aberrations, and AFSCs lines with specific monogenic disease mutations can provide very powerful tools for disease modeling in future research.

7 Limitations and Challenges

Despite relative advantages of AFSCs over other types of stem cells in basic research and regenerative medicine, the diverse subpopulations of multipotent cells in amniotic fluid which differ in marker expression, morphology, and growth kinetics might introduce a bias toward producing particular subpopulations of cells during culture and isolation. The lack of agreement on isolation and culture method in directing AFSCs differentiation toward particular cell fates is one limit on AFSCs application as the differences in isolation and culture method make it uncertain if the same populations of cells are being compared across experiments. The low rate of differentiation toward a desired phenotype prior to AFSCs transplantation may decrease the chance of engraftment in some tissues. Additionally, the gestational stage at which AF is collected and the passage number of the cultured cells exacerbate this situation. On the other hand, although AFSCs reportedly possess low immunogenicity which can survive transplantation into xenogeneic or allogeneic hosts, there is evidence indicating that AF cells were rejected upon transplantation into immunocompetent animals due to the recruitment of host immune system (Rennie et al. 2012).

8 Conclusion and Future Remarks

Since the identification of AFSCs, the knowledge about these cells has increased dramatically so that they are widely accepted as a new powerful tool for basic research as well as for the establishment of new stem cell-based therapy concepts. Generation of monoclonal genomically stable AFSCs lines possessing high proliferative potential and establishment of international registries of cell lines derived from amniotic fluid are promising approaches in basic and clinical studies. Aside from the proposed approach reducing methodological variations, the creation of a library of information pertaining to the research and preclinical study on AFSCs would allow researchers to choose the most appropriate cell line for a particular application in regenerative therapies. Additionally, in order to study the role of signaling pathways in human, as a hallmark in many human cancers studies, application of siRNA-mediated prolonged gene silenced AFSCs may be an efficient approach. In this regard one of the most important pathways involved in different cellular function is mTOR pathway so that the mutations in the mTOR pathway component genes trigger the development of many human genetic syndromes. It is strongly believed that the approach of siRNA-mediated knockdown of endogenous gene expression in monoclonal human AFSCs lines is a very powerful tool for future projects dealing with the molecular regulation of differentiation (Rosner et al. 2012). Another interesting aspect for future basic research is the banking of AFS cell lines carrying naturally occurring mutations, which are of relevance for certain human pathological phenotypes (Rosner et al. 2012). In modeling human diseases, AFSCs might be useful in generating disease-specific stem cell lines and can replace human pluripotent stem cells like ESCs and iPS cells and eliminate the concerns associated with their shortage. Furthermore, since AFSCs can be easily reprogrammed, they are able to erase the epigenetic memory after reprogramming and can be easily differentiated into different cell types (Antonucci et al. 2016). AFSCs may also represent an alternative to iPS for drug discovery and safety assays. In drug development and toxicity drug screening, the poor predictability of the preclinical studies carried out on animal models, which are mined by significant species-specific differences, justifies the application of in vitro cell culture systems including AFSCs which provides an efficient assay to individuate a selected list of promising drug candidates for further studies (Antonucci et al. 2016). Large-scale amplification of AFSCs in three-dimensional (3D) chimeric organoids (Antonucci et al. 2016), which enables the 3D growth of cells, can represent a very interesting physiologically relevant novel system to adequately test pharmaceutical agents. AFSCs may also be applicable as a model of germ cell precursors which allow investigating the mechanisms underlying drug-induced effects in gametogenesis and screening either natural or synthetic compounds potentially useful in fertility preservation. As recently demonstrated, AFSCs have some features of primordial germ cells and thus may be useful as an efficient in vitro model to study human gametogenesis and help to get insight on germ-line diseases and diseases caused by epigenetic alterations (Antonucci et al. 2016). Furthermore, since AFSCs are highly proliferative and less differentiated, they are good candidate for gene therapy and have great potential in cell therapies and regenerative medicine. However, to transit from bench to beside and clinical application of AFSCs, safety of AMSCs administration should be assessed in long-term preclinical and clinical studies.

Study Finding/Competing Interest The authors indicate no potential conflict of interest.

References

Antonucci I, Provenzano M, Rodrigues M, Pantalone A, Salini V, Ballerini P et al (2016) Amniotic fluid stem cells: a novel source for modeling of human genetic diseases. Int J Mol Sci 17(4):607

Bollini S, Cheung KK, Riegler J, Dong X, Smart N, Ghionzoli M et al (2011) Amniotic fluid stem cells are cardioprotective following acute myocardial infarction. Stem Cells Dev 20(11):1985–1994, Epub 2011/05/04. eng

Cannon TW, Lee JY, Somogyi G, Pruchnic R, Smith CP, Huard J et al (2003) Improved sphincter contractility after allogenic muscle-derived progenitor cell injection into the denervated rat urethra. Urology 62(5):958–963

Carraro G, Perin L, Sedrakyan S, Giuliani S, Tiozzo C, Lee J et al (2008) Human amniotic fluid stem cells can integrate and differentiate into epithelial lung lineages. Stem Cells 26(11):2 902–2911, Pubmed Central PMCID: Pmc3174105. Epub 2008/08/23. eng

Castellani C, Vescovo G, Ravara B, Franzin C, Pozzobon M, Tavano R et al (2013) The contribution of stem cell therapy to skeletal muscle remodeling in heart failure. Int J Cardiol 168(3):2014–2021, Epub 2013/03/05. eng

Cheng F-C, Tai M-H, Sheu M-L, Chen C-J, Yang D-Y, Su H-L et al (2010) Enhancement of regeneration with glia cell line-derived neurotrophic factor-transduced human amniotic fluid mesenchymal stem cells after sciatic nerve crush injury [RETRACTED] Laboratory investigation. J Neurosurg 112(4):868–879

Chiavegato A, Bollini S, Pozzobon M, Callegari A, Gasparotto L, Taiani J et al (2007) Human amniotic fluid-derived stem cells are rejected after transplantation in the myocardium of normal, ischemic, immuno-suppressed or immuno-deficient rat. J Mol Cell Cardiol 42(4):746–759, Epub 2007/02/16. eng

Choi JY, Chun SY, Kim BS, Kim HT, Yoo ES, Shon Y-H et al (2015) Pre-clinical efficacy and safety evaluation of human amniotic fluid-derived stem cell injection in a mouse model of urinary incontinence. Yonsei Med J 56(3):648–657

Chun SY, Cho DH, Chae SY, Choi KH, Lim HJ, Yoon GS et al (2012) Human amniotic fluid stem cell-derived muscle progenitor cell therapy for stress urinary incontinence. J Korean Med Sci 27(11):1300–1307, Pubmed Central PMCID: Pmc3492662. Epub 2012/11/21. eng

Chun SY, Kwon JB, Chae SY, Lee JK, Bae JS, Kim BS et al (2014) Combined injection of three different lineages of early-differentiating human amniotic fluid-derived cells restores urethral sphincter function in urinary incontinence. BJU Int 114(5):770–783, Epub 2014/05/21. eng

Chun SY, Mack DL, Moorefield E, Oh SH, Kwon TG, Pettenati MJ et al (2015) Pdx1 and controlled culture conditions induced differentiation of human amniotic fluid-derived stem cells to insulin-producing clusters. J Tissue Eng Regen Med 9(5):540–549

Colosimo A, Curini V, Russo V, Mauro A, Bernabo N, Marchisio M et al (2013) Characterization, GFP gene Nucleofection, and allotransplantation in injured tendons of ovine amniotic fluid-derived stem cells. Cell Transplant 22(1):99–117, Epub 2012/04/18. eng

Corcos J, Beaulieu S, Donovan J, Naughton M, Gotoh M (2002) Quality of life assessment in men and women with urinary incontinence. J Urol 168(3):896–905, Epub 2002/08/21. eng

De Coppi P, Bartsch G, Siddiqui MM, Xu T, Santos CC, Perin L et al (2007) Isolation of amniotic stem cell lines with potential for therapy. Nat Biotechnol 25(1):100–106

Deng DY (2011) Urinary incontinence in women. Med Clin N Am 95(1):101–109

Dziadosz M, Basch RS, Young BK (2016) Human amniotic fluid: a source of stem cells for possible therapeutic use. Am J Obstet Gynecol 214(3):321–327

Edwards RG, Hollands P (2007) Will stem cells in cord blood, amniotic fluid, bone marrow and peripheral blood soon be unnecessary in transplantation? Reprod Biomed Online 14(3):396–401

Eslaminejad MB, Jahangir S (2012) Amniotic fluid stem cells and their application in cell-based tissue regeneration. Int J Fertil Steril 6(3):147

Gholizadeh-Ghalehaziz S, Farahzadi R, Fathi E, Pashaiasl M (2015) A mini overview of isolation, characterization and application of amniotic fluid stem cells. Int J Stem Cells 8(2):115

Guan X, Delo DM, Atala A, Soker S (2011) In vitro cardiomyogenic potential of human amniotic fluid stem cells. J Tissue Eng Regen Med 5(3):220–228, Pubmed Central PMCID: Pmc2975013. Epub 2010/08/06. eng

Gundacker C, Dolznig H, Mikula M, Rosner M, Brandau O, Hengstschläger M (2012) Amniotic fluid stem cell-based models to study the effects of gene mutations and toxicants on male germ cell formation. Asian J Androl 14(2):247–250

Hauser PV, De Fazio R, Bruno S, Sdei S, Grange C, Bussolati B et al (2010) Stem cells derived from human amniotic fluid contribute to acute kidney injury recovery. Am J Pathol 177(4):2011–2021, Pubmed Central PMCID: Pmc2947295. Epub 2010/08/21. eng

Iordache F, Constantinescu A, Andrei E, Amuzescu B, Halitzchi F, Savu L, et al (2016) Electrophysiology, immunophenotype, and gene expression characterization of senescent and cryopreserved human amniotic fluid stem cells. J Physiol Sci 1–14

Joo S, Ko IK, Atala A, Yoo JJ, Lee SJ (2012) Amniotic fluid-derived stem cells in regenerative medicine research. Arch Pharm Res 35(2):271–280

Kim BS, Chun SY, Lee JK, Lim HJ, Bae J-s, Chung H-Y et al (2012) Human amniotic fluid stem cell injection therapy for urethral sphincter regeneration in an animal model. BMC Med 10(1):94

Kim EY, Lee K-B, Kim MK (2014) The potential of mesenchymal stem cells derived from amniotic membrane and amniotic fluid for neuronal regenerative therapy. BMB Rep 47(3):135–140

Kobold S, Guhr A, Kurtz A, Löser P (2015) Human embryonic and induced pluripotent stem cell research trends: complementation and diversification of the field. Stem Cell Rep 4(5):914–925

Koike C, Zhou K, Takeda Y, Fathy M, Okabe M, Yoshida T et al (2014) Characterization of amniotic stem cells. Cell Reprogram 16(4):298–305

Kunisaki SM, Freedman DA, Fauza DO (2006) Fetal tracheal reconstruction with cartilaginous grafts engineered from mesenchymal amniocytes. J Pediatr Surg 41(4):675–682

Liu YW, Roan JN, Wang SP, Hwang SM, Tsai MS, Chen JH et al (2013) Xenografted human amniotic fluid-derived stem cell as a cell source in therapeutic angiogenesis. Int J Cardiol 168(1):66–75, Epub 2012/10/11. eng

Loukogeorgakis SP, De Coppi P (2016) Stem cells from amniotic fluid—potential for regenerative medicine. Best Pract Res Clin Obstet Gynaecol 31:45–57

Maraldi T, Riccio M, Resca E, Pisciotta A, La Sala GB, Ferrari A et al (2011) Human amniotic fluid stem cells seeded in fibroin scaffold produce in vivo mineralized matrix. Tissue Eng Part A 17(21–22):2833–2843

Mareschi K, Rustichelli D, Comunanza V, De Fazio R, Cravero C, Morterra G et al (2009) Multipotent mesenchymal stem cells from amniotic fluid originate neural precursors with functional voltage-gated sodium channels. Cytotherapy 11(5):534–547

Martinelli D, Pereira RC, Mogni M, Benelli R, Mastrogiacomo M, Coviello D et al (2016) A humanized system to expand in vitro amniotic fluid-derived stem cells intended for clinical application. Cytotherapy 18(3):438–451

Murphy SV, Atala A (2013) Amniotic fluid and placental membranes: unexpected sources of highly multipotent cells. Semin Reprod Med 31(1):62–68

Odibo AO, Gray DL, Dicke JM, Stamilio DM, Macones GA, Crane JP (2008) Revisiting the fetal loss rate after second-trimester genetic amniocentesis: a single center's 16-year experience. Obstet Gynecol 111(3):589–595

Ohi Y, Qin H, Hong C, Blouin L, Polo JM, Guo T et al (2011) Incomplete DNA methylation underlies a transcriptional memory of somatic cells in human iPS cells. Nat Cell Biol 13(5):541–549

Pan H-C, Cheng F-C, Chen C-J, Lai S-Z, Lee C-W, Yang D-Y et al (2007) Post-injury regeneration in rat sciatic nerve facilitated by neurotrophic factors secreted by amniotic fluid mesenchymal stem cells. J Clin Neurosci 14(11):1089–1098

Pederiva F, Ghionzoli M, Pierro A, De Coppi P, Tovar J (2013) Amniotic fluid stem cells rescue both in vitro and in vivo growth, innervation, and motility in nitrofen-exposed hypoplastic rat lungs through paracrine effects. Cell Transplant 22(9):1683–1694

Perin L, Sedrakyan S, Giuliani S, Da Sacco S, Carraro G, Shiri L et al (2010) Protective effect of human amniotic fluid stem cells in an immunodeficient mouse model of acute tubular necrosis. PLoS One 5(2):e9357, Pubmed Central PMCID: Pmc2827539. Epub 2010/03/03. eng

Piccoli M, Franzin C, Bertin E, Urbani L, Blaauw B, Repele A et al (2012) Amniotic fluid stem cells restore the muscle cell niche in a HSA-Cre, Smn(F7/F7) mouse model. Stem Cells 30(8):1675–1684, Epub 2012/05/31. eng

Pozzobon M, Piccoli M, De Coppi P (2014) Stem cells from fetal membranes and amniotic fluid: markers for cell isolation and therapy. Cell Tissue Bank 15(2):199–211

Prasongchean W, Bagni M, Calzarossa C, De Coppi P, Ferretti P (2011) Amniotic fluid stem cells increase embryo survival following injury. Stem Cells Dev 21(5):675–688

Ramachandra DL, Shaw SS, Shangaris P, Loukogeorgakis S, Guillot PV, Coppi PD et al (2014) In utero therapy for congenital disorders using amniotic fluid stem cells. Front Pharmacol 5:270

Rennie K, Gruslin A, Hengstschläger M, Pei D, Cai J, Nikaido T et al (2012) Applications of amniotic membrane and fluid in stem cell biology and regenerative medicine. Stem Cells Int 2012:721538

Rennie K, Haukenfrers J, Ribecco-Lutkiewicz M, Ly D, Jezierski A, Smith B et al (2013) Therapeutic potential of amniotic fluid-derived cells for treating the injured nervous system. Biochem Cell Biol 91(5):271–286

Riccio M, Maraldi T, Pisciotta A, La Sala GB, Ferrari A, Bruzzesi G et al (2012) Fibroin scaffold repairs critical-size bone defects in vivo supported by human amniotic fluid and dental pulp stem cells. Tissue Eng Part A 18(9–10):1006–1013

Rodrigues MT, Lee SJ, Gomes ME, Reis RL, Atala A, Yoo JJ (2012) Bilayered constructs aimed at osteochondral strategies: the influence of medium supplements in the osteogenic and chondrogenic differentiation of amniotic fluid-derived stem cells. Acta Biomater 8(7):2795–2806

Rortveit G, Hannestad YS, Daltveit AK, Hunskaar S (2001) Age- and type-dependent effects of parity on urinary incontinence: the Norwegian EPINCONT study. Obstet Gynecol 98(6):1004–1010, Epub 2002/01/05. eng

Rosner M, Siegel N, Fuchs C, Slabina N, Dolznig H, Hengstschläger M (2010) Efficient siRNA-mediated prolonged gene silencing in human amniotic fluid stem cells. Nat Protoc 5(6):1081–1095

Rosner M, Schipany K, Shanmugasundaram B, Lubec G, Hengstschläger M (2012) Amniotic fluid stem cells: future perspectives. Stem Cells Int 2012:1–6

Rosner M, Schipany K, Hengstschläger M (2014) The decision on the"optimal" human pluripotent stem cell. Stem Cells Transl Med 3(5):553–559

Rota C, Imberti B, Pozzobon M, Piccoli M, De Coppi P, Atala A et al (2012) Human amniotic fluid stem cell preconditioning improves their regenerative potential. Stem Cells Dev 21(11):1911–1923, Pubmed Central PMCID: Pmc3396139. Epub 2011/11/10. eng

Roubelakis MG, Trohatou O, Anagnou NP (2012) Amniotic fluid and amniotic membrane stem cells: marker discovery. Stem Cells Int 2012:1–9

Schmidt D, Achermann J, Odermatt B, Breymann C, Mol A, Genoni M et al (2007) Prenatally fabricated autologous human living heart valves based on amniotic fluid derived progenitor cells as single cell source. Circulation 116(11 Suppl):I64–I70, Epub 2007/09/14. eng

Sedrakyan S, Da Sacco S, Milanesi A, Shiri L, Petrosyan A, Varimezova R et al (2012) Injection of amniotic fluid stem cells delays progression of renal fibrosis. J Am Soc Nephrol 23(4):661–673, Pubmed Central PMCID: Pmc3312511. Epub 2012/02/04. eng

Sessarego N, Parodi A, Podestà M, Benvenuto F, Mogni M, Raviolo V et al (2008) Multipotent mesenchymal stromal cells from amniotic fluid: solid perspectives for clinical application. Haematologica 93(3):339–346

Shaw S-WS (2014) Amniotic fluid stem cells for minimally invasive prenatal cell therapy. Gynecol Minimally Invasive Ther 3(1):1–6

Streubel B, Martucci-Ivessa G, Fleck T, Bittner R (1995) In vitro transformation of amniotic cells to muscle cells—background and outlook. Wien Med Wochenschr 146(9–10):216–217

Sun Q, Li F, Li H, Chen RH, Gu YZ, Chen Y et al (2015) Amniotic fluid stem cells provide considerable advantages in epidermal regeneration: B7H4 creates a moderate inflammation microenvironment to promote wound repair. Sci Rep 5:11560, Pubmed Central PMCID: Pmc4477371. Epub 2015/06/24. eng

Torricelli F, Brizzi L, Bernabei P, Gheri G, Di Lollo S, Nutini L et al (1992) Identification of hematopoietic progenitor cells in human amniotic fluid before the 12th week of gestation. Ital J Anat Embryol 98(2):119–126

Tsai M-S, Hwang S-M, Tsai Y-L, Cheng F-C, Lee J-L, Chang Y-J (2006) Clonal amniotic fluid-derived stem cells express characteristics of both mesenchymal and neural stem cells. Biol Reprod 74(3):545–551

Vadasz S, Jensen T, Moncada C, Girard E, Zhang F, Blanchette A et al (2014) Second and third trimester amniotic fluid mesenchymal stem cells can repopulate a de-cellularized lung scaffold and express lung markers. J Pediatr Surg 49(11):1554–1563

Xinaris C, Benedetti V, Novelli R, Abbate M, Rizzo P, Conti S et al (2016) Functional human podocytes generated in organoids from amniotic fluid stem cells. J Am Soc Nephrol 27(5):1400–1411, Pubmed Central PMCID: Pmc4849826. Epub 2015/10/31. eng

Yeh YC, Wei HJ, Lee WY, Yu CL, Chang Y, Hsu LW et al (2010) Cellular cardiomyoplasty with human amniotic fluid stem cells: in vitro and in vivo studies. Tissue Eng Part A 16(6):1925–1936, Epub 2010/01/14. eng

Yokoyama T, Huard J, Chancellor MB (2000) Myoblast therapy for stress urinary incontinence and bladder dysfunction. World J Urol 18(1):56–61

Yoon BS, Moon JH, Jun EK, Kim J, Maeng I, Kim JS et al (2010) Secretory profiles and wound healing effects of human amniotic fluid-derived mesenchymal stem cells. Stem Cells Dev 19(6):887–902, Epub 2009/08/19. eng

Zhu H, Lensch MW, Cahan P, Daley GQ (2011) Investigating monogenic and complex diseases with pluripotent stem cells. Nat Rev Genet 12(4):266–275

GMP-Compliant Perinatal Tissue-Derived Stem Cells

Babak Arjmand, Parisa Goodarzi, Khadijeh Falahzadeh, Hamid Reza Aghayan, Fakher Rahim, Fereshteh Mohamadi-Jahani, and Bagher Larijani

1 Introduction

Regenerative medicine is a new field which uses biological substitutes like stem cell-based therapies to repair, replace, and enhance the lost function of an organ or tissue (Riazi et al. 2009; Baghbaderani et al. 2015). Based on potential abilities of stem cells, they have a crucial role in regenerative medicine (Preynat-Seauve and Krause 2011; Ding and Schultz 2004). Accordingly, various sources of stem cells have been used for cell therapy, tissue engineering, and regenerative medicine including adult, embryonic, fetal, and perinatal tissue-derived stem cells. In recent years, perinatal stem cells as a promising cell source have been suggested for regenerative medicine. In comparison to other counterparts, these cells have great advantages such as

B. Arjmand, M.D., Ph.D. (✉)
Endocrinology and Metabolism Research Center, Endocrinology and Metabolism Clinical Sciences Institute,Tehran University of Medical Sciences, Tehran, Iran

Brain and Spinal Cord Injury Research Center, Tehran University of Medical Sciences, Tehran, Iran
e-mail: b_arjmand@farabi.tums.ac.ir

P. Goodarzi, M.Sc. • F. Mohamadi-Jahani, M.Sc.
Brain and Spinal Cord Injury Research Center, Tehran University of Medical Sciences, Tehran, Iran

K. Falahzadeh, M.Sc. • H.R. Aghayan, M.D., Ph.D.
Chronic Diseases Research Center, Endocrinology and Metabolism Population Sciences Institute, Tehran University of Medical Sciences, Tehran, Iran

F. Rahim, Ph.D.
Institute of Health Research, Thalassemia and Hemoglobinopathies Research Center, Ahvaz Jundishapur University of Medical Sciences, Ahvaz, Iran

B. Larijani, M.D.
Endocrinology and Metabolism Research Center, Endocrinology and Metabolism Clinical Sciences Institute, Tehran University of Medical Sciences, Tehran, Iran

B. Arjmand (ed.), *Perinatal Tissue-Derived Stem Cells*, Stem Cell Biology and Regenerative Medicine, DOI 10.1007/978-3-319-46410-7_9

convenient accessibility, low immunogenicity, high proliferation capacity, and minimal ethical limitations. On the other hand, they have some pluripotency properties like to embryonic stem cells (Witkowska-Zimny and Wrobel 2011b). However, good manufacturing practice (GMP) is needed to provide cells with defined quality, safety, and efficacy for clinical application (Unger et al. 2008). Today, producing of GMP-compliant stem cells is feasible and available for therapeutic uses. Like other types of stem cells, perinatal stem cells manufacturing in accordance with GMP regulations and standards could provide a valuable source of stem cells for cell therapy and regenerative medicine (Brooke et al. 2009; Barlow et al. 2008).

2 Sources of Stem Cells for Cell Therapy

2.1 Adult Stem Cells

Adult stem cells are undifferentiated cells from various tissues (Crisan et al. 2008). Mesenchymal stem cells (MSCs) are multipotent stromal cells which are the most common stem cells used for clinical cell transplantation trials. They can be differentiated into mesodermal as well as nonmesodermal cell types (Pittenger et al. 1999; Da Silva Meirelles et al. 2006; Ferrari et al. 1998; Makino et al. 1999; Zhao et al. 2011). Regarding to the immunomodulatory characteristics of MSCs they have been candidate as an immunosuppressive adjuvant therapy (Engela et al. 2012; Rojewski et al. 2008; Bieback 2008; Schafer and Northoff 2008; Abdi et al. 2008; Rasmusson 2006). First time, isolation of MSCs was reported from human bone marrow (BM) which is the common source for clinical trials (Pittenger et al. 1999; Mackay et al. 1998; Lennon and Caplan 2006). Although, BM is the most common source the number of its MSCs and their differentiation potential decrease with age and also isolation of MSCs from BM is invasive and painful. Therefore, scientists tried to find and develop an alternative source of MSCs. Therefore, researchers try to find other sources for MSCs (Bieback et al. 2008b; Aghayan et al. 2015). Accordingly, some other tissues have been introduced including adipose tissue, the gut, peripheral blood, the lung, perinatal tissues, etc. (Kern et al. 2006; Rotter et al. 2008; Lanzoni et al. 2009; Cao et al. 2005; Griffiths et al. 2005; Beltrami et al. 2003).

2.2 Human Embryonic Stem Cells

Embryonic stem cells (ESCs) were first isolated from mouse embryos in 1981 and then were isolated from inner cell mass (ICM) of human blastocysts that resulted in successful establishment of hESCs in 1998 by Thomson's group (Moon et al. 2006). Although ESCs with high proliferative capacity appeared as a suitable alternative for adult MSCs, their use in cellular therapy has raised some concerns which limit the application of hESCs in clinical trials (Jung 2009). According to pluripotent nature of hESCs, they are prone to form teratoma after transplantation (Moon et al. 2006).

From a clinical aspect, the potential of teratoma formation of hESCs should be tested before their therapeutic applications. This is considered as a safety assay which determines the efficacy of hESCs transplantation (Prokhorova et al. 2009). Moreover, it has been demonstrated that transplanted cells need long-term survival to regenerate injured tissues (Wingard et al. 2011; Hagell and Brundin 2001). In other words, immune privileged cells such as hESCs can be used for regenerative medicine without any immune rejection concerns (Charron et al. 2009; Saric et al. 2008). But, their use is still arguable because of their tumorigenicity potential and also some ethical concerns (Grinnemo et al. 2006, 2008; Drukker et al. 2006; Moon et al. 2006).

2.3 *Perinatal and Fetal Tissue-Derived Stem Cells*

Trying to find an alternative for hESCs has led to the use of perinatal tissues such as umbilical cord and placenta for cellular therapy since 1989 (Gluckman et al. 1989). Perinatal tissue-derived stem cells were isolated from human cord blood (Prindull et al. 1978). Since then, some other perinatal tissues such as placenta, cord blood tissue, chorion villi, fetal membranes, and amniotic fluid were introduced as alternative sources for cellular therapy and regenerative medicine (Bieback et al. 2004; Parolini et al. 2008; Guillot et al. 2008; Piskorska-Jasiulewicz and Witkowska-Zimny 2015; Witkowska-Zimny and Wrobel 2011a). Additionally, human fetal tissues including liver, lung, and brain were used to harvest different types of stem cells (Larijani et al. 2015a; Ghodsi et al. 2012). In spite of several advantages, fetal tissue-derived stem cells have raised significant ethical concerns. Therefore, perinatal tissue-derived stem cells have been increasingly suggested to use in regenerative medicine. One of the most common perinatal tissues used for cell therapy is umbilical cord tissue (Troyer and Weiss 2008). In addition, MSCs could be isolated from Wharton's jelly as well as the perivascular region (Seshareddy et al. 2008). The perinatal stem cell collection and processing is easy and convenient, with no risk for both mother and newborn. Perinatal tissues could provide a large number of stem cells (Bieback and Brinkmann 2010). Furthermore, in comparison with adult stem cells, perinatal tissue-derived stem cells exhibit more telomerase activity and more expression of the pluripotency genes which makes them more effective source with immune privileged and nontumorigenic properties (Bieback and Brinkmann 2010). It has been indicated that placenta-derived MSCs have a greater proliferative potential than other counterparts. Moreover, placenta-derived MSCs represented more immunosuppressive effect on T cell in comparison with BM-derived MSCs which makes them one of the most promising source for clinical applications (Heazlewood and Atkinson 2013).

3 Stem Cells and Good Manufacturing Practice

The main step in stem cells manufacturing is the isolation and in vitro culture and expansion which performs usually under research-grade conditions (Wang et al. 2012b). Clinical applications of stem cells need cell manufacturing according to current regulations, including good tissue practice (GTP) and GMP (Bieback et al. 2008b). These regulations include all steps and components involved in cell production in order to ensure the safety and quality of final cell products. Although in a cell-based manufacturing system, conversion of cell production under research-grade conditions to the GMP grade one is necessary, the majority of cellular therapy centers are not GMP certified. Since there are no quality control standard and a worldwide uniform procedure for in vitro culture and expansion of cells, findings are not in accordance with the international standards and principles (Wang et al. 2012b). According to current reports, there are various protocols for preparation of (stem) cells due to different methods used for cell preparation including culture method, growth factors, supplements, cell seeding density, subculture, cryopreservation, route of transplantation, and injection vehicle. These parameters have dramatic effects on the quality and safety of the final cell-based products (Ikebe and Suzuki 2014). So, the most appropriate and valid protocols must be standardized and developed in order to clinical- and GMP-compliant processes. Accordingly, development of a quality assurance system is required to avoid deviations from applicable regulations and current standards for cellular therapy (Wang et al. 2012b). Therefore, using GMP-grade strategies in cell-based products manufacturing enhances the efficacy and safety of perinatal stem cells.

4 GMP-Grade Raw Materials

According to regulation No. [EC] 1394/2007 of the European Parliament established by the European Medicines Agency (EMA), MSCs are included as advanced therapy medicinal products (ATMPs). Therefore, producing MSCs adhering GMP standards is necessary to ensure the safety, quality, sterility, and reproducibility of final product (Fekete et al. 2012b). The quality and safety of raw materials is critical to ensure GMP-compliant stem cells. Accordingly, clinical-grade materials should be replaced with research grade one, if possible (Bieback et al. 2011). A number of important research-grade and animal-derived reagents currently used in cellular therapy centers are discussed here.

4.1 Serum Supplements

4.1.1 Fetal Bovine Serum and Human Serum

Among various available protocols for isolation and expansion of stem cells, the most commonly used culture media is Dulbecco's Modified Eagle's Medium (DMEM) or alpha-minimum essential medium (alpha-MEM) supplemented with serum, often fetal bovine serum (FBS) (Bieback et al. 2011). FBS as the most widely used serum supplement in cell culture methods is a complex mixture of a variety of biomolecules which is used for stimulating and accelerating cell growth and proliferation (Rauch et al. 2011; Herrera and Inman 2009; Muramatsu et al. 1995; Gstraunthaler 2003). As FBS is an animal origin reagent and rich in uncharacterized components, its use is a major limitation of clinical-grade cell manufacturing (Lindroos et al. 2011; Larijani et al. 2015a). FBS contains xenogeneic proteins with potential risk for viral and prion diseases transmission and also immunological reactions that raised concerns about its use in cell manufacturing (Bieback et al. 2009, 2010; Kocaoemer et al. 2007; Lepperdinger et al. 2008; Martin et al. 2005). The etiology of transmissible spongiform encephalopathy as a fatal neurodegenerative disease is a prion which can be transmitted using animal origin materials such as FBS. It has been reported that some countries, notably Australia and New Zealand, are free of bovine spongiform encephalitis. Therefore, the Australian Therapeutic Goods Authority which has a position equal to FDA in the USA has no problem to allow FBS obtained from cattle of these countries and its use in clinic (Brooke et al. 2009). However, FBS is not a xeno-free supplement and includes xenogeneic reagents such as ruminant proteins that may cause immunogenic response and serious adverse reactions in recipients. Accordingly, the availability of clinical-grade FBS does not completely eliminate concerns about the risk for zoonotic diseases and reactions (Bieback et al. 2008b; Horwitz et al. 2002; Sundin et al. 2007). Therefore, to avoid any negative effect of FBS, it is strongly recommended to replace it with a safer xeno-free supplement. For instance, the use of human serum has been suggested as an alternative to FBS with its own advantages and disadvantages compared to FBS (Kocaoemer et al. 2007; Bieback et al. 2009, 2012; Fekete et al. 2014; Tateishi et al. 2008; Aghayan et al. 2012). According to current reports, the use of autologous human serum has led to elimination of negative effects of FBS. This replacement sounds a highly controversial topic regarding to both proliferative and differentiation capacity (Stute et al. 2004; Goodarzi et al. 2014). There are several studies demonstrating an increased proliferative capacity of MSCs using autologous human serum (Shahdadfar et al. 2005; Shigeno and Ashton 1995; Koller et al. 1998; Nimura et al. 2008; Dahl et al. 2008; Stute et al. 2004), whereas some other reports depicted no significant difference between autologous human serum and FBS in terms of proliferation rate (Spees et al. 2004; Anselme et al. 2002; Yamamoto et al. 2003). Moreover, there are some reports on positive effects of autologous human serum including stability in long-term culture, ability to differentiate into multilineage cells, high cell motility and variability in DNA methylation, antiapoptotic activity, and angiogenic effects (Shahdadfar et al. 2005; Heiskanen et al. 2007). But, comparative studies have demonstrated conflicting results regarding to osteogenesis and adipogenesis potential of MSCs (Yamamoto et al. 2003; Oreffo and Triffitt 1999). On the other hand, due to variability observed in different batches of autologous human serum, isolated serum from different patients was pooled to solve this problem. Pooling of sera resulted in reduced colony formation and proliferation with swift senescence in cell culture (Shahdadfar et al. 2005). These limitations of autologous human serum have been considered as a serious problem to develop the clinical applications of human serum (Nimura et al. 2008; Brinchmann 2008; Bieback et al. 2009; Stute et al. 2004).

4.1.2 Platelet Lysate

Other reports have revealed that human platelet lysate (HPL) can be a safe and effective substitute for FBS (Schallmoser et al. 2007, 2009; Schallmoser and Strunk 2013; Fekete et al. 2012a; Iudicone et al. 2014). HPL is a xeno-free serum substitute which contains a large amount of growth factors (Bernardi et al. 2013). HPL can increase expansion capacity, differentiation potential, and immunomodulatory effects of MSCs (Muller et al. 2006; Gottipamula et al. 2012; Doucet et al. 2005). The same as autologous human serum, HPL is a variable supplement and to reduce this variability, it is obtained from different donors after pooling (Bieback 2013; Wuchter et al. 2015). Moreover, its storage at −20 °C for a long period of time is possible with no effect on the growth factor content (Kocaoemer et al. 2007; Fekete et al. 2012a; Rauch et al. 2011). Obviously, HPL could be added to the media until the end of its shelf life which is around 4–6 days following blood donation (Bieback 2013). Body of literature has revealed that HPL could be used as a GMP-compliant substitute for FBS (Avanzini et al. 2009; Bernardo et al. 2007; Bieback et al. 2009; Crespo-Diaz et al. 2011; Lange et al. 2007; Schallmoser et al. 2007, 2009; Griffiths et al. 2013; Castiglia et al. 2014). According to microarray analyses, it has been demonstrated that cell culture conditions have profound effects on the gene expression profile in terms of differentiation, development, the interaction between cells, adhesion to the extracellular matrix, TGF-β signaling, apoptosis, cell cycle, DNA replication, and purine metabolism (Lange et al. 2007; Bieback et al. 2010). Fekete et al. indicated that both blood-derived pooled and apheresis-derived platelet concentrates could be used as the clinical-grade supplement for isolation and expansion of MSCs (Fekete et al. 2012a). Both kinds of HPL included the same cytokines such as basic fibroblast growth factor (bFGF), sCD40L, platelet-derived growth factor AA (PDGF-AA), PDGF-AB/AA, sVCAM-1, sICAM-1, RANTES, and TGF-β1 which led to increased proliferation of MSCs (Kinzebach and Bieback 2013). GMP certified alternatives for FBS including HPL have represented different characteristics regarding to proliferative capacity and gene expression profile. The comparative studies intended to clarify effects of human serum on cell culture system, demonstrated that HPL have resulted in accelerated proliferative activity without any effect on chromosomal stability compared to traditional FBS (Crespo-Diaz et al. 2011; Dahl et al. 2008). From a clinical point of view, care should be taken to assess the potential effects of these changes on cell culture system in order to reduce the risk for using HPL in clinical applications. Because of using serum in cell culture gives rise to high lot-to-lot variability in cellular growth rate and the probability of disease transmission, another alternative including human serum called serum-free (SF) media was represented (Tonti and Mannello 2008).

4.1.3 Serum-Free Medium

Novel serum substitutes as serum-free media proposed to decrease the risks and limitations of conventional serum supplemented media (Koller et al. 1998; Parker et al. 2007; Lian et al. 2007; Meuleman et al. 2006; Chase et al. 2010). There is promising evidence for replacement of autologous human serum added media by serum-free media indicating higher proliferative capacity and multipotent state maintenance (Lindroos et al. 2009; Rajala et al. 2010; Parker et al. 2007). Hartmann et al. has successfully reported the isolation and expansion of umbilical cord tissue-derived MSCs in a serum- and xeno-free medium {Hartmann, 2010 #44}. Although they used first StemPro MSC serum-free media supplemented with 2 % GMP-grade human serum, they succeeded to develop their protocol to a completely serum- and xeno-free conditions. Additionally, they have demonstrated significant increase in proliferative capacity and viability of MSCs cultured in xeno-free and serum-free GMP-compliant manner (Hartmann et al. 2010). There are various studies demonstrating MSCs isolation and expansion in xeno- and serum-free culture system as an essential requirement for translation from the basic to the clinic (Patrikoski et al. 2013). Swamynathan et al. have reported the isolation and large-scale expansion of MSCs derived from Wharton jelly in a xeno-free and serum-free condition (Swamynathan et al. 2014). Therefore, the optimizing of serum-free media could be considered as a safer and more efficient method for manufacturing of MSCs for clinical applications (Chase et al. 2010; Gottipamula et al. 2013). Despite several advantages of this newly introduced method, it needs to be optimized for large-scale expansion of perinatal tissue-derived stem cells in accordance with GMP regulations.

4.2 Enzymes

From the clinical perspective, optimal GMP compliance needs a completely xeno-free condition for the isolation, expansion, and cryopreservation of cells (Hartmann et al. 2010). Some studies have revealed adverse reactions including anaphylaxis and immune reactions caused by cell processing by animal origin reagents and supplements (Mackensen et al. 2000; Selvaggi et al. 1997). Furthermore, the potential risk for transmission of viral and bacterial infections, prions, and other types of transmissible diseases between animals and humans should not be neglected (Selvaggi et al. 1997; Will et al. 1996). Ficoll and animal origin reagents and enzymes such as collagenase, dispase, and trypsin, which are commonly used in research-based cell culture, are not GMP or clinical grade and therefore not appropriate for therapeutic use. Thus, safer alternatives should be used in clinical intended uses (Bergstrom et al. 2011; Swamynathan et al. 2014; Arjmand and Aghayan 2014; Ilic et al. 2011). Enzymatic dissociation is an important step in stem cell isolation and the most common used enzymes such as collagenase contains animal origin components, therefore it should be replaced by a GMP-compliant alternative in clinical trials. A number of GMP approved collagenase including CLSAFA (Worthington, Lakewood, NJ), NB-6 GMP grade (Serva, Heidelberg, Germany), and Liberase MTF-S GMP grade (Roche Diagnostics, Basel, Switzerland) are available (Crook et al. 2007; Szot et al. 2009; Aghayan et al. 2015; Arjmand and Aghayan 2014; Carvalho et al. 2013). Carvalho et al. have identified that the use of animal origin free and also GMP-grade collagenase have been led to the same differentiation potential and expression of cell surface markers of adipose-derived stem cells. Moreover, the use of clinical-grade collagenase had no negative effect on the yield and functional properties of human adipose-derived stem cells and it could be considered as a safe alternative (Carvalho et al. 2013). The most commonly available trypsin is derived from porcine that should be substituted by its xeno-free counterparts such as TrypLE Select (Invitrogen, Carlsbad, CA) and TrypZean (Sigma-Aldrich, St. Louis, MO) (Arjmand and Aghayan 2014). TrypLESelect has been represented as a recombinant animal and human component-free trypsin. This proteolytic enzyme is a safe clinical-grade enzyme that could be used in clinical trials (Bergstrom et al. 2011; Swamynathan et al. 2014; Chen et al. 2013; Larijani et al. 2015b). Additionally, GMP-grade Ficoll-Paque PREMIUM (GE Healthcare Life Sciences, USA), as a substitute for routinely used Ficoll, has been producing as a high-performance, animal-origin-free reagent (Arjmand and Aghayan 2014; Ilic et al. 2011; Arjmand et al. 2012). Recent advancements in regenerative medicine give rise to commercially available xeno-free reagents which are compatible with GMP regulations.

5 Long Storage and Cryopreservation

It has been demonstrated that clinical use of manufactured MSCs needs their long-term storage (Cooper and Viswanathan 2011; Gong et al. 2012; Thirumala et al. 2009). Thus, the final cell-based product should be cryopreserved for long-term storage. According to various reports, MSCs maintain their properties and functions after freezing and thawing (Todorov et al. 2010; Gordon et al. 2001; Pal et al. 2008; Bruder et al. 1997). Additionally, due to the complexity and low probability of human leukocyte antigen (HLA) matching process, cryopreservation and storage of perinatal stem cells from different tissues such as cord blood, placenta, and umbilical cord could be a potential valuable source for treatment of different disorders (Wang et al. 2012b). Therefore, cryopreservation and long storage of MSCs according to GMP conditions is an important issue in clinical-grade manufacturing of perinatal tissue-derived stem cells. Cryopreservation makes it possible to transport cell-based products across international borders (Wang et al. 2012b). In order to long term storage, liquid nitrogen or the gas phase above the liquid phase is used (Mccullough et al. 2010).

5.1 Cross-contamination

Since stored products in liquid nitrogen are prone to cross-contamination, it is suggested to use a secondary system which protects the primary cryovials containing cellular product if possible. The potential risk for contamination of stored cells in gas phase of liquid nitrogen is not considerable (Hunt 2011). Contrary to the GMP regulations, the most commonly used cryovials could not be properly sealed leading to a leakage of liquid nitrogen into the cryovials during immersion, over time. The use of heat-sealable membrane to cover the cap of vials has been employed to prevent liquid nitrogen infiltration into the cryovials with no effect on the viability of postthawed cells (Chen et al. 2006a). More recently, another type of cryovials suitable for pharmaceutical applications has been developed to cryopreserve stem cells manufactured for clinical applications (Woods et al. 2010). Therefore, implementing an efficient quality management system ensures the safety and quality of final product avoiding possible cross-contamination.

5.2 Cryoprotectants

During freezing and thawing processes, moving between −15 and −60 °C is the most critical point which contributes to cellular damage (Mazur 1988). Both slow and rapid cooling of cells has their own destructive effects. Slow cooling leads to increased osmolality in the cell's extracellular environment due to ice crystal formation. This state results in passing of water from the cell membrane and dehydration, in order to equalize osmolality. Rapid cooling has been shown to cause the ice formation inside the cells (Li and Ma 2012). Since all present freezing and thawing protocols destroy the cell structure, it is essential to use safer cryoprotectants (Fuller 2004; Janz Fde et al. 2012). A variety of cryoprotectants including low molecular weight compounds such as dimethylsulfoxide (DMSO), ethylene glycol (EG), and propylene glycol (PG) permeate the cells and prevent intracellular ice formation. Sugars such as sucrose and trehalose, and high molecular weight polymers like polyvinylpyrrolidone (PVP) and hydroxyl-ethyl starch (HES) with different functions are also available (Fuller 2004; Meryman 2007). The most routinely used cryopreservation medium for MSCs generally contains FBS and DMSO (Liu et al. 2010). In accordance with GMP-compliant protocols, the use of FBS and DMSO should be limited. Toxicity of DMSO is influenced by different factors such as time, temperature, concentration, and cell type. Some clinical trials have reported the adverse reactions related to DMSO in patients (Berz et al. 2007; Galvao et al. 2014). To eliminate toxic effects of DMSO, washout process using Dextran, as an osmotic buffer, compensates imbalance osmolality after removal of DMSO. Also, some automated cell washing methods have been developed (Berz et al. 2007; Rodriguez et al. 2004). Another way to decrease toxicity and side effects of DMSO is using of reduced concentration of DMSO for cryopreservation. It has been revealed that the use of 5 % DMSO (instead of 10 %) can reduce adverse reactions (Rubinstein et al. 1995; Woods et al. 2003; Morris et al. 2014). Berz et al. have reported the successful use of 2 % DMSO (Berz et al. 2007). The reduced concentration of DMSO from 2 to 0 % had lethal effects on adipose-derived stem cells that highlighted the critical concentration of DMSO for cell cryopreservation (Thirumala et al. 2009). However, since DMSO is a toxic agent, it is highly recommended to replace it with a safer cryoprotectant with similar efficacy. Using of hydroxyethyl starch and trehalose as alternatives for DMSO needs further clarification and optimization (Hayakawa et al. 2010; Mccullough et al. 2010; Buchanan et al. 2004; Stolzing et al. 2012; Motta et al. 2014). There is a variety of cryoprotectants which achieved European Conformity (CE) certification. CE-marked cryoprotectants are introduced as GMP-compliant agents and are available for therapeutic applications (Hunt 2011). Thus, the use of commercially available cryoprotectants with validation for sterility and bacterial endotoxin test is infinitely preferable to the homemade one. Furthermore, in order to manufacturing of hESCs and induced pluripotent stem (iPS) cells, successful use of commercially produced cryopreservation media and wash solutions has been reported (Holm et al. 2010). Although, the components of commercially available cryoprotectant are disclosed as a combination of DMSO, glucose, and a high molecular weight polymer in phosphate-buffered saline (PBS), the characteristic of high molecular weight polymer is not revealed. These solutions show high efficiency recovery of cryopreserved stem cells, which represent an appropriate postthawing method. On the other hand, there are some concerns about uncharacterized properties and potential effects of high molecular weight polymer used in cryoprotectants which should not be ignored (Hunt 2011).

6 GMP Facility and Staff Training

As in GMP certified cell manufacturing centers all procedures should be performed in accordance with GMP regulations, the processes lead to the reproducible results with the highest standards of safety and quality for final product (Unger et al. 2008). Prolonged cell culture may cause contamination and since the sterilization of cell-based products inevitably leads to a reduction of their biological functions, the final products of a cell manufacturing process should be a sterilized product resulted from aseptic processing. Furthermore, due to the lack of a specific test for detecting all probable contaminations, it is essential to use sterile raw materials and adhere to the aseptic techniques during cell manufacturing. Therefore, controlling the environment of cell manufacturing facility in addition to adhering to all components of GMP guideline can provide a condition to avoid exogenous contaminations of cell-based products (Arjmand et al. 2012).

6.1 GMP Facility

A GMP facility should be designed by a group of highly experienced architects and also GMP experts with a compliance to the GMP regulations (Burger 2000; Arjmand et al. 2012). In international standard organization (ISO) 14644-1, a GMP facility (clean room) is described as "a room in which the concentration of airborne particles is controlled, and which is constructed and used in a manner to minimize the introduction, generation, and retention of particles inside the room, and in which other relevant parameters, e.g., temperature, humidity, and pressure, are controlled as necessary" (Arjmand et al. 2012). Some noteworthy elements in GMP facility are controlled temperature and humidity, easily cleanable work surfaces and walls and

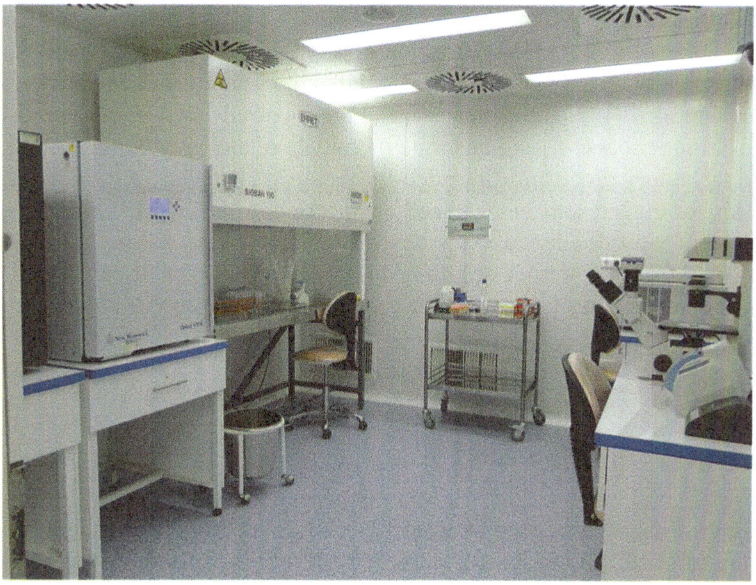

Fig. 1 GMP-compliant cell manufacturing facility

ceilings, high efficiency particulate air (HEPA) filtration system under positive pressure, airborne particle monitoring in clean room environments (Fig. 1), routine cleaning, and disinfection and decontamination solutions (Burger 2003). In order to clean room validation, it is essential to qualify and test procedures performing in clean room. Three main stages of testing a clean room are as-built (phase I), at-rest (phase II), and operational (phase II) testing. Phase I testing is intended for installation qualification after fully installed equipment to validate the correct installation of equipment. Phase II should be done for operational qualification during production to confirm that all equipments operate as desired. And finally, operational testing is performed when the personnel are working in accordance to prove that the whole process of cell manufacturing is performed according to GMP regulations at the presence of the staff who are involved in cellular manufacturing (O'Donoghue 2011). In addition, the location of a GMP facility is very important. An ideal situation would be the proximity of hospitals to the clinical-grade cell manufacturing centers. All GMP facilities need initial budget evaluation by a GMP specialist to improve and ensure the quality and safety of final product as the main objective of GMP guidelines (Burger 2000).

6.2 *Staff Training*

Development of a GMP facility will be achieved by the GMP experts in cooperation with the quality assurance (QA) personnel. All staff working in GMP areas should pass GMP training courses which are in accordance with QMS standards. A combined QMS and GMP approach could give rise to a translation to GMP adherent laboratory (Ilic et al. 2011). It is noteworthy that application of QMSs such as ISO 9001 and ISO 13485 will help the GMP facility performance and improvement (Arjmand et al. 2012). The personnel who are involved in GMP-grade cell-based manufacturing process should be aware of guidelines and specific regulations relevant to laboratory procedures intended for therapeutic applications. In addition to the needs for trained staff in GMP cell engineering skills, a GMP facility requires support staff who developed skills beyond those acquired at the bench (Burger 2000). Therefore, the presence of trained and expert staff in well-engineered clean rooms is considered as key points for cell manufacturing in GMP-compliant cell production centers.

7 Large-Scale Expansion

The fast growing advancement in regenerative medicine leads to commercial manufacturing of stem cells in recent years. A high level of quantity of stem cells is required for cellular therapy which highlights the necessity for large-scale manufacturing of cell-based products. It should be considered that successful commercialization of cell-based products requires a high level of compliance to the GMP regulations. The prevalent approach to stem cell culture is the use of cell culture flasks and containers that compared to newly introduced bioreactors is a time- and money-consuming method for mass production (Schallmoser et al. 2008; Sensebe 2008; Chen et al. 2006b; Gastens et al. 2007). As GMP-compliant cell manufacturing requires safe and viable cells under controlled conditions, implementation of bioreactors will facilitate the large-scale and well-controlled stem cell production. The disposable bioreactor systems could provide a safe and time-efficient tool with the elimination of possible risk of cross-contamination (Schallmoser et al. 2008; Sensebe 2008; dos Santos et al. 2014; Elseberg et al. 2015). Besides the limitations of available bioreactors their suppliers and manufacturers try to do their best for development of better defined conditions of effective bioreactors for stem cell expansion in accordance with GMP regulations. Since stem cell therapies need a large number of safe and efficient stem cells manufactured under GMP-compliant culture systems, the use of bioreactors is considered as a cost-effective tool to respond clinical-grade stem cell market demand (dos Santos et al. 2013, 2014; Elseberg et al. 2015). The large-scale production of stem cells requires the replacement of conventional cell culture system with automated devices (Eibes et al. 2010; Thomas et al. 2009a, b). Recently, scale-up expansion of hESCs in a stirred micro carrier system was reported as a method to improve the yield of cell-based product. This improvement is probably resulted from improved oxygen and nutrient diffusion rates and advanced mass transport of metabolites parallel to reduced toxic effects (Fernandes et al. 2009). In this system adherent cells could be cultivated in suspension leading to larger amount of manufactured cells in comparison with conventional methods (Nie et al. 2009). Although all bioreactor systems as a cost-effective method with increased postthawing recovery are not GMP certified, scientist try to develop micro carrier system for large-scale expansion of different types of cells in accordance with GMP regulations for therapeutic applications (Carmelo et al. 2015a, b; Eibes et al. 2010; Nie et al. 2009; Thomas et al. 2009b).

8 Discussion and Conclusion

The recent advancement in regenerative medicine is offering novel promising clinical applications for stem cells from different tissues. Therapeutic applications require the compliance of cell manufacturing to GMP standards in order to optimum assurance of produced cells for clinical applications (George 2011). In recent years, among different types of stem cells, MSCs are the most commonly used cells for cellular therapy due to their regenerative potential (Lazarus et al. 1995; Horwitz et al. 2002; Garcia-Olmo et al. 2005; Le Blanc et al. 2004; Wei et al. 2013; Lewis and Suzuki 2014; Wang et al. 2012a). MSCs have been used for treatment of a variety of diseases such as blood disease, acute respiratory distress syndrome, spinal cord injury, liver injury, autoimmune diseases, bone disorders, chronic myocardial infarction, critical limb ischemia, etc. (Hayes et al. 2012; Ishikane et al. 2008; Nakajima et al. 2012; Puglisi et al. 2011; Hass et al. 2011; Wang et al. 2012a). The considerable advantages of MSCs which make them an interesting candidate for cell therapy is easy accessibility, availability in different adult tissues such as BM, adipose tissue, and even fetal tissues, as well as perinatal tissues (Kern et al. 2006; Bieback et al. 2008a; Bartmann et al. 2007; Reinisch et al. 2007; Piskorska-Jasiulewicz and Witkowska-Zimny 2015; Wouters et al. 2007). Perinatal tissues are often discarded as medical wastes, while they can solve the ethical problems related to fetal tissues as an alternative source for cell therapy and regenerative medicine. Although regenerative medicine has revealed the amazing versatility of stem cells, in the majority of cellular therapy centers cell manufacturing process is not performed in accordance with GMP guidelines (Bieback et al. 2008b). GMP guidelines cover quality and safety standards for manufacturing cell-based products for clinical transplantation. GMP regulations include all aspects of cell manufacturing process including donor eligibility, ethical concerns, controlled environment, equipments, staff training, raw material and reagents, storage conditions, and all steps of manufacturing processes to ensure the quality, purity, reproducibility, and safety of final products. Since the progress in regenerative medicine happen rapidly, worldwide development of GMP facilities should be accelerated. In terms of raw materials required for cell processing, an ideal condition would be the use of xeno-free or serum-free reagents. One of the problems caused by xenogeneic reagents in clinical trials is the probability of transferring animal derivatives into the patient's body resulting in immune reaction as a consequence of cell transplantation (Spees et al. 2004; Heiskanen et al. 2007; Martin et al. 2005; Lepperdinger et al. 2008). Moreover, the strong possibility of viral or bacterial infections transmission to the recipients is another serious concern (Selvaggi et al. 1997; Will et al. 1996; Lepperdinger et al.

2008). Serum proteins deliver important nutrients and attachment factors to cells, therefore it is essential for cell culture procedure. The most commonly used serum supplement is animal-derived FBS that according to GMP regulations should be replaced by a xeno-free substitute. Autologous or allogeneic human serum or platelet-derived factors contain a mixture of required factors—including some unknown molecules—for stem cell culture that should be considered as an appropriate and safe alternative for FBS (Bieback et al. 2008b, 2009; Mannello and Tonti 2007; Kocaoemer et al. 2007; Bieback 2013). On the other hand, human serum shows high variability and negative effects on proliferative capacity of cells have limited its use in regenerative medicine (Tapp et al. 2009; Herrera and Inman 2009; Gstraunthaler 2003; Nimura et al. 2008; Su et al. 2009; Luttun et al. 2006; Caterson et al. 2002; Frechette et al. 2005; Johansson et al. 2003; Salvade et al. 2010; Witzeneder et al. 2013). Moreover, disease transmission possibility adding to human serum limitations and problems have strongly proposed the development of serum-free or xeno-free cell manufacturing methods to eliminate adverse effects of serum (Tonti and Mannello 2008). There is promising evidences for replacement of serum-based media by serum-free or xeno-free media (Lindroos et al. 2009; Rajala et al. 2010; Parker et al. 2007; Chase et al. 2010; Gottipamula et al. 2013). Development of serum-free media with greater proliferation effects represents a potential GMP grade alternative for traditional serum supplemented media. Since the use of serum-free media is a growing area of stem cell research, it requires further investigations to clarify the possible effects of this media on the safety and quality of cell-based products. In order to achieve a high level of compliance to the GMP standards, other animal-derived reagents like trypsin and collagenase should be replaced by clinical-grade alternatives such as TrypLE Select and GMP grade collagenase, respectively (Aghayan et al. 2015; Carvalho et al. 2013). Moreover, cryoprotectant agents have an important role in cryopreservation efficacy and postthawing viability of cells. Cryopreservation of human stem cells is one of the critical steps required to provide stocks of stored cells (Li and Ma 2012). The current reagent used for laboratory cryopreservation is 10% DMSO whose toxicity limits its clinical applications (Wang et al. 2012b; Hunt 2011). To alleviate the problems caused by DMSO, the use of reduced concentrations of DMSO (2%), CE-marked DMSO, and commercially available hydroxy ethyl starch and trehalose are suggested (Hunt 2011). According to GMP guidelines, all operational process should be performed in a GMP facility (clean room) with controlled temperature, air filtration, and sterility by qualified staff with a comprehensive understanding of GMP regulations and guidelines. Clean rooms should be engineered by expert architects in collaboration with biologists and GMP experts following regular validation tests to ensure adherence of staff, facility, equipments, and procedures to GMP principles. Today GMP-compliant stem cells including perinatal tissue-derived stem cells can be manufactured in accordance with GMP regulations to provide a valuable source for cell therapy and regenerative medicine (Brooke et al. 2009; Barlow et al. 2008). In summary, the use of stem cells for cellular therapy and regenerative medicine needs to be guaranteed for safety and quality. Accordingly, all aspects of the cell-based products manufacturing from the donor to the recipient should be compatible with GMP regulations to ensure the safety and efficacy of final product. Therefore, cellular therapy centers should not only obtain general quality assurance programs including ISO 9001 and ISO 13485 but also should meet GMP requirements.

Acknowledgements The authors would like to acknowledge Dr. Bahram Moazami, Maryam moghadari, and Maryam sadat Gousheh for their kind support in procurement of the donated tissues. We also like to thank Dr. Mohamad Vasei, Dr. Nasser Ahmadbeigi, Dr. Yousof Gheisari, Dr. Hossein Adibi, Shokouh Salimi, Hanieh Rostamabadi, and Azam Ranjbar.

References

Abdi R, Fiorina P, Adra CN, Atkinson M, Sayegh MH (2008) Immunomodulation by mesenchymal stem cells: a potential therapeutic strategy for type 1 diabetes. Diabetes 57:1759–1767

Aghayan H-R, Arjmand B, Norouzi-Javidan A, Saberi H, Soleimani M, Tavakoli SA-H, Khodadadi A, Tirgar N, Mohammadi-Jahani F (2012) Clinical grade cultivation of human Schwann cell, by the using of human autologous serum instead of fetal bovine serum and without growth factors. Cell Tissue Bank 13:281–285

Aghayan HR, Goodarzi P, Larijani B, Mohamadi-Jahani F, Norouzi-Javidan A, Dehpour AR, Fallahzadeh K, Sayahpour FA, Arjmand B (2015) Clinical grade human adipose tissue-derived mesenchymal stem cell banking. Acta Med Iran 53(9):540–546

Anselme K, Broux O, Noel B, Bouxin B, Bascoulergue G, Dudermel AF, Bianchi F, Jeanfils J, Hardouin P (2002) In vitro control of human bone marrow stromal cells for bone tissue engineering. Tissue Eng 8:941–953

Arjmand B, Aghayan HR (2014) Cell manufacturing for clinical applications. Stem Cells 32:2557–2558

Arjmand B, Emami-Razavi SH, Larijani B, Norouzi-Javidan A, Aghayan HR (2012) The implementation of tissue banking experiences for setting up a cGMP cell manufacturing facility. Cell Tissue Bank 13:587–596

Avanzini MA, Bernardo ME, Cometa AM, Perotti C, Zaffaroni N, Novara F, Visai L, Moretta A, Del Fante C, Villa R, Ball LM, Fibbe WE, Maccario R, Locatelli F (2009) Generation of mesenchymal stromal cells in the presence of platelet lysate: a phenotypic and functional comparison of umbilical cord blood- and bone marrow-derived progenitors. Haematologica 94:1649–1660

Baghbaderani BA, Tian X, Neo BH, Burkall A, Dimezzo T, Sierra G, Zeng X, Warren K, Kovarcik DP, Fellner T (2015) cGMP-manufactured human induced pluripotent stem cells are available for pre-clinical and clinical applications. Stem cell Rep 5:647–659

Barlow S, Brooke G, Chatterjee K, Price G, Pelekanos R, Rossetti T, Doody M, Venter D, Pain S, Gilshenan K, Atkinson K (2008) Comparison of human placenta- and bone marrow-derived multipotent mesenchymal stem cells. Stem Cells Dev 17:1095–1107

Bartmann C, Rohde E, Schallmoser K, Purstner P, Lanzer G, Linkesch W, Strunk D (2007) Two steps to functional mesenchymal stromal cells for clinical application. Transfusion 47:1426–1435

Beltrami AP, Barlucchi L, Torella D, Baker M, Limana F, Chimenti S, Kasahara H, Rota M, Musso E, Urbanek K, Leri A, Kajstura J, Nadal-Ginard B, Anversa P (2003) Adult cardiac stem cells are multipotent and support myocardial regeneration. Cell 114:763–776

Bergstrom R, Strom S, Holm F, Feki A, Hovatta O (2011) Xeno-free culture of human pluripotent stem cells. Methods Mol Biol 767:125–136

Bernardi M, Albiero E, Alghisi A, Chieregato K, Lievore C, Madeo D, Rodeghiero F, Astori G (2013) Production of human platelet lysate by use of ultrasound for ex vivo expansion of human bone marrow-derived mesenchymal stromal cells. Cytotherapy 15:920–929

Bernardo ME, Avanzini MA, Perotti C, Cometa AM, Moretta A, Lenta E, Del Fante C, Novara F, de Silvestri A, Amendola G, Zuffardi O, Maccario R, Locatelli F (2007) Optimization of in vitro expansion of human multipotent mesenchymal stromal cells for cell-therapy approaches: further insights in the search for a fetal calf serum substitute. J Cell Physiol 211:121–130

Berz D, Mccormack EM, Winer ES, Colvin GA, Quesenberry PJ (2007) Cryopreservation of hematopoietic stem cells. Am J Hematol 82:463–472

Bieback K (2008) Basic biology of mesenchymal stem cells. Transfus Med Hemother 35:151–152

Bieback K (2013) Platelet lysate as replacement for fetal bovine serum in mesenchymal stromal cell cultures. Transfus Med Hemother 40:326–335

Bieback K, Brinkmann I (2010) Mesenchymal stromal cells from human perinatal tissues: from biology to cell therapy. World J Stem Cells 2:81–92

Bieback K, Kern S, Kluter H, Eichler H (2004) Critical parameters for the isolation of mesenchymal stem cells from umbilical cord blood. Stem Cells 22:625–634

Bieback K, Kern S, Kocaomer A, Ferlik K, Bugert P (2008a) Comparing mesenchymal stromal cells from different human tissues: bone marrow, adipose tissue and umbilical cord blood. Biomed Mater Eng 18:S71–S76

Bieback K, Schallmoser K, Kluter H, Strunk D (2008b) Clinical protocols for the isolation and expansion of mesenchymal stromal cells. Transfus Med Hemother 35:286–294

Bieback K, Hecker A, Kocaomer A, Lannert H, Schallmoser K, Strunk D, Kluter H (2009) Human alternatives to fetal bovine serum for the expansion of mesenchymal stromal cells from bone marrow. Stem Cells 27:2331–2341

Bieback K, Ha VA, Hecker A, Grassl M, Kinzebach S, Solz H, Sticht C, Kluter H, Bugert P (2010) Altered gene expression in human adipose stem cells cultured with fetal bovine serum compared to human supplements. Tissue Eng Part A 16:3467–3484

Bieback K, Kinzebach S, Karagianni M (2011) Translating research into clinical scale manufacturing of mesenchymal stromal cells. Stem Cells Int 2010:193519

Bieback K, Hecker A, Schlechter T, Hofmann I, Brousos N, Redmer T, Besser D, Kluter H, Muller AM, Becker M (2012) Replicative aging and differentiation potential of human adipose tissue-derived mesenchymal stromal cells expanded in pooled human or fetal bovine serum. Cytotherapy 14:570–583

Brinchmann JE (2008) Expanding autologous multipotent mesenchymal bone marrow stromal cells. J Neurol Sci 265:127–130

Brooke G, Rossetti T, Pelekanos R, Ilic N, Murray P, Hancock S, Antonenas V, Huang G, Gottlieb D, Bradstock K, Atkinson K (2009) Manufacturing of human placenta-derived mesenchymal stem cells for clinical trials. Br J Haematol 144:571–579

Bruder SP, Jaiswal N, Haynesworth SE (1997) Growth kinetics, self-renewal, and the osteogenic potential of purified human mesenchymal stem cells during extensive subcultivation and following cryopreservation. J Cell Biochem 64:278–294

Buchanan SS, Gross SA, Acker JP, Toner M, Carpenter JF, Pyatt DW (2004) Cryopreservation of stem cells using trehalose: evaluation of the method using a human hematopoietic cell line. Stem Cells Dev 13:295–305

Burger SR (2000) Design and operation of a current good manufacturing practices cell-engineering laboratory. Cytotherapy 2:111–122

Burger SR (2003) Current regulatory issues in cell and tissue therapy. Cytotherapy 5:289–298

Cao C, Dong Y, Dong Y (2005) [Study on culture and in vitro osteogenesis of blood-derived human mesenchymal stem cells]. Zhongguo Xiu Fu Chong Jian Wai Ke Za Zhi 19:642–647

Carmelo JG, Fernandes-Platzgummer A, Cabral JM, da Silva CL (2015a) Scalable ex vivo expansion of human mesenchymal stem/stromal cells in microcarrier-based stirred culture systems. Methods Mol Biol 1283:147–159

Carmelo JG, Fernandes-Platzgummer A, Diogo MM, da Silva CL, Cabral JM (2015b) A xeno-free microcarrier-based stirred culture system for the scalable expansion of human mesenchymal stem/stromal cells isolated from bone marrow and adipose tissue. Biotechnol J 10:1235–1247

Carvalho PP, Gimble JM, Dias IR, Gomes ME, Reis RL (2013) Xenofree enzymatic products for the isolation of human adipose-derived stromal/stem cells. Tissue Eng Part C Methods 19:473–478

Castiglia S, Mareschi K, Labanca L, Lucania G, Leone M, Sanavio F, Castello L, Rustichelli D, Signorino E, Gunetti M, Bergallo M, Bordiga AM, Ferrero I, Fagioli F (2014) Inactivated human platelet lysate with psoralen: a new perspective for mesenchymal stromal cell production in Good Manufacturing Practice conditions. Cytotherapy 16:750–763

Caterson EJ, Nesti LJ, Danielson KG, Tuan RS (2002) Human marrow-derived mesenchymal progenitor cells: isolation, culture expansion, and analysis of differentiation. Mol Biotechnol 20:245–256

Charron D, Suberbielle-Boissel C, Al-Daccak R (2009) Immunogenicity and allogenicity: a challenge of stem cell therapy. J Cardiovasc Transl Res 2:130–138

Chase LG, Lakshmipathy U, Solchaga LA, Rao MS, Vemuri MC (2010) A novel serum-free medium for the expansion of human mesenchymal stem cells. Stem Cell Res Ther 1:8

Chen HI, Tsai CD, Wang HT, Hwang SM (2006a) Cryovial with partial membrane sealing can prevent liquid nitrogen penetration in submerged storage. Cryobiology 53:283–287

Chen X, Xu H, Wan C, Mccaigue M, Li G (2006b) Bioreactor expansion of human adult bone marrow-derived mesenchymal stem cells. Stem Cells 24:2052–2059

Chen AK, Reuveny S, Oh SK (2013) Application of human mesenchymal and pluripotent stem cell microcarrier cultures in cellular therapy: achievements and future direction. Biotechnol Adv 31:1032–1046

Cooper K, Viswanathan C (2011) Establishment of a mesenchymal stem cell bank. Stem Cells Int 2011:905621

Crespo-Diaz R, Behfar A, Butler GW, Padley DJ, Sarr MG, Bartunek J, Dietz AB, Terzic A (2011) Platelet lysate consisting of a natural repair proteome supports human mesenchymal stem cell proliferation and chromosomal stability. Cell Transplant 20:797–811

Crisan M, Yap S, Casteilla L, Chen CW, Corselli M, Park TS, Andriolo G, Sun B, Zheng B, Zhang L, Norotte C, Teng PN, Traas J, Schugar R, Deasy BM, Badylak S, Buhring HJ, Giacobino JP, Lazzari L, Huard J, Peault B (2008) A perivascular origin for mesenchymal stem cells in multiple human organs. Cell Stem Cell 3:301–313

Crook JM, Peura TT, Kravets L, Bosman AG, Buzzard JJ, Horne R, Hentze H, Dunn NR, Zweigerdt R, Chua F, Upshall A, Colman A (2007) The generation of six clinical-grade human embryonic stem cell lines. Cell Stem Cell 1:490–494

Da Silva Meirelles L, Chagastelles PC, Nardi NB (2006) Mesenchymal stem cells reside in virtually all post-natal organs and tissues. J Cell Sci 119:2204–2213

Dahl JA, Duggal S, Coulston N, Millar D, Melki J, Shahdadfar A, Brinchmann JE, Collas P (2008) Genetic and epigenetic instability of human bone marrow mesenchymal stem cells expanded in autologous serum or fetal bovine serum. Int J Dev Biol 52:1033–1042

Ding S, Schultz PG (2004) A role for chemistry in stem cell biology. Nat Biotechnol 22:833–840

dos Santos FF, Andrade PZ, da Silva CL, Cabral JM (2013) Bioreactor design for clinical-grade expansion of stem cells. Biotechnol J 8:644–654

dos Santos F, Campbell A, Fernandes-Platzgummer A, Andrade PZ, Gimble JM, Wen Y, Boucher S, Vemuri MC, da Silva CL, Cabral JM (2014) A xenogeneic-free bioreactor system for the clinical-scale expansion of human mesenchymal stem/stromal cells. Biotechnol Bioeng 111:1116–1127

Doucet C, Ernou I, Zhang Y, Llense JR, Begot L, Holy X, Lataillade JJ (2005) Platelet lysates promote mesenchymal stem cell expansion: a safety substitute for animal serum in cell-based therapy applications. J Cell Physiol 205:228–236

Drukker M, Katchman H, Katz G, Even-Tov Friedman S, Shezen E, Hornstein E, Mandelboim O, Reisner Y, Benvenisty N (2006) Human embryonic stem cells and their differentiated derivatives are less susceptible to immune rejection than adult cells. Stem Cells 24:221–229

Eibes G, dos Santos F, Andrade PZ, Boura JS, Abecasis MM, da Silva CL, Cabral JM (2010) Maximizing the ex vivo expansion of human mesenchymal stem cells using a microcarrier-based stirred culture system. J Biotechnol 146:194–197

Elseberg CL, Salzig D, Czermak P (2015) Bioreactor expansion of human mesenchymal stem cells according to GMP requirements. Methods Mol Biol 1283:199–218

Engela AU, Baan CC, Dor FJ, Weimar W, Hoogduijn MJ (2012) On the interactions between mesenchymal stem cells and regulatory T cells for immunomodulation in transplantation. Front Immunol 3:126

Fekete N, Gadelorge M, Furst D, Maurer C, Dausend J, Fleury-Cappellesso S, Mailander V, Lotfi R, Ignatius A, Sensebe L, Bourin P, Schrezenmeier H, Rojewski MT (2012a) Platelet lysate from whole blood-derived pooled platelet concentrates and apheresis-derived platelet concentrates for the isolation and expansion of human bone marrow mesenchymal stromal cells: production process, content and identification of active components. Cytotherapy 14:540–554

Fekete N, Rojewski MT, Furst D, Kreja L, Ignatius A, Dausend J, Schrezenmeier H (2012b) GMP-compliant isolation and large-scale expansion of bone marrow-derived MSC. PLoS One 7:e43255

Fekete N, Rojewski MT, Lotfi R, Schrezenmeier H (2014) Essential components for ex vivo proliferation of mesenchymal stromal cells. Tissue Eng Part C Methods 20:129–139

Fernandes AM, Marinho PA, Sartore RC, Paulsen BS, Mariante RM, Castilho LR, Rehen SK (2009) Successful scale-up of human embryonic stem cell production in a stirred microcarrier culture system. Braz J Med Biol Res 42:515–522

Ferrari G, Cusella-De Angelis G, Coletta M, Paolucci E, Stornaiuolo A, Cossu G, Mavilio F (1998) Muscle regeneration by bone marrow-derived myogenic progenitors. Science 279:1528–1530

Frechette JP, Martineau I, Gagnon G (2005) Platelet-rich plasmas: growth factor content and roles in wound healing. J Dent Res 84:434–439

Fuller BJ (2004) Cryoprotectants: the essential antifreezes to protect life in the frozen state. Cryo Letters 25:375–388

Galvao J, Davis B, Tilley M, Normando E, Duchen MR, Cordeiro MF (2014) Unexpected low-dose toxicity of the universal solvent DMSO. FASEB J 28:1317–1330

Garcia-Olmo D, Garcia-Arranz M, Herreros D, Pascual I, Peiro C, Rodriguez-Montes JA (2005) A phase I clinical trial of the treatment of Crohn's fistula by adipose mesenchymal stem cell transplantation. Dis Colon Rectum 48:1416–1423

Gastens MH, Goltry K, Prohaska W, Tschope D, Stratmann B, Lammers D, Kirana S, Gotting C, Kleesiek K (2007) Good manufacturing practice-compliant expansion of marrow-derived stem and progenitor cells for cell therapy. Cell Transplant 16:685–696

George B (2011) Regulations and guidelines governing stem cell based products: clinical considerations. Perspect Clin Res 2:94–99

Ghodsi M, Heshmat R, Amoli M, Keshtkar A-A, Arjmand B, Aghayan H, Hosseini P, Sharifi AM, Larijani B (2012) The effect of fetal liver-derived cell suspension allotransplantation on patients with diabetes: first year of follow-up. Acta Med Iran 50:541

Gluckman E, Broxmeyer HA, Auerbach AD, Friedman HS, Douglas GW, Devergie A, Esperou H, Thierry D, Socie G, Lehn P et al (1989) Hematopoietic reconstitution in a patient with Fanconi's anemia by means of umbilical-cord blood from an HLA-identical sibling. N Engl J Med 321:1174–1178

Gong W, Han Z, Zhao H, Wang Y, Wang J, Zhong J, Wang B, Wang S, Wang Y, Sun L, Han Z (2012) Banking human umbilical cord-derived mesenchymal stromal cells for clinical use. Cell Transplant 21:207–216

Goodarzi P, Arjmand B, Emami-Razavi SH, Soleimani M, Khodadadi A, Mohamadi-Jahani F, Aghayan HR (2014) Human autologous serum as a substitute for fetal bovine serum in human Schwann cell culture. Acta Med Iran 52:241

Gordon SL, Oppenheimer SR, Mackay AM, Brunnabend J, Puhlev I, Levine F (2001) Recovery of human mesenchymal stem cells following dehydration and rehydration. Cryobiology 43:182–187

Gottipamula S, Sharma A, Krishnamurthy S, Majumdar AS, Seetharam RN (2012) Human platelet lysate is an alternative to fetal bovine serum for large-scale expansion of bone marrow-derived mesenchymal stromal cells. Biotechnol Lett 34:1367–1374

Gottipamula S, Muttigi MS, Kolkundkar U, Seetharam RN (2013) Serum-free media for the production of human mesenchymal stromal cells: a review. Cell Prolif 46:608–627

Griffiths MJ, Bonnet D, Janes SM (2005) Stem cells of the alveolar epithelium. Lancet 366:249–260

Griffiths S, Baraniak PR, Copland IB, Nerem RM, Mcdevitt TC (2013) Human platelet lysate stimulates high-passage and senescent human multipotent mesenchymal stromal cell growth and rejuvenation in vitro. Cytotherapy 15:1469–1483

Grinnemo KH, Kumagai-Braesch M, Mansson-Broberg A, Skottman H, Hao X, Siddiqui A, Andersson A, Stromberg AM, Lahesmaa R, Hovatta O, Sylven C, Corbascio M, Dellgren G (2006) Human embryonic stem cells are immunogenic in allogeneic and xenogeneic settings. Reprod Biomed Online 13:712–724

Grinnemo KH, Sylven C, Hovatta O, Dellgren G, Corbascio M (2008) Immunogenicity of human embryonic stem cells. Cell Tissue Res 331:67–78

Gstraunthaler G (2003) Alternatives to the use of fetal bovine serum: serum-free cell culture. ALTEX 20:275–281

Guillot PV, de Bari C, Dell'accio F, Kurata H, Polak J, Fisk NM (2008) Comparative osteogenic transcription profiling of various fetal and adult mesenchymal stem cell sources. Differentiation 76:946–957

Hagell P, Brundin P (2001) Cell survival and clinical outcome following intrastriatal transplantation in Parkinson disease. J Neuropathol Exp Neurol 60:741–752

Hartmann I, Hollweck T, Haffner S, Krebs M, Meiser B, Reichart B, Eissner G (2010) Umbilical cord tissue-derived mesenchymal stem cells grow best under GMP-compliant culture conditions and maintain their phenotypic and functional properties. J Immunol Methods 363:80–89

Hass R, Kasper C, Bohm S, Jacobs R (2011) Different populations and sources of human mesenchymal stem cells (MSC): a comparison of adult and neonatal tissue-derived MSC. Cell Commun Signal 9:12

Hayakawa J, Joyal EG, Gildner JF, Washington KN, Phang OA, Uchida N, Hsieh MM, Tisdale JF (2010) 5% dimethyl sulfoxide (DMSO) and pentastarch improves cryopreservation of cord blood cells over 10% DMSO. Transfusion 50:2158–2166

Hayes M, Curley G, Laffey JG (2012) Mesenchymal stem cells—a promising therapy for Acute Respiratory Distress Syndrome. F1000 Med Rep 4:2

Heazlewood C, Atkinson K (2013) Optimal tissue sources of mesenchymal stromal cells for clinical applications. In: Chase LG, Vemuri MC (eds) Mesenchymal stem cell therapy. Humana Press, New York

Heiskanen A, Satomaa T, Tiitinen S, Laitinen A, Mannelin S, Impola U, Mikkola M, Olsson C, Miller-Podraza H, Blomqvist M, Olonen A, Salo H, Lehenkari P, Tuuri T, Otonkoski T, Natunen J, Saarinen J, Laine J (2007) N-glycolylneuraminic acid xenoantigen contamination of human embryonic and mesenchymal stem cells is substantially reversible. Stem Cells 25:197–202

Herrera B, Inman GJ (2009) A rapid and sensitive bioassay for the simultaneous measurement of multiple bone morphogenetic proteins. Identification and quantification of BMP4, BMP6 and BMP9 in bovine and human serum. BMC Cell Biol 10:20

Holm F, Strom S, Inzunza J, Baker D, Stromberg AM, Rozell B, Feki A, Bergstrom R, Hovatta O (2010) An effective serum- and xeno-free chemically defined freezing procedure for human embryonic and induced pluripotent stem cells. Hum Reprod 25:1271–1279

Horwitz EM, Gordon PL, Koo WK, Marx JC, Neel MD, Mcnall RY, Muul L, Hofmann T (2002) Isolated allogeneic bone marrow-derived mesenchymal cells engraft and stimulate growth in children with osteogenesis imperfecta: implications for cell therapy of bone. Proc Natl Acad Sci U S A 99:8932–8937

Hunt CJ (2011) Cryopreservation of human stem cells for clinical application: a review. Transfus Med Hemother 38:107–123

Ikebe C, Suzuki K (2014) Mesenchymal stem cells for regenerative therapy: optimization of cell preparation protocols. Biomed Res Int 2014:951512

Ilic N, Brooke G, Murray P, Barlow S, Rossetti T, Pelekanos R, Hancock S, Atkinson K (2011) Manufacture of clinical grade human placenta-derived multipotent mesenchymal stromal cells. Methods Mol Biol 698:89–106

Ishikane S, Ohnishi S, Yamahara K, Sada M, Harada K, Mishima K, Iwasaki K, Fujiwara M, Kitamura S, Nagaya N, Ikeda T (2008) Allogeneic injection of fetal membrane-derived mesenchymal stem cells induces therapeutic angiogenesis in a rat model of hind limb ischemia. Stem Cells 26:2625–2633

Iudicone P, Fioravanti D, Bonanno G, Miceli M, Lavorino C, Totta P, Frati L, Nuti M, Pierelli L (2014) Pathogen-free, plasma-poor platelet lysate and expansion of human mesenchymal stem cells. J Transl Med 12:28

Janz Fde L, Debes Ade A, Cavaglieri Rde C, Duarte SA, Romao CM, Moron AF, Zugaib M, Bydlowski SP (2012) Evaluation of distinct freezing methods and cryoprotectants for human amniotic fluid stem cells cryopreservation. J Biomed Biotechnol 2012:649353

Johansson L, Klinth J, Holmqvist O, Ohlson S (2003) Platelet lysate: a replacement for fetal bovine serum in animal cell culture? Cytotechnology 42:67–74

Jung KW (2009) Perspectives on human stem cell research. J Cell Physiol 220:535–537

Kern S, Eichler H, Stoeve J, Kluter H, Bieback K (2006) Comparative analysis of mesenchymal stem cells from bone marrow, umbilical cord blood, or adipose tissue. Stem Cells 24:1294–1301

Kinzebach S, Bieback K (2013) Expansion of Mesenchymal Stem/Stromal cells under xenogenic-free culture conditions. Adv Biochem Eng Biotechnol 129:33–57

Kocaoemer A, Kern S, Kluter H, Bieback K (2007) Human AB serum and thrombin-activated platelet-rich plasma are suitable alternatives to fetal calf serum for the expansion of mesenchymal stem cells from adipose tissue. Stem Cells 25:1270–1278

Koller MR, Maher RJ, Manchel I, Oxender M, Smith AK (1998) Alternatives to animal sera for human bone marrow cell expansion: human serum and serum-free media. J Hematother 7:413–423

Lange C, Cakiroglu F, Spiess AN, Cappallo-Obermann H, Dierlamm J, Zander AR (2007) Accelerated and safe expansion of human mesenchymal stromal cells in animal serum-free medium for transplantation and regenerative medicine. J Cell Physiol 213:18–26

Lanzoni G, Alviano F, Marchionni C, Bonsi L, Costa R, Foroni L, Roda G, Belluzzi A, Caponi A, Ricci F, Luigi Tazzari P, Pagliaro P, Rizzo R, Lanza F, Roberto Baricordi O, Pasquinelli G, Roda E, Paolo Bagnara G (2009) Isolation of stem cell populations with trophic and immunoregulatory functions from human intestinal tissues: potential for cell therapy in inflammatory bowel disease. Cytotherapy 11:1020–1031

Larijani B, Aghayan H-R, Goodarzi P, Arjmand B (2015a) GMP-grade human fetal liver-derived mesenchymal stem cells for clinical transplantation. Stem Cells Good Manuf Pract Methods Protoc Regul 1283:123–136

Larijani B, Arjmand B, Ahmadbeigi N, Falahzadeh K, Soleimani M, Sayahpour FA, Aghayan HR (2015b) A simple and cost-effective method for isolation and expansion of human fetal pancreas derived mesenchymal stem cells. Arch Iran Med 18:770–775

Lazarus HM, Haynesworth SE, Gerson SL, Rosenthal NS, Caplan AI (1995) Ex vivo expansion and subsequent infusion of human bone marrow-derived stromal progenitor cells (mesenchymal progenitor cells): implications for therapeutic use. Bone Marrow Transplant 16:557–564

Le Blanc K, Rasmusson I, Sundberg B, Gotherstrom C, Hassan M, Uzunel M, Ringden O (2004) Treatment of severe acute graft-versus-host disease with third party haploidentical mesenchymal stem cells. Lancet 363:1439–1441

Lennon DP, Caplan AI (2006) Isolation of human marrow-derived mesenchymal stem cells. Exp Hematol 34:1604–1605

Lepperdinger G, Brunauer R, Jamnig A, Laschober G, Kassem M (2008) Controversial issue: is it safe to employ mesenchymal stem cells in cell-based therapies? Exp Gerontol 43:1018–1023

Lewis CM, Suzuki M (2014) Therapeutic applications of mesenchymal stem cells for amyotrophic lateral sclerosis. Stem Cell Res Ther 5:32

Li Y, Ma T (2012) Bioprocessing of cryopreservation for large-scale banking of human pluripotent stem cells. Biores Open Access 1:205–214

Lian Q, Lye E, Suan Yeo K, Khia Way Tan E, Salto-Tellez M, Liu TM, Palanisamy N, El Oakley RM, Lee EH, Lim B, Lim SK (2007) Derivation of clinically compliant MSCs from CD105+, CD24- differentiated human ESCs. Stem Cells 25:425–436

Lindroos B, Boucher S, Chase L, Kuokkanen H, Huhtala H, Haataja R, Vemuri M, Suuronen R, Miettinen S (2009) Serum-free, xeno-free culture media maintain the proliferation rate and multipotentiality of adipose stem cells in vitro. Cytotherapy 11:958–972

Lindroos B, Suuronen R, Miettinen S (2011) The potential of adipose stem cells in regenerative medicine. Stem Cell Rev 7:269–291

Liu Y, Xu X, Ma X, Martin-Rendon E, Watt S, Cui Z (2010) Cryopreservation of human bone marrow-derived mesenchymal stem cells with reduced dimethylsulfoxide and well-defined freezing solutions. Biotechnol Prog 26:1635–1643

Luttun A, Ross JJ, Verfaillie C, Aranguren XL, Prosper F (2006) Differentiation of multipotent adult progenitor cells into functional endothelial and smooth muscle cells. Curr Protoc Immunol Chapter 22:Unit 22F.9

Mackay AM, Beck SC, Murphy JM, Barry FP, Chichester CO, Pittenger MF (1998) Chondrogenic differentiation of cultured human mesenchymal stem cells from marrow. Tissue Eng 4:415–428

Mackensen A, Drager R, Schlesier M, Mertelsmann R, Lindemann A (2000) Presence of IgE antibodies to bovine serum albumin in a patient developing anaphylaxis after vaccination with human peptide-pulsed dendritic cells. Cancer Immunol Immunother 49:152–156

Makino S, Fukuda K, Miyoshi S, Konishi F, Kodama H, Pan J, Sano M, Takahashi T, Hori S, Abe H, Hata J, Umezawa A, Ogawa S (1999) Cardiomyocytes can be generated from marrow stromal cells in vitro. J Clin Invest 103:697–705

Mannello F, Tonti GA (2007) Concise review: no breakthroughs for human mesenchymal and embryonic stem cell culture: conditioned medium, feeder layer, or feeder-free; medium with fetal calf serum, human serum, or enriched plasma; serum-free, serum replacement nonconditioned medium, or ad hoc formula? All that glitters is not gold! Stem Cells 25:1603–1609

Martin MJ, Muotri A, Gage F, Varki A (2005) Human embryonic stem cells express an immunogenic nonhuman sialic acid. Nat Med 11:228–232

Mazur P (1988) Stopping biological time. The freezing of living cells. Ann N Y Acad Sci 541:514–531

Mccullough J, Haley R, Clay M, Hubel A, Lindgren B, Moroff G (2010) Long-term storage of peripheral blood stem cells frozen and stored with a conventional liquid nitrogen technique compared with cells frozen and stored in a mechanical freezer. Transfusion 50:808–819

Meryman HT (2007) Cryopreservation of living cells: principles and practice. Transfusion 47:935–945

Meuleman N, Tondreau T, Delforge A, Dejeneffe M, Massy M, Libertalis M, Bron D, Lagneaux L (2006) Human marrow mesenchymal stem cell culture: serum-free medium allows better expansion than classical alpha-MEM medium. Eur J Haematol 76:309–316

Moon SY, Park YB, Kim DS, Oh SK, Kim DW (2006) Generation, culture, and differentiation of human embryonic stem cells for therapeutic applications. Mol Ther 13:5–14

Morris C, de Wreede L, Scholten M, Brand R, van Biezen A, Sureda A, Dickmeiss E, Trneny M, Apperley J, Chiusolo P, van Imhoff GW, Lenhoff S, Martinelli G, Hentrich M, Pabst T, Onida F, Quinn M, Kroger N, de Witte T, Ruutu T (2014) Should the standard dimethyl sulfoxide concentration be reduced? Results of a European Group for Blood and Marrow Transplantation prospective noninterventional study on usage and side effects of dimethyl sulfoxide. Transfusion 54:2514–2522

Motta JP, Paraguassu-Braga FH, Bouzas LF, Porto LC (2014) Evaluation of intracellular and extracellular trehalose as a cryoprotectant of stem cells obtained from umbilical cord blood. Cryobiology 68:343–348

Muller I, Kordowich S, Holzwarth C, Spano C, Isensee G, Staiber A, Viebahn S, Gieseke F, Langer H, Gawaz MP, Horwitz EM, Conte P, Handgretinger R, Dominici M (2006) Animal serum-free culture conditions for isolation and expansion of multipotent mesenchymal stromal cells from human BM. Cytotherapy 8:437–444

Muramatsu T, Pinontoan R, Okumura J (1995) Biopotency of fetal bovine serum, and insulin and insulin-like growth factors I and II in enhancing whole-body protein synthesis of chicken embryos cultured in vitro. Comp Biochem Physiol C Pharmacol Toxicol Endocrinol 111:281–286

Nakajima H, Uchida K, Guerrero AR, Watanabe S, Sugita D, Takeura N, Yoshida A, Long G, Wright KT, Johnson WE, Baba H (2012) Transplantation of mesenchymal stem cells promotes

an alternative pathway of macrophage activation and functional recovery after spinal cord injury. J Neurotrauma 29:1614–1625
Nie Y, Bergendahl V, Hei DJ, Jones JM, Palecek SP (2009) Scalable culture and cryopreservation of human embryonic stem cells on microcarriers. Biotechnol Prog 25:20–31
Nimura A, Muneta T, Koga H, Mochizuki T, Suzuki K, Makino H, Umezawa A, Sekiya I (2008) Increased proliferation of human synovial mesenchymal stem cells with autologous human serum: comparisons with bone marrow mesenchymal stem cells and with fetal bovine serum. Arthritis Rheum 58:501–510
O'Donoghue K (2011) Validating and monitoring the cleanroom. In: Kanegsberg B, Kanegsberg E (eds) Handbook for critical cleaning: applications, processes, and controls, 2nd edn. CRC Press, Boca Raton
Oreffo RO, Triffitt JT (1999) Future potentials for using osteogenic stem cells and biomaterials in orthopedics. Bone 25:5s–9s
Pal R, Hanwate M, Totey SM (2008) Effect of holding time, temperature and different parenteral solutions on viability and functionality of adult bone marrow-derived mesenchymal stem cells before transplantation. J Tissue Eng Regen Med 2:436–444
Parker A, Shang H, Khurgel M, Katz A (2007) Low serum and serum-free culture of multipotential human adipose stem cells. Cytotherapy 9:637–646
Parolini O, Alviano F, Bagnara GP, Bilic G, Buhring HJ, Evangelista M, Hennerbichler S, Liu B, Magatti M, Mao N, Miki T, Marongiu F, Nakajima H, Nikaido T, Portmann-Lanz CB, Sankar V, Soncini M, Stadler G, Surbek D, Takahashi TA, Redl H, Sakuragawa N, Wolbank S, Zeisberger S, Zisch A, Strom SC (2008) Concise review: isolation and characterization of cells from human term placenta: outcome of the first international Workshop on Placenta Derived Stem Cells. Stem Cells 26:300–311
Patrikoski M, Juntunen M, Boucher S, Campbell A, Vemuri MC, Mannerstrom B, Miettinen S (2013) Development of fully defined xeno-free culture system for the preparation and propagation of cell therapy-compliant human adipose stem cells. Stem Cell Res Ther 4:27
Piskorska-Jasiulewicz MM, Witkowska-Zimny M (2015) [Perinatal sources of stem cells]. Postepy Hig Med Dosw (Online) 69:327–334
Pittenger MF, Mackay AM, Beck SC, Jaiswal RK, Douglas R, Mosca JD, Moorman MA, Simonetti DW, Craig S, Marshak DR (1999) Multilineage potential of adult human mesenchymal stem cells. Science 284:143–147
Preynat-Seauve O, Krause KH (2011) Stem cell sources for regenerative medicine: the immunological point of view. Semin Immunopathol 33:519–524
Prindull G, Prindull B, Meulen N (1978) Haematopoietic stem cells (CFUc) in human cord blood. Acta Paediatr Scand 67:413–416
Prokhorova TA, Harkness LM, Frandsen U, Ditzel N, Schroder HD, Burns JS, Kassem M (2009) Teratoma formation by human embryonic stem cells is site dependent and enhanced by the presence of Matrigel. Stem Cells Dev 18:47–54
Puglisi MA, Tesori V, Lattanzi W, Piscaglia AC, Gasbarrini GB, D'ugo DM, Gasbarrini A (2011) Therapeutic implications of mesenchymal stem cells in liver injury. J Biomed Biotechnol 2011:860578
Rajala K, Lindroos B, Hussein SM, Lappalainen RS, Pekkanen-Mattila M, Inzunza J, Rozell B, Miettinen S, Narkilahti S, Kerkela E, Aalto-Setala K, Otonkoski T, Suuronen R, Hovatta O, Skottman H (2010) A defined and xeno-free culture method enabling the establishment of clinical-grade human embryonic, induced pluripotent and adipose stem cells. PLoS One 5:e10246
Rasmusson I (2006) Immune modulation by mesenchymal stem cells. Exp Cell Res 312:2169–2179
Rauch C, Feifel E, Amann EM, Spotl HP, Schennach H, Pfaller W, Gstraunthaler G (2011) Alternatives to the use of fetal bovine serum: human platelet lysates as a serum substitute in cell culture media. ALTEX 28:305–316
Reinisch A, Bartmann C, Rohde E, Schallmoser K, Bjelic-Radisic V, Lanzer G, Linkesch W, Strunk D (2007) Humanized system to propagate cord blood-derived multipotent mesenchymal stromal cells for clinical application. Regen Med 2:371–382

Riazi AM, Kwon SY, Stanford WL (2009) Stem cell sources for regenerative medicine. Methods Mol Biol 482:55–90

Rodriguez L, Azqueta C, Azzalin S, Garcia J, Querol S (2004) Washing of cord blood grafts after thawing: high cell recovery using an automated and closed system. Vox Sang 87:165–172

Rojewski MT, Weber BM, Schrezenmeier H (2008) Phenotypic characterization of mesenchymal stem cells from various tissues. Transfus Med Hemother 35:168–184

Rotter N, Oder J, Schlenke P, Lindner U, Bohrnsen F, Kramer J, Rohwedel J, Huss R, Brandau S, Wollenberg B, Lang S (2008) Isolation and characterization of adult stem cells from human salivary glands. Stem Cells Dev 17:509–518

Rubinstein P, Dobrila L, Rosenfield RE, Adamson JW, Migliaccio G, Migliaccio AR, Taylor PE, Stevens CE (1995) Processing and cryopreservation of placental/umbilical cord blood for unrelated bone marrow reconstitution. Proc Natl Acad Sci U S A 92:10119–10122

Salvade A, Della Mina P, Gaddi D, Gatto F, Villa A, Bigoni M, Perseghin P, Serafini M, Zatti G, Biondi A, Biagi E (2010) Characterization of platelet lysate cultured mesenchymal stromal cells and their potential use in tissue-engineered osteogenic devices for the treatment of bone defects. Tissue Eng Part C Methods 16:201–214

Saric T, Frenzel LP, Hescheler J (2008) Immunological barriers to embryonic stem cell-derived therapies. Cells Tissues Organs 188:78–90

Schafer R, Northoff H (2008) Characteristics of mesenchymal stem cells—new stars in regenerative medicine or unrecognized old fellows in autologous regeneration? Transfus Med Hemother 35:154–159

Schallmoser K, Strunk D (2013) Generation of a pool of human platelet lysate and efficient use in cell culture. Methods Mol Biol 946:349–362

Schallmoser K, Bartmann C, Rohde E, Reinisch A, Kashofer K, Stadelmeyer E, Drexler C, Lanzer G, Linkesch W, Strunk D (2007) Human platelet lysate can replace fetal bovine serum for clinical-scale expansion of functional mesenchymal stromal cells. Transfusion 47:1436–1446

Schallmoser K, Rohde E, Reinisch A, Bartmann C, Thaler D, Drexler C, Obenauf AC, Lanzer G, Linkesch W, Strunk D (2008) Rapid large-scale expansion of functional mesenchymal stem cells from unmanipulated bone marrow without animal serum. Tissue Eng Part C Methods 14:185–196

Schallmoser K, Rohde E, Bartmann C, Obenauf AC, Reinisch A, Strunk D (2009) Platelet-derived growth factors for GMP-compliant propagation of mesenchymal stromal cells. Biomed Mater Eng 19:271–276

Selvaggi TA, Walker RE, Fleisher TA (1997) Development of antibodies to fetal calf serum with arthus-like reactions in human immunodeficiency virus-infected patients given syngeneic lymphocyte infusions. Blood 89:776–779

Sensebe L (2008) Clinical grade production of mesenchymal stem cells. Biomed Mater Eng 18:S3–S10

Seshareddy K, Troyer D, Weiss ML (2008) Method to isolate mesenchymal-like cells from Wharton's Jelly of umbilical cord. Methods Cell Biol 86:101–119

Shahdadfar A, Fronsdal K, Haug T, Reinholt FP, Brinchmann JE (2005) In vitro expansion of human mesenchymal stem cells: choice of serum is a determinant of cell proliferation, differentiation, gene expression, and transcriptome stability. Stem Cells 23:1357–1366

Shigeno Y, Ashton BA (1995) Human bone-cell proliferation in vitro decreases with human donor age. J Bone Joint Surg Br 77:139–142

Spees JL, Gregory CA, Singh H, Tucker HA, Peister A, Lynch PJ, Hsu SC, Smith J, Prockop DJ (2004) Internalized antigens must be removed to prepare hypoimmunogenic mesenchymal stem cells for cell and gene therapy. Mol Ther 9:747–756

Stolzing A, Naaldijk Y, Fedorova V, Sethe S (2012) Hydroxyethylstarch in cryopreservation—mechanisms, benefits and problems. Transfus Apher Sci 46:137–147

Stute N, Holtz K, Bubenheim M, Lange C, Blake F, Zander AR (2004) Autologous serum for isolation and expansion of human mesenchymal stem cells for clinical use. Exp Hematol 32:1212–1225

Su CY, Kuo YP, Lin YC, Huang CT, Tseng YH, Burnouf T (2009) A virally inactivated functional growth factor preparation from human platelet concentrates. Vox Sang 97:119–128

Sundin M, Ringden O, Sundberg B, Nava S, Gotherstrom C, Le Blanc K (2007) No alloantibodies against mesenchymal stromal cells, but presence of anti-fetal calf serum antibodies, after transplantation in allogeneic hematopoietic stem cell recipients. Haematologica 92:1208–1215

Swamynathan P, Venugopal P, Kannan S, Thej C, Kolkundar U, Bhagwat S, Ta M, Majumdar AS, Balasubramanian S (2014) Are serum-free and xeno-free culture conditions ideal for large scale clinical grade expansion of Wharton's jelly derived mesenchymal stem cells? A comparative study. Stem Cell Res Ther 5:88

Szot GL, Lee MR, Tavakol MM, Lang J, Dekovic F, Kerlan RK, Stock PG, Posselt AM (2009) Successful clinical islet isolation using a GMP-manufactured collagenase and neutral protease. Transplantation 88:753–756

Tapp H, Hanley EN Jr, Patt JC, Gruber HE (2009) Adipose-derived stem cells: characterization and current application in orthopaedic tissue repair. Exp Biol Med (Maywood) 234:1–9

Tateishi K, Ando W, Higuchi C, Hart DA, Hashimoto J, Nakata K, Yoshikawa H, Nakamura N (2008) Comparison of human serum with fetal bovine serum for expansion and differentiation of human synovial MSC: potential feasibility for clinical applications. Cell Transplant 17:549–557

Thirumala S, Goebel WS, Woods EJ (2009) Clinical grade adult stem cell banking. Organogenesis 5:143–154

Thomas RJ, Anderson D, Chandra A, Smith NM, Young LE, Williams D, Denning C (2009a) Automated, scalable culture of human embryonic stem cells in feeder-free conditions. Biotechnol Bioeng 102:1636–1644

Thomas RJ, Hope AD, Hourd P, Baradez M, Miljan EA, Sinden JD, Williams DJ (2009b) Automated, serum-free production of CTX0E03: a therapeutic clinical grade human neural stem cell line. Biotechnol Lett 31:1167–1172

Todorov P, Hristova E, Konakchieva R, Michova A, Dimitrov J (2010) Comparative studies of different cryopreservation methods for mesenchymal stem cells derived from human fetal liver. Cell Biol Int 34:455–462

Tonti GA, Mannello F (2008) From bone marrow to therapeutic applications: different behaviour and genetic/epigenetic stability during mesenchymal stem cell expansion in autologous and foetal bovine sera? Int J Dev Biol 52:1023–1032

Troyer DL, Weiss ML (2008) Wharton's jelly-derived cells are a primitive stromal cell population. Stem Cells 26:591–599

Unger C, Skottman H, Blomberg P, Dilber MS, Hovatta O (2008) Good manufacturing practice and clinical-grade human embryonic stem cell lines. Hum Mol Genet 17:R48–R53

Wang S, Qu X, Zhao RC (2012a) Clinical applications of mesenchymal stem cells. J Hematol Oncol 5:19

Wang Y, Han ZB, Song YP, Han ZC (2012b) Safety of mesenchymal stem cells for clinical application. Stem Cells Int 2012:652034

Wei X, Yang X, Han ZP, Qu FF, Shao L, Shi YF (2013) Mesenchymal stem cells: a new trend for cell therapy. Acta Pharmacol Sin 34:747–754

Will RG, Ironside JW, Zeidler M, Cousens SN, Estibeiro K, Alperovitch A, Poser S, Pocchiari M, Hofman A, Smith PG (1996) A new variant of Creutzfeldt-Jakob disease in the UK. Lancet 347:921–925

Wingard JR, Majhail NS, Brazauskas R, Wang Z, Sobocinski KA, Jacobsohn D, Sorror ML, Horowitz MM, Bolwell B, Rizzo JD, Socie G (2011) Long-term survival and late deaths after allogeneic hematopoietic cell transplantation. J Clin Oncol 29:2230–2239

Witkowska-Zimny M, Wrobel E (2011a) Perinatal sources of mesenchymal stem cells: Wharton's jelly, amnion and chorion. Cell Mol Biol Lett 16:493–514

Witkowska-Zimny M, Wrobel E (2011b) Perinatal sources of mesenchymal stem cells: Wharton's jelly, amnion and chorion. Cell Mol Biol Lett 16:493–514

Witzeneder K, Lindenmair A, Gabriel C, Holler K, Theiss D, Redl H, Hennerbichler S (2013) Human-derived alternatives to fetal bovine serum in cell culture. Transfus Med Hemother 40:417–423

Woods EJ, Liu J, Pollok K, Hartwell J, Smith FO, Williams DA, Yoder MC, Critser JK (2003) A theoretically optimized method for cord blood stem cell cryopreservation. J Hematother Stem Cell Res 12:341–350

Woods EJ, Bagchi A, Goebel WS, Vilivalam VD, Vilivalam VD (2010) Container system for enabling commercial production of cryopreserved cell therapy products. Regen Med 5:659–667

Wouters G, Grossi S, Mesoraca A, Bizzoco D, Mobili L, Cignini P, Giorlandino C (2007) Isolation of amniotic fluid-derived mesenchymal stem cells. J Prenat Med 1:39–40

Wuchter P, Bieback K, Schrezenmeier H, Bornhauser M, Muller LP, Bonig H, Wagner W, Meisel R, Pavel P, Tonn T, Lang P, Muller I, Renner M, Malcherek G, Saffrich R, Buss EC, Horn P, Rojewski M, Schmitt A, Ho AD, Sanzenbacher R, Schmitt M (2015) Standardization of Good Manufacturing Practice-compliant production of bone marrow-derived human mesenchymal stromal cells for immunotherapeutic applications. Cytotherapy 17:128–139

Yamamoto N, Isobe M, Negishi A, Yoshimasu H, Shimokawa H, Ohya K, Amagasa T, Kasugai S (2003) Effects of autologous serum on osteoblastic differentiation in human bone marrow cells. J Med Dent Sci 50:63–69

Zhao JW, Gao ZL, Mei H, Li YL, Wang Y (2011) Differentiation of human mesenchymal stem cells: the potential mechanism for estrogen-induced preferential osteoblast versus adipocyte differentiation. Am J Med Sci 341:460–468

GMP Facilities for Clinical Cell Therapy Product Manufacturing: A Brief Review of Requirements and Design Considerations

Hamid Reza Aghayan, Babak Arjmand, and Scott R. Burger

1 Introduction

The field of cell therapy is evolving quickly, with potentially transformational new treatment modalities generating great excitement in the scientific and clinical communities, among patients, and in the biopharmaceutical industry (Dawson et al. 2003). In the decade 2000–2010, cell therapy products accounted for over 2700 clinical trials (Culme-Seymour et al. 2012) with the aim of addressing unmet medical needs (Hampson et al. 2008). Approved, marketed cell therapy products include expanded autologous chondrocytes, fibroblasts, keratinocytes, dendritic cells, limbal stem cells, and tissue-engineered skin substitutes (Bersenev 2011, 2012).

Cell therapy products present numerous challenges, including complex manufacturing processes involving a high degree of manual operations that require rigorous control, manufacturing environments in which multiple products are processed simultaneously, complex raw materials that may or may not be part of the final product, and products that cannot be fully characterized (US Pharmacopoeia 2012). Major risks associated with cell therapy products include microbiological contamination, loss of cell function, cell transformation malignancies, immunogenicity, and ectopic engraftment (Giancola et al. 2012).

H.R. Aghayan, M.D. Ph.D.
Chronic Diseases Research Center, Endocrinology and Metabolism Population Sciences Institute, Tehran University of Medical Sciences, Tehran, Iran

B. Arjmand, M.D. Ph.D.
Endocrinology and Metabolism Research Center, Endocrinology and Metabolism Clinical Sciences Institute, Tehran University of Medical Sciences, Tehran, Iran

Brain and Spinal Cord Injury Research Center, Tehran University of Medical Sciences, Tehran, Iran

S.R. Burger, M.D. (✉)
Advanced Cell & Gene Therapy, LLC, Chapel Hill, North Carolina, USA
e-mail: sburger@ac-gt.com

B. Arjmand (ed.), *Perinatal Tissue-Derived Stem Cells*, Stem Cell Biology and Regenerative Medicine, DOI 10.1007/978-3-319-46410-7_10

Regulatory agencies have responded to these concerns by establishing risk-based regulatory structures, in which more rigorous controls are required for products that have been more extensively manipulated or are thought to pose increased risk in other respects (Burger 2003). In recent years, numerous regulations, standards, and guidance documents about cell therapy product manufacturing have been published, but current Good Manufacturing Practices (cGMPs, often abbreviated GMPs) are the most fundamental.

The definition of GMPs is similar worldwide. The European Medicines Agency (EMA), for example, describes GMP is "part of quality assurance which ensures that products are consistently produced and controlled to the quality standards appropriate to their intended use." The International Society for Pharmaceutical Engineering (ISPE) defines GMP as: "A system for ensuring that products are consistently produced and controlled according to quality standards. It is designed to minimize the risks involved in any pharmaceutical production that cannot be eliminated through testing the final product."

GMPs are composed of multiple elements covering all aspects of production. Although each element is equally important, the GMP facility is the most obvious and tangible aspect (Arjmand et al. 2012; Burger 2003). A GMP facility represents a large capital cost and fixed investment, and design and construction must be rigorously planned, with input from multiple disciplines and much technical information (Signore and Jacobs 2005). This chapter summarizes basic requirements for cell therapy GMP facilities and provides an overview of facility planning and design, cleanroom classifications, and operating procedures.

2 Regulation of Cell Therapy Product Manufacturing

Manufacturing cell therapy products often requires complex procedures, such as cell isolation/selection, ex vivo expansion, differentiation, activation, gene modification, and encapsulation. Since the final product is living cells, terminal sterilization or removal/inactivation of microbial contaminants is not possible. Rigorously controlled manufacturing, including qualified starting materials, validated aseptic manufacturing processes, and appropriate testing are critical factors to ensure safety and consistency of the product (Giancola et al. 2012; Brandenberger et al. 2011; Bosse et al. 2000). For these reasons, current regulations require GMP-compliant manufacturing of advanced cell therapy products used in clinical studies (Arjmand and Aghayan 2014).

Regulatory requirements for cell therapy products follow a risk-based, data-driven approach (Burger 2003). These regulations are codified in the United States by the U.S. Food and Drug Administration (FDA), and by The European Medicines Agency (EMA) in the European Union (Ährlund-Richter et al. 2009). Complex, extensive manufacturing procedures involve greater risk than simple, brief processes, and so cell-based products that are extensively processed are assigned a higher risk category. To classify the associated risks and to define the level of oversight, the FDA describes these types of products as "more-than-minimally manipulated," and the EMA as "substantial manipulation." Nearly all advanced cell therapies would be considered more-than-minimally manipulated, and hence require a higher degree of process control and laboratory sophistication. Manufacturing processes involving only minimal manipulation, such as cryopreservation of autologous peripheral blood progenitor cells, are required to comply only with Good Tissue Practices (GTPs), rather than the more stringent and extensive GMPs (Burger 2000, 2003).

3 GMP Facilities

Annex 1 of the Pharmaceutical Inspection Co-operation Scheme (PIC/S) guide to GMP states, "The manufacture of sterile products is subject to special requirements in order to minimize risks of microbiological contamination, and of particulate and pyrogen contamination." According to this guide, "The manufacture of sterile products should be carried out in clean areas entry to which should be through airlocks for personnel and/or for equipment and materials. Clean areas should be maintained to an appropriate cleanliness standard and supplied with air which has passed through filters of an appropriate efficiency" (PIC/S Secretariat 2015). This facility must be capable of supporting the manufacturing process, as well as characterization testing, and must incorporate the quality systems and infrastructure required for GMP compliance (Burger 2009).

Manufacturing areas, designated Class A, B, C, or D depending on air quality (Table 1), cover the processing requirements of a GMP facility (Arjmand et al. 2012; Giancola et al. 2012). Although the roots of cleanroom design and management go back more than 100 years, when they were used within the hospital environment to decrease the risk of spreading infection, the need for a clean environment for industrial manufacturing is a requirement of modern society (Whyte 2001). The controlled environment of a carefully designed, constructed, validated, and maintained cleanroom minimizes the risks of environmental contamination during aseptic processing and decreases the possibility of cross-contamination between patient-specific products (US Pharmacopoeia 2008). Cleanroom areas have special

Table 1 Cleanroom classification based on GMP guidelines

Cleanroom grade	Maximum number of particles/m^3 equal to or greater than the tabulated size			
	At rest		In operation	
	0.5 μm	5 μm	0.5 μm	5 μm
A	3520	20	3520	20
B	3520	29	352,000	2900
C	352,000	2900	3,520,000	29,000
D	3,520,000	29,000	Not classified	

meaning that goes beyond simply clean space. The Federal Standard 209E defines it as "A room in which the concentration of airborne particles is controlled and which contains one or more clean zones." In ISO 14644-1 definition, a cleanroom is "A room in which the concentration of airborne particles is controlled, and which is constructed and used in a manner to minimize the introduction, generation, and retention of particles inside the room and in which other relevant parameters, e.g. temperature, humidity, and pressure, are controlled as necessary" (Whyte 2001; ISO 1999). The Federal Standard 209E cleanliness classification is convenient but generally inadequate by itself to describe a facility used for biopharmaceutical manufacturing. The presence of viable particles, which are not considered in this standard, may affect operations of these cleanrooms. A measure of both viable and nonviable particles is required to provide sufficient information regarding the suitability of the cleanroom for its intended purpose. GMP guidelines emphasize that cleanrooms and clean air devices should be classified in accordance with ISO 14644-1. The maximum permitted airborne particle concentration for each grade is shown in Table 1. Since the particle concentration is dependent on the particle-generating activities going on in the room, cleanroom classification should be carried out when the room is as build, at rest, and operational (Whyte 1999). Almost all emerging cell therapy manufacturing processes start as research laboratory procedures in which controlling, monitoring, and evaluating impact of key parameters on target cells is difficult (Kirouac and Zandstra 2008). These types of procedures typically involve open process steps, which can expose cells to the external environment, rather than enclosed, sterile process systems which incorporate aseptic access ports. Based on the GMP guideline, open procedures for aseptically prepared cell-based products must be performed in a Class A environment—usually a biological safety cabinet housed within a Class B process room (Sensebé et al. 2010).

4 GMP Facility Design

Given their cost and complexity, cell therapy GMP facilities must be designed with flexibility and varied applications in mind because the more adaptable the laboratory, the longer it is likely to remain useful. There is no a single "right" way to construct a GMP facility or cleanroom, as each should be designed to address end-user requirements. Centers looking to establish or expand facilities would benefit from discussions with regulatory experts and persons with experience setting up cell therapy GMP facilities. Active communication, both formal and informal, with regulatory bodies is also invaluable in the planning phase. Preventing design and construction mistakes likely will far outweigh the costs obtaining consultations and visiting other centers.

GMP guidelines for cell therapy manufacturing are still evolving, but are based principally on existing regulations for medicinal products. Considering GMP requirements from the outset of facility design ensure compliance with regulatory requirements. Facility designs that poorly address GMP requirements can be expected to encounter regulatory difficulties and may not be licensed without significant changes (Signore and Jacobs 2005).

Table 2 Guidance documents for cleanroom design

Document	Title	Publication date
ISO 14644-4	Cleanrooms and associated controlled environments—design, construction, and startup	2015
PIC/S GMP guide	Annex 1 (Manufacture of sterile medicinal products)	2015
21 CFR Part 211	Current good manufacturing practice for finished pharmaceuticals—subpart c: buildings and facilities	2015
United States Pharmacopoeia	Chapter <1116> Microbiological control and monitoring of aseptic processing environments	2015
ISPE Baseline Guide	Biopharmaceutical manufacturing facilities	2013
ISPE Baseline Guide	Sterile product manufacturing facilities	2011
WHO Technical Report Series, No. 961	Annex 6—WHO good manufacturing practices for sterile pharmaceutical products	2011
ISPE Good Practice Guide	Heating, Ventilation, and Air Conditioning (HVAC)	2009
EU GMP	Annex 1 (Manufacture of sterile medicinal products)	2008
FDA Guidance for industry	Sterile drug products produced by aseptic processing—current good manufacturing practice	2004
ISO 14698-1	Biocontamination control—general principles and methods	2003

Several guidance documents address aspects of facility and cleanroom design apart from GMPs and are listed in Table 2. These guidelines use very general terms, which must be interpreted with regard to specific applications, but do describe how a facility can enhance process control. For example, 21 CFR 211 Part C states that "Any such building shall have adequate space for the orderly placement of equipment and materials to prevent mix-ups between different components, drug product containers, closures, labeling, in-process materials, or drug products, and to prevent contamination."

Establishing a new GMP facility can be divided into several phases including planning, design, construction, commissioning, operation, and qualification.

- *Planning Phase*: As a first step, it is important to assemble a project team whose responsibility will be to identify the scope of manufacturing and its related process maps. The number of people, their backgrounds and expertise, their availability, and their responsibilities play a very important role in project execution and success (Odum 2004). Although some GMP cell manufacturing knowledge and skills may be available within the institution, in most cases outside help is required. In addition to architects, engineers, and project managers—with extensive experience in GMP facility design—persons with knowledge and experience in GMP cell therapy manufacturing, regulatory requirements, development and execution of validation programs, and standard operating procedure (SOP) development will be needed (Burger 2000). This project team should define the user requirements specifications (URS) for the equipment, utilities, and rooms.

- A comprehensive review of the preliminary design should be performed to ensure that the user requirements are met, and that the design complies with GMP requirements. ISO Standard 14644-4 is also a useful reference for preliminary design review (White 2009; Dietz et al. 2007).
- The project team is also responsible for finding the appropriate location for building the facility. Few projects begin as a "greenfield" site with unlimited building area. The limitations of the project site should be thoroughly understood, and reasonable decisions in fitting the process and operational requirements into the existing site should be made. If the facility will be located in an existing building, the project team must determine whether the proposed location is feasible for renovation to house a GMP facility. For example, ceiling space available in a conventionally constructed building may not be adequate to accommodate the large air-handling systems necessary for aseptic processing areas (Burger 2000). In this scenario, modular cleanrooms or other temporary structures might be better design options. These options operationally provide aseptic processing cleanrooms within a conventional room. These arrangements are often temporary and limited in size, suitable only for early phase clinical trials (Dietz et al. 2007).
- In the planning phase, the role of the facility in the commercialization process should be defined. As cell therapy clinical trials move from early (Phase I/II) to late-stage clinical development (Phase III), the level of process control will be intensified. A facility designed for early phase clinical trials may not be adequate for late clinical development or licensed cell therapy manufacturing. For facilities which support Phase I/II clinical trials at a single center, validation, though still rigorous, will not be as stringent as for facilities involved in Phase III trials.
- People are the main source of cleanroom contamination. It is therefore useful to know the number of people expected to work in the cleanrooms during planning phase. This parameter influences calculations on the quantity of supplied air in each room, and consequently the specifications of HVAC system. Air handling units should be designed, constructed, and maintained to minimize the risk of cross-contamination between different manufacturing areas and may need to be specific for an area. Some general areas such as offices, rest room, conference room, file storage, central supply room (CSR), quality control laboratory, and support rooms should be in mind. The space limitations could be compensated by outsourcing of some services. For example, it is wise to purchase sterile supplies and reagents rather than providing in-house sterilization services (Gee 2009a).
- *Design Phase*: This phase begins with evaluation of predefined processes and product requirements. The designer should first gain an understanding of product and process requirements and use this information to develop a conceptual layout. This layout should be appraised by the project team and subsequently enhanced and refined to produce final layout of facility and equipment. Thinking about contamination and mix-up prevention starts with drawing the layout of work areas and flow diagrams. Process flow diagrams are used to plan movements of materials, equipment, personnel, products, and wastes into the cleanrooms. These flow diagrams are also applied in developing the site Validation Master Plan (VMP), regulatory submissions, and SOPs. Whenever possible, flow should be unidirec-

tional, to prevent cross-contamination of different areas within the facility. Some GMP facilities employ single-pass patterns in which the cleanrooms are situated between clean and dirty corridors. Staff, materials, and reagents move in a unidirectional pattern from the clean corridor, to the manufacturing areas and from there through the dirty corridor to de-gowning area. In single-corridor design, there is multidirectional traffic, but differential pressures are set to protect the manufacturing areas from the corridors or general areas. The unidirectional pattern provides the highest degree of stringency but is expensive to maintain and requires the largest footprint due to the double-corridor design (Gee 2009b). The layout of cleanrooms should facilitate maintenance, pressure differentials, and temperature/humidity control by isolating critical spaces and by excluding non-clean operations (Burger 2000; ISPE Baseline Guide 2011). An example of materials, waste, and personnel flows is shown in Fig. 1.

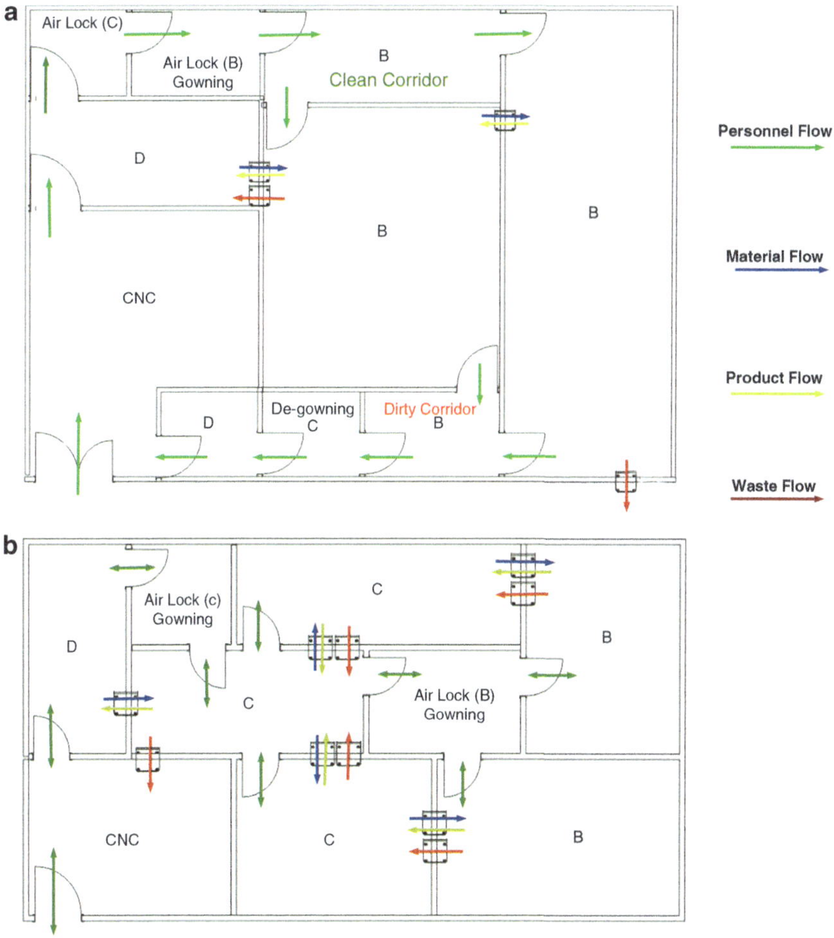

Fig. 1 An example of materials, waste, and personnel flows in unidirectional (**a**) and single-corridor (**b**) cleanrooms

- Cleanroom monitoring, and mechanisms to maintain facility integrity in the case of power failures or other emergencies, should be considered. Examples include easily observed monitoring systems (for air pressure, temperatures, particle counts, and humidity) and backup power supply for critical equipment (Dietz et al. 2007; Larijani et al. 2015a).
- The origin of cells can also influence facility design. Allogeneic stem cells (SCs) have been successfully isolated from different adult, fetal, and perinatal tissues. These cells have been used in different clinical trials and recent studies have suggested that allogeneic SCs from healthy donor could be a more suitable source to trigger tissue regeneration (Zhang et al. 2015; Larijani et al. 2015b). If a facility aims to prepare the therapeutic batches of allogeneic SCs, a scale-up approach should be considered, in which multiple doses of cell-based products are manufactured from a single donor. Therefore, a few large cleanrooms could be appropriate for this. By contrast, autologous (patient-specific) cells are not amenable to a scale-up approach. In this case, the facility should be suitable for scale-out manufacturing process, which involves concurrent manufacturing of multiple products at a one product per patient scale. A facility with multiple, smaller cleanrooms would be more suitable for this type of manufacturing (Hampson et al. 2008; Eaker et al. 2013).
- When using open systems, each cleanroom generally must be dedicated to manufacturing a single cell product. Closed-system processing permits efficient use of manufacturing space because it provides process and product isolation, and thus can eliminate the need to dedicate a room to one product for one patient (Burger 2002).
- The air pressure of cleanroom should be positive to all surrounding zones of lower classification. It is normal practice to "cascade" air quality from higher to lower quality levels. It means that the critical zone should be surrounded by areas of lower classifications, which eventually lead to a controlled not classified (CNC) area (Fig. 2). The minimum value of differential pressure should be 10–15 Pa (ISPE Baseline Guide 2011).
- *Construction Phase*: All components used in the construction of a cleanroom should comply with the relevant local regulations and national laws.
- General principles of design should be followed in building cleanrooms or an aseptic processing facility. Many points to consider can be found in ISO 14644-4, the FDA aseptic processing guideline, and in EU Annex 1. For example, in the EU-GMP the requirements for the quality of surface have been described as: "surfaces to be smooth, impervious and without sharp edges, free of pores, abrasion resistant, unbroken, easy to clean, as well as resistant to cleaning agents and disinfectants." In general, floors, walls, and ceilings should be finished with smooth, nonporous surfaces. Resin-based wall- and floor-coverings with sealed seams often are used. To increase the workspace flexibility and to facilitate the cleaning procedure, it is wise to avoid installing permanent equipment (Burger 2000).
- The air-handling system may recirculate air, but single-pass (fully exhausted) air is preferable when there is a potential source of contamination or active biologics in the room. If recirculated air is used, potential sources of cross-contamination should be considered.

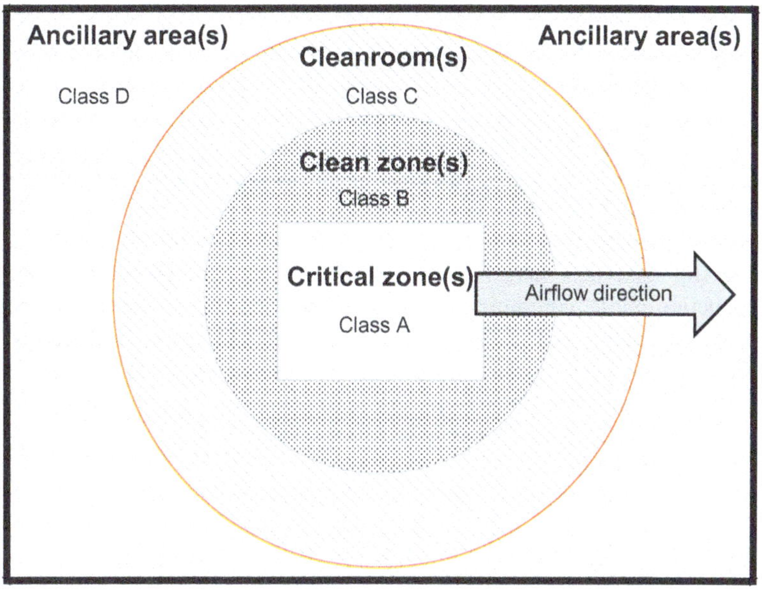

Fig. 2 Cleanroom cleanliness cascade (adapted from ISO 14644-4). In this configuration, the clean zone would be regarded as a more stringently controlled area

- Airlocks or anterooms help maintain pressurization differentials between spaces of different classifications. The at-rest grade of an airlock should be the same grade as the area into which it leads. It provides physical barrier to people, materials, and air. This barrier is logical location to enforce gowning requirements for controlled area. When the risk of cross-contamination within the cleanrooms is high, the use of separate changing rooms for entering and leaving clean areas is desirable. Certain airlocks may be designated as an equipment or material airlock and provide a space to remove clean equipment or materials before they are introduced into the cleanroom. Pass-through chambers should be installed to allow the flow of small articles, raw materials, test objects, and products through the clean zones without requiring personnel to exit higher controlled areas.
- Sinks and drains are not recommended in cleanrooms and should be prohibited in grade A/B areas used for aseptic manufacturing. Windows are recommended in cleanrooms to facilitate supervision and for safety, unless prohibited by the facility protocol for security reasons (Dietz et al. 2007). Use of closed-circuit television cameras (CCTVs) is an effective tool for surveillance system. The CCTVs should be strategically placed to cover all critical areas. It is better to consider high resolution, low light/night vision camera with recording capabilities. Access to the manufacturing facility must be restricted to authorized persons only, using some form of personal identification badges (Odum 2004).
- *Commissioning Phase*: In this phase, the equipment, utilities, and facility are tested to ensure they meet design specifications and user requirements. Commissioning ensures that any system operates properly and all necessary procedures are in place to ensure consistent operation. In the commissioning phase,

a less formal change management system is in place, allowing changes to be made and documented with lower levels of approval than would be necessary during the qualification phase (White 2009; Odum 2004).

- *Qualification Phase*: The qualification phase begins through aseptic process simulations and process validation runs. Process simulation should be performed as initial validation with three consecutive satisfactory simulations per shift. Process simulation should be repeated at defined intervals and after any significant modification to the HVAC system, equipment, and process. Some installation qualification (IQ) activities may begin in the commissioning phase or even in the construction phase, continuing into the qualification phase. This phase includes operational qualification (OQ) and performance qualification (PQ) activities. If no changes are made to the equipment, some OQ activities may be performed during commissioning. Cleanroom certification activities are typically performed in the qualification phase (White 2009; Sharp 2005).

5 GMP Facility Maintenance and Operations

Professional standards and regulatory documents require that the facility and equipment be maintained in a clean condition. Therefore, the organization should develop and implement effective cleaning methods, procedures to monitor cleaning efficiency, and documentation of cleaning procedures. In addition to normal cleaning requirements, a program of environmental monitoring should be in place to assure that the facility is consistently meeting its classification. The facility must have written SOPs for cleaning procedure and its monitoring. The SOPs should be simple and easy to understand because it may be read and used by staff who are not scientists. It should contain clear instructions on cleaning agent selection, preparation and use, the areas that should be cleaned, the cleaning intervals in each zone, and the documentation and record keeping of the cleaning procedure. Frequency of cleaning and environmental monitoring is determined by a number of factors including the types and number of products, cleanliness classification, and changeover procedures (Lindblad 2009). Cleaning processes can be classified in various ways, such as routine cleaning in which the floor and benches are cleaned, versus complete cleaning, which also includes walls, windows, and ceilings. Personnel working in a cleanroom may also carry out cleaning resulting from the process (Ramstorp 2008).

To prevent cross-contamination between products, cleaning should ensure that all traces of a product are removed from a manufacturing site before a different product is introduced (Gee and Lyon 2009). US Pharmacopeia recommendations on sampling frequency for each cleanroom grade are shown in Table 3.

Most academic cell manufacturing facilities are not in continuous operation, hence each facility need to develop a monitoring plan that reflects the manufacturing environmental conditions over a specified period. Appropriate microbial monitoring should include quantitation of the microbial content of room air, surfaces, equipment, floors, walls, and personnel garments (US Pharmacopeia 2008).

Table 3 Suggested frequency of sampling on the basis of criticality of controlled environment

Sampling area	Frequency of sampling
ISO 5 (formerly Class 100) or better room designations	Each operating shift
Supporting areas immediately adjacent to ISO 5 (e.g., ISO 7, formerly Class 10,000)	Each operating shift
Other support areas (ISO 8, formerly Class 100,000)	Twice per week
Potential product/container contact areas	Twice per week
Other support areas to aseptic processing areas but nonproduct contact (ISO 8 or lower)	Once per week

Access to the manufacturing site should be via a gowning area and should be restricted to authorized personnel. To eliminate particle shedding from the personnel, use of special clothing, known as cleanroom garments, is necessary. Gowning may take place in up to three incrementally classified areas (Alici and Blomberg 2010). Cleanroom garments act as filters and retain any generated particles until the clothing is removed in an area where this poses no risk to the production process. Depending on the nature of the manufacturing environment, different types of gowning practices (from laboratory coats to full sterile gowning) might be used (Gee 2009a). The ambient temperature and humidity should not be uncomfortably high because of the nature of the garments.

6 Summary

The essential concept of GMP is to ensure that a production process is reliable, reproducible, and transferrable (Ährlund-Richter et al. 2009). The facility is not the sole component of GMP and GMP-compliant cell manufacturing would not be achieved by transfer of current methodology to cleanrooms (Arjmand et al. 2012). Naturally, moving to cleanrooms will decrease the risk of contaminations but more importantly, GMP implementation will lead to development of validated SOPs for the entire process. It is a misconception that implementing GMP by itself will yield products of the highest quality, and which are most efficient for a certain application. The biggest benefit of GMP is actually the reproducible manufacturing of products, which will ensure traceability of the process and clinical safety (Alici and Blomberg 2010). The integration of GMP into a cell therapy manufacturing facility is challenging, but it should be remembered that most of the GMP elements (SOPs, laboratory controls, training, and documentation) already exist in some form in most cell-processing laboratories (Davis-Sproul 1999). Based on our previous experiences, the implementation of general quality management systems (such as ISO 9001 and ISO 13485) could facilitate this transition (Arjmand et al. 2012). Appropriate regulation of cell therapy products is essential to ensure public safety and trust while minimizing unnecessary barriers to product development (von Tigerstrom 2008). Regulatory requirements should be considered when planning and building a GMP facility for cell therapy manufacturing. Long-term planning is crucial to facilitate suitability of the facility for future use (Alici and Blomberg 2010). It is far too easy to decide to construct a GMP facility based on the perception that any advanced academic center should have one. Some centers undoubtedly do require on-site facilities. Others, however, may find it more cost effective to collaborate with an existing GMP facility. It could be beneficial for small centers with early stage clinical development to use the services of contract manufacturing organizations (CMOs), which can facilitate clinical trial operations by outsourcing the actual manufacturing steps, managing regulatory requirements, conserving capital, and speeding up the development cycle (Fitzpatrick 2008).

References

Ährlund-Richter L, De Luca M, Marshak DR, Munsie M, Veiga A, Rao M (2009) Isolation and production of cells suitable for human therapy: challenges ahead. Cell Stem Cell 4(1):20–26

Alici E, Blomberg P (2010) GMP facilities for manufacturing of advanced therapy medicinal products for clinical trials: an overview for clinical researchers. Curr Gene Ther 10(6):508–515

Arjmand B, Aghayan HR (2014) Cell manufacturing for clinical applications. Stem Cells 32(9):2557–2558. doi:10.1002/stem.1751

Arjmand B, Emami-Razavi SH, Larijani B, Norouzi-Javidan A, Aghayan HR (2012) The implementation of tissue banking experiences for setting up a cGMP cell manufacturing facility. Cell Tissue Bank 13(4):587–596. doi:10.1007/s10561-011-9276-y

Bersenev A (2011) RegenMed and cell therapeutic products available on the market. Stem Cell Assays blog. http://stemcellassays.com/2011/04/regenmed-and-cell-therapeutic-products--available-on-the-market/. Accessed 13 April 2011

Bersenev A (2012) Stem cell therapeutic products on the market. Stem Cell Assays blog. http://stemcellassays.com/2012/02/stem-cell-therapeutic-products-market/. Accessed 29 Feb 2012

Bosse R, Kulmburg P, Von Kalle C, Engelhardt M, Dwenger A, Rosenthal F, Schulz G (2000) Production of stem-cell transplants according to good manufacturing practice. Ann Hematol 79(9):469–476

Brandenberger R, Burger S, Campbell A, Fong T, Lapinskas E, Rowley JA (2011) Cell therapy bioprocessing. BioProcess Int 9(Suppl 1):30–37

Burger SR (2000) Design and operation of a current good manufacturing practices cell-engineering laboratory. Cytotherapy 2(2):111–122

Burger S (2002) Advanced cell and gene therapies: translational development and GMP production. BioProcessing 1:19–23

Burger S (2003) CTP/GMP cell engineering for cell and gene therapies. Bioprocessing J 2:1–4

Burger SR (2009) Commercial manufacturing of cell therapy products—considering the options. World Stem Cell Report. Genetics Policy Institute, pp 149-153

Culme-Seymour EJ, Davie NL, Brindley DA, Edwards-Parton S, Mason C (2012) A decade of cell therapy clinical trials (2000–2010). Regen Med 7(4):455–462

Davis-Sproul J (1999) cGMP and cell processing. Cytotherapy 1(6):431–432

Dawson L, Bateman-House AS, Agnew DM, Bok H, Brock DW, Chakravarti A, Greene M, King PA, O'Brien SJ, Sachs DH (2003) Safety issues in cell-based intervention trials. Fertil Steril 80(5):1077–1085

Dietz AB, Padley D, Gastineau D (2007) Infrastructure development for human cell therapy translation. Clin Pharmacol Ther 82(3):320–324

Eaker S, Armant M, Brandwein H, Burger S, Campbell A, Carpenito C, Clarke D, Fong T, Karnieli O, Niss K (2013) Concise review: guidance in developing commercializable autologous/patient-specific cell therapy manufacturing. Stem Cells Transl Med 2(11):871–883

Fitzpatrick I (2008) Cellular therapy success through integrated automation. BioProcess Int 6(9)

Gee A (2009a) Baylor College of Medicine—Center for Cell and Gene Therapy (CAGT). In: Gee A (ed) Cell therapy: cGMP facilities and manufacturing. Springer, Boston, pp 67–77

Gee A (2009b) Design of a new GMP facility—lessons learned. In: Gee A (ed) Cell therapy: cGMP facilities and manufacturing. Springer, Boston, pp 79–84

Gee A, Lyon DL (2009) Cleaning procedures. In: Gee A (ed) Cell therapy: cGMP facilities and manufacturing. Springer, Boston, pp 135–144

Giancola R, Bonfini T, Iacone A (2012) Cell therapy: cGMP facilities and manufacturing. Muscles Ligaments Tendons J 2(3):243–247

Hampson B, Rowley J, Venturi N (2008) Manufacturing patient-specific cell therapy products. BioProcess Int 6(8)

International Organization for Standardization (1999) ISO 14644-1, Cleanrooms and associated controlled environments—part 1: classification of air cleanliness

ISPE Baseline Guide (2011) Sterile product manufacturing facilities, 1st edn. ISPE, Tampa

Kirouac DC, Zandstra PW (2008) The systematic production of cells for cell therapies. Cell Stem Cell 3(4):369–381

Larijani B, Aghayan HR, Goodarzi P, Arjmand B (2015a) GMP-grade human fetal liver-derived mesenchymal stem cells for clinical transplantation. Methods Mol Biol 1283:123–136

Larijani B, Arjmand B, Ahmadbeigi N, Falahzadeh K, Soleimani M, Sayahpour FA, Aghayan HR (2015b) A simple and cost-effective method for isolation and expansion of human fetal pancreas derived mesenchymal stem cells. Arch Iran Med 18(11):770–775

Lindblad RW (2009) Regulation of cell product manufacturing and delivery: a United States perspective. In: Gee A (ed) Cell therapy: cGMP facilities and manufacturing. Springer, Boston, pp 3–25

Odum JN (2004) Sterile product facility design and project management, 2nd edn. CRC Press, Boca Raton

US Pharmacopoeia (2008) Chapter <1116> Microbiological evaluation of clean rooms and other controlled environments

US Pharmacopoeia (2012) Chapter <1046> Cellular and tissue-based products

PIC/S Secretariat (2015) Guide to good manufacturing practice for medicinal products—Annex 1

Ramstorp M (2008) Introduction to contamination control and cleanroom technology. Wiley, New York

Sensebé L, Bourin P, Tarte K (2010) Good manufacturing practices production of mesenchymal stem/stromal cells. Hum Gene Ther 22(1):19–26

Sharp J (2005) Good pharmaceutical manufacturing practice: rationale and compliance. CRC Press, Boca Raton

Signore AA, Jacobs T (2005) Good design practices for GMP pharmaceutical facilities. Taylor & Francis Group, Boca Raton

von Tigerstrom BJ (2008) The challenges of regulating stem cell-based products. Trends Biotechnol 26(12):653–658

White E (2009) Cleanroom design, construction, and qualification. J Validat Technol 15(4):30

Whyte W (1999) Cleanroom design, 2nd edn. Wiley Online Library, Chichester, pp 21–49

Whyte W (2001) Cleanroom technology: fundamentals of design, testing and operation, 2nd edn. Wiley Online Library, Chichester, pp 1–20

Zhang J, Huang X, Wang H, Liu X, Zhang T, Wang Y, Hu D (2015) The challenges and promises of allogeneic mesenchymal stem cells for use as a cell-based therapy. Stem Cell Res Ther 6(1):1–7

Ethical Issues in Perinatal Tissue Derivation and Regenerative Medicine

Leila Afshar

1 Introduction

Regenerative medicine and Tissue Engineering are fields of research which are rapidly developing in every different branches of medicine. The methods and techniques that are involved in tissue engineering are multitude. And so the scope of regenerative medicine could not be described easily. Some commentators define tissue engineering as "an interdisciplinary field that applies the principles of engineering and life sciences toward the development of biological substitutes that restore, maintain or improve tissue function or a whole organ" (Langer and Vacanti 1993).

This definition includes different scientific and therapeutic purposes in regenerative medicine; from developing a cell line in laboratory to construction of whole autologous organs that could be a solution for scare donor organs, and also an answer to the problem of adverse host response and minimize the risk of rejection. This wide scope and field of area means that the ethical aspect of tissue engineering and regenerative medicine is complex and multidimensional. Based on the techniques, kind and source of the subjects, and the goal of the procedures, there would be different ethical considerations. Furthermore, the issue of social justice is a major challenge in this regard. In some cases such as embryonic stem cell research or enhancement, the discussions are highly controversial. This article try to present the most important ethical aspects of the issue in order to identify the good and the worthwhile and to guide the direction process of this new field in order to find the most acceptable means and ways.

L. Afshar, M.D., Ph.D. (✉)
Medical Ethics Department, Shahid Beheshti University of Medical Sciences, Tehran, Iran
e-mail: leilaafshar@sbmu.ac.ir

B. Arjmand (ed.), *Perinatal Tissue-Derived Stem Cells*, Stem Cell Biology and Regenerative Medicine, DOI 10.1007/978-3-319-46410-7_11

2 Criteria for Ethical Judgment

Every human action is characterized by two aspects, each of which is important for its ethical evaluation. Teleologically an action could be regarded in terms of its externality (*what* the agent does) and its internality (*why* he/she is doing it) (Schockenhoff 2009; Veatch 2012). Although the intention of the agent and the justification of the objectives and goals play important role for ethical judgment of actions these are not sufficient. Rather, the ethical evaluation of actions must also include a judgment of the means by which we wish to attain our goals. In addition, good motivation and the intent to achieve high-ranking goals relate only to one aspect of our actions, which needs to be supplemented by another one.

In ethical assessment of actions what we do in pursuing our goals, it is important, because the others' rights will not be nullified by our high desirable goals. In this regard, besides justifying our goals, we need to legitimate the mean or our selected method to exclude those ones that violate the human dignity or fundamental human rights. One aspect of violating the dignity of a human being is to utilize him/her as a mere object of someone else's will and to use him/her solely for achieving extraneous purposes irrelative to his/her existence that is a form of instrumentalization. In relation to the regenerative medicine, the question must accordingly be raised of whether the various forms of obtaining human stem cells make an embryo into a mere object at any phase of the production process and reduce its existence to an instrument?

In addition to justification of the goals and the scrutiny of the means, the third criterion of ethical judgment is to take required responsibility for the foreseeable consequences of an action, especially not intended and perhaps harmful effects that should be incorporated in the overall judgment of an action.

This model of comprehensive ethical evaluation could be applied for both individual actions in the interpersonal relationship and for the collective actions of social institutions and therefore could be used for the activities of the scientific community and also in the field of stem cell and regenerative medicine.

It is scientists' task to justify the goals of their actions, scrutinize the appropriate means for achieving them, and take responsibility on the consequences of their actions.

Application of this approach to research projects and therapeutic models of the tissue engineering and regenerative medicine leads to a comprehensive analysis of the ethical aspect of the issue.

3 Main Ethical Challenges

3.1 The Kind and the Origin of the Subject

For some tissue engineering projects human embryonic cells are demanding resources for new biotechnological developments. Their pluripotential characteristics make them desirable to use. But the human embryo is a special and precious entity. These cases stimulate ethical controversies in human embryonic experiments and also in regenerative medicine. The ontological controversy about the nature of the human embryo (is it a person, a thing, or an entity with certain moral standing) and the ambivalent moral status of it makes it difficult to justify destroying an embryo for use as a source of the raw materials for tissue engineering and regenerative medicine (Afshar 2015).

On the other hand, because of the limitation of the embryonic stem cell resources and the ethical challenges of their use we can obtain usable pluripotential cells for tissue engineering from other sources besides the fetus, these postembryonic, perinatal sources including the amniotic fluid and membrane, placenta, Wharton's jelly, and umbilical cord blood. Since these tissues are discarded at the time of birth, could be a simple and safe means for stem cell deprivation. In addition, peripheral maternal blood can be a source of fetal cells (Si et al. 2015).

Because of their least invasiveness, the amniotic fluid and the placenta are clinically most appealing ones. And most importantly, these cells effectively avoid ethical issue involving the moral status of the embryo.

3.2 *The Intended Goal of Regenerative Medicine, Its Means and Methods*

The ethical discussions about the use of bioengineering and regenerative medicine begin with questioning the therapeutic goals of medicine versus cosmetic and enhancement ones. The justifying arguments for the use of these techniques are mainly based on its goal, the treatment. If the goal is to cure or treat human disease, then the benefits will outweigh the burdens, the cost, and difficulty. In this regard, tissue engineering ought not to be used in a wasteful manner (Geron Ethics Advisory Board 1999). Indeed, the question is can we mean the therapeutic purpose of regenerative medicine in terms of restoring the normal function of the tissues, organs, and humans as a whole? And concluding that it is necessary and also possible to restrict the technique based on it? In this regard, for example, using neurons of another person for a Parkinson patient may raise the question of identity and its nature. This means that drawing a distinctive line between different goals of regenerative medicine would be difficult. Yet it is more unclear that who should decide or evaluate the issue?

One may say that the utilizing therapeutic options of regenerative medicine are still not available, but there are many researches in this area that may lead to new treatments. This may generate a basic question; is it possible that the regenerative medicine alter the definition of health? And therefore the enhancement therapies one day became part of the medicine. And may be this change will impact on the medical profession's self-image, focusing on optimizing the human nature or on the antiaging medicine than on healing the diseases?

One of the main ethical challenges in using new techniques is that everything which is technically feasible may not be ethically acceptable. This notes the potentially harmful consequences of the new scientific achievements and also their undesirable methods and means. Overcoming this obstacle needs to draw distinctive lines between ethically permissible, still acceptable, and unacceptable methods. In tissue engineering and regenerative medicine, this question can be translated to different questions. Whether the cellular biology and tissue development researches should be performed only on adult stem cells or it may also have recourse to embryonic stem cells? Whether we can use the cloned embryo for depriving their stem cells? These are philosophical questions of value judgment which the scientific criteria may not apply to assess it.

And by setting aside these ontological questions, there are still ethical considerations in tissue engineering researches that also presented in other areas of clinical research. For example, do we have a favorable balance of risks and benefits? However, in the context of regenerative medicine issues are more problematic because of the high degree of uncertainty. These uncertainties are about the nature of the risks, the probability of them, and the absence of an acceptable standard for the assessment. Therefore, a simple and general risk assessment is not possible and also favorable.

One of the risk indicators of the tissue engineering is the degree of invasiveness of the methods. An in vitro experiment on cells is nearly riskless, but the implantation of a cultivated cluster of cells implies the risks of contamination, incompatibility, and errors. The more modification in cells will increase the risk of side effects and more surgical procedures add the risk of invasive surgery. Furthermore, there is risk due to the applied substances and their interactions. If the rationale of the tissue engineering and regenerative medicine is to get nearer to physiological process of the body, the risks of the methods should be evaluated based on their effect on the body's own physiological process.

Another factor that should be mentioned in assessing the risk–benefit ratio is the intended function of the engineered tissue. As much as the underlying disease be serious, the producing risks would be acceptable. However, a life-threatening disease does not necessarily justify any kind of dangerous therapy. This means that "anything is better than this state" is not a careful and sufficient way of risk assessment.

Other risks due to the tissue engineering are associated to the scientific risks of different methods that are impossible to rule out them. However, there are important source of risk that should be considered such as infection, modification of the cells, mistaken identity of the cells and delivery of unwanted cells (Sutherland and Mayer 2003), and also the risk of induced cancer. Evaluation of risk–benefit ratio needs a careful analysis of possible alternative treatments, which are usually neglected. It is somehow different from situations that there is a gold standard of therapy and therefore needs an individualistic approach for each patient based on her medical and personal preferences. Finally, there are unexpected risks in any new experimental therapies, the important issue is to not ignore them and general considerations of these risks should be mentioned as one of the important aspects of the informed consent.

3.3 *The Issue of Social Justice*

Tissue engineering is a complex procedure still in experimental stages. Yet to be an ethical technology, it must be directed toward accessibility, just distribution, and efficacy (Zoloth 2014). In the context of widespread healthcare disparity, justice is a problem which will raise some ethical questions.

First, in a world that the majority of the people suffer from easily preventable or treatable infectious diseases such as tuberculosis, malaria, diarrhea, and AIDS, can tissue engineering and regenerative medicine be justly a promotion in the face of the health?

If we consider the widespread use of tissue engineering as its goal, which surely needs careful support and monitoring, we can justify the procedure. However, the question of how to achieve this goal has not yet been solved.

This perspective needs a careful attention to how the market may drive technology toward specific research goals (framed by value of profit) rather than the humanistic ones such as healing, and solidarity, which itself create the possibility for serious conflicts of interest.

4 Conclusion

Like any other field of scientific activities in modern medicine, the stem cell biology and regenerative medicine should be valued only within established ethical frameworks (Zacharias et al. 2012). By reviewing the ethical challenges of tissue engineering and regenerative medicine we find that like any other new technologies and scientific achievements, this field is also facing some important issues, from ontological to methodological and social ones. The controversy surrounding human embryo destruction to generate stem cells could be bypassed by using the perinatal stem cells. However, some other important challenges such as safety, and appropriate clinical application, still remain. Confronting these challenges need a regulatory system. To have an effective regulatory system for prohibiting these ethical considerations we need set of standards, for safety, for efficacy, and for fair and just use of regenerative medicine, in addition to guidelines for clinical trials and clinical use, and protocols for tissue stability and purity. All these require a regulatory structure including local committees, institutional, and national reviewing boards. Constitutions that play a great role in policy writing, addressing the issues involving the process of tissue engineering, applying in clinical level for regenerative purposes, in order to protect humanistic values of medicine and science.

However, it is also important to remember that ethics is not a barrier to restrict the scientific progress. The function of ethics is to lead the scientific progress in accepted directions and to emphasize the most acceptable means on this way.

References

Afshar L (2015) Moral Status of the human embryo; Applying philosophical consideration in medical decision making. Shahid Beheshti University of Medical Sciences Publication, Tehran (in Persian)

Geron Ethics Advisory Board (1999) Research with human embryonic stem cells: ethical considerations. Hastings Cent Rep 29:31e6

Langer R, Vacanti JP (1993) Tissue engineering. Science 260:920–926

Schockenhoff E (2009) Tissue engineering and regenerative medicine. Their goals, their methods and their consequences from an ethical viewpoint. In: Meyer U et al (eds) Fundamentals of tissue engineering and regenerative medicine. Springer, Berlin

Si JW, Wang XD, Shen S (2015) Perinatal stem cells: a promising cell resource for tissue engineering of craniofacial bone. World J Stem Cells 7(1):149–159

Sutherland FWH, Mayer JE Jr (2003) Ethical and regulatory issues concerning engineered tissues for congenital heart repair. Semin Thorac Cardiovasc Surg Pediatr Card Surg Annu 6:152–163

Veatch RM (2012) The basics of bioethics, 3rd edn. Pearson Education, Boston

Zacharias DG, Nelson TJ, Mueller PS, Hook CC (2012) Impedance of novel therapeutic technologies: the case of stem cells. Clin Transl Sci 5:422–427

Zoloth L (2014) Ethical issues. In: Lanza R, Langer R, Vacanti J (eds) Principles of tissue engineering. Academic Press of Elsevier, San Diego

ERRATUM

Perinatal Tissue-Derived Stem Cells

Alternative Sources of Fetal Stem Cells

Babak Arjmand

B. Arjmand (ed.), *Perinatal Tissue-Derived Stem Cells*, Stem Cell Biology and Regenerative Medicine, DOI 10.1007/978-3-319-46410-7

DOI 10.1007/978-3-319-46410-7_12

In Chapter 2 titled "Immunomodulatory Properties of Perinatal Tissue-Derived Mesenchymal Stem Cells" corrections have been made, to typesetting errors, throughout the chapter.

In Chapter 6 titled "Current Understanding Realities of Umbilical Cord Stem Cells Biology and Future Perspectives in Clinical Application" the order of authors has been corrected to Somayeh Ebrahimi-Barough, Reza Rahbarghazi, Zohreh Bagher, Jafar Ai, Elham Hoveizi.

The Front Matter has been updated to reflect the corrections to Chapter 2 and Chapter 6.

The updated online version of the original book can be found at http://dx.doi.org/10.1007/978-3-319-46410-7

B. Arjmand (ed.), *Perinatal Tissue-Derived Stem Cells*, Stem Cell Biology and Regenerative Medicine, DOI 10.1007/978-3-319-46410-7_12

Index

B. Arjmand (ed.), *Perinatal Tissue-Derived Stem Cells*, Stem Cell Biology and Regenerative Medicine, DOI 10.1007/978-3-319-46410-7

The manufacturer's authorised representative in the EU is Springer Nature Customer Service Centre GmbH, Europaplatz 3, 69115 Heidelberg, Germany. If you have any concerns regarding our products, please contact ProductSafety@springernature.com

Printed and bound by CPI Group (UK) Ltd, Croydon, CR0 4YY
22/07/2026
02174142-0003